ERRATA

In Chapter 1, "Mutagenesis from a Chemical Perspective," by B. Singer, a number of illustrations have been misplaced:

The figure on page 7 is Fig. 6 and should appear on page 15
The figure on page 8 is Fig. 2 and should appear on page 7
The figure on page 9 is Fig. 3 and should appear on page 8
The figure on page 12 is Fig. 4 and should appear on page 9
The figure on page 15 is Fig. 5 and should appear on page 12

MOLECULAR AND CELLULAR MECHANISMS OF MUTAGENESIS, 0-306-41006-0
edited by J. F. Lemontt and W. M. Generoso

MOLECULAR AND CELLULAR MECHANISMS OF MUTAGENESIS

BASIC LIFE SCIENCES

Alexander Hollaender, General Editor

Associated Universities, Inc.
Washington, D.C.

Recent volumes in the series:

Volume 11 • **PHOTOSYNTHETIC CARBON ASSIMILATION**
Edited by Harold W. Siegelman and Geoffrey Hind

Volume 12 • **GENETIC MOSAICS AND CHIMERAS IN MAMMALS**
Edited by Liane B. Russell

Volume 13 • **POLYPLOIDY: Biological Relevance**
Edited by Walter H. Lewis

Volume 14 • **GENETIC ENGINEERING OF OSMOREGULATION**
Impact on Plant Productivity for Food, Chemicals, and Energy
Edited by D. W. Rains, R. C. Valentine, and Alexander Hollaender

Volume 15 • **DNA REPAIR AND MUTAGENESIS IN EUKARYOTES**
Edited by W. M. Generoso, M.D. Shelby, and F. J. deSerres

Volume 16 • **DEVELOPMENT AND NEUROBIOLOGY OF DROSOPHILA**
Edited by O. Siddiqi, P. Babu, Linda M. Hall, and Jeffrey C. Hall

Volume 17 • **GENETIC ENGINEERING OF SYMBIOTIC NITROGEN FIXATION**
Edited by J. M. Lyons, R. C. Valentine, D. A. Phillips, D. W. Rains, and R. C. Huffaker

Volume 18 • **TRENDS IN THE BIOLOGY OF FERMENTATIONS FOR FUELS AND CHEMICALS**
Edited by Alexander Hollaender, Robert Rabson, Palmer Rogers, Anthony San Pietro, Raymond Valentine, and Ralph Wolfe

Volume 19 • **GENETIC ENGINEERING OF MICROORGANISMS FOR CHEMICALS**
Edited by Alexander Hollaender, Ralph D. DeMoss, Samuel Kaplan, Jordan Konisky, Dwayne Savage, and Ralph S. Wolfe

Volume 20 • **MOLECULAR AND CELLULAR MECHANISMS OF MUTAGENESIS**
Edited by J. F. Lemontt and W. M. Generoso

A Continuation Order Plan is available for this series. A continuation order will bring delivery of each new volume immediately upon publication. Volumes are billed only upon actual shipment. For further information please contact the publisher.

MOLECULAR AND CELLULAR MECHANISMS OF MUTAGENESIS

Edited by

J. F. Lemontt
and
W. M. Generoso

Oak Ridge National Laboratory
Oak Ridge, Tennessee

PLENUM PRESS • NEW YORK AND LONDON

Library of Congress Cataloging in Publication Data

Symposium on Molecular and Cellular Mechanisms of Mutagenesis (1981: Gatlinburg, Tenn.)
Molecular and cellular mechanisms of mutagenesis.

(Basic life sciences; n. 20)
"Proceedings of a Symposium on Molecular and Cellular Mechanisms of Mutagenesis held April 5–9, 1981, in Gatlinburg, Tennessee"—T.p. verso.
Bibliography: p.
Includes index.
1. Mutagenesis—Congresses. I. Lemontt, J. F., 1944– . II. Generoso, W. M. III. Title. IV. Series.
QH460.S95 1981   574.87'322   82-5300
ISBN 0-306-41006-0   AACR2

41,899

Proceedings of a symposium on Molecular and Cellular Mechanisms of Mutagenesis, held April 5–9, 1981, in Gatlinburg, Tennessee

©1982 Plenum Press, New York
A Division of Plenum Publishing Corporation
233 Spring Street, New York, N.Y. 10013

All rights reserved

No part of this book may be reproduced, stored in a retrieval system, or transmitted in any form or by any means, electronic, mechanical, photocopying, microfilming, recording, or otherwise, without written permission from the Publisher

Printed in the United States of America

FOREWORD

It has been nearly 35 years since the peacetime Biology Division of Oak Ridge National Laboratory was started, born of rather inauspicious conditions. Virtually no facilities were available and most of the wartime scientists had left. So, when we started out, it was obvious to me that something had to be done to reestablish research. Even more, because Oak Ridge was not known at that time for its biological work but rather for the separation of Uranium 235, nuclear reactor development, and radioisotope production, a new tradition had to be promoted. Although good biological work had been done at Oak Ridge during the war to protect the workers and the results of this work were quite excellent, very few installations remained.

When we started the work of the Biology Division, it became essential to make it part of the flow of modern biology all over the world. As Director, I had to do more than just attract promising scientists. We created an atmosphere conducive to creative research and nurtured all of the other aspects of a productive laboratory. Of course, we carefully prepared the results of our work in publishable form. We made a sincere effort to invite seminar speakers and lecturers to come to Oak Ridge despite the sacrifices this presented to our early budget. We also had to do something more, and here I "cashed in" on my experience of the previous 15 years.

In the early 1930s, I was in close contact with our colleagues at Cold Spring Harbor Laboratory. Having spent a few summers in Woods Hole, I worked with a number of physiologists, geneticists, and others. My interest in always knowing what was going on at Cold Spring Harbor served me very well as I began to develop the Biology Division at Oak Ridge. I had actually attended one of the first symposia that was held under Dr. Reginald Harris, and my close contact with Cold Spring Harbor increased through my friendship with Milislav Demerec, who later became the director of the biology work there. I came to know Dr. Berwind Kaufmann, a cytologist on the staff; we became very close and exchanged assistants with our laboratory at the National Institutes of Health in Bethesda. I was especially interested in the way the programs of their annual symposia were organized. Not only did the Laboratory organize the proceedings, but they also published most conferences. I saw that this was a very good way to break the isolation of a not-so-accessible laboratory. My close interest in the developments at Cold Spring Harbor has not diminished over the years and I am still a Trustee Emeritus.

In those early days at Oak Ridge following the war, we met for our first symposium in a wooden structure in a section of Oak Ridge that was a shopping center situated near the only restaurant in town. I, of course, concentrated my efforts on the areas of genetics, biochemistry, and biophysics, in which I was most interested, and steered the laboratory along those lines.

Our first symposium was held in 1948, and I had learned how to organize these symposia since I had worked in close cooperation with Demerec and had attended many of the symposia at Cold Spring Harbor. I don't think I missed many of these lectures. I chose the topic "Radiation Genetics," in which I had a strong interest and I knew many of the outstanding working scientists.

How then did we get from Oak Ridge to Gatlinburg? Very simple. We planned a symposium on photosynthesis and we wanted to invite Linus Pauling, who was at that time persona non grata to the Atomic Energy Commission. No objection was made to our inviting him, but some officials did not want him around the Oak Ridge area because it was still under heavy security at that time. In order to accommodate our guest speaker, our workshop was held in nearby Gatlinburg at the Mountain View Hotel. With no formal meeting room, the proprietors simply screened across the entrance hall and we sat around the fireplace to begin our first workshop there. In fact, I was told this was the first time that anyone had held a scientific meeting in Gatlinburg!

In the early 1950s, we moved to Gatlinburg, and it became our headquarters for symposia since housing facilities, except for a rather rustic guest house, were nonexistent in Oak Ridge. Thus the tradition was set to conduct scientific meetings in Gatlinburg. It took the hotel owners a few years to catch on that this could be an important source of income to them before the start of the tourist season. Gatlinburg was at that time a rather sleepy town without the number of tourists that now pass through its borders, so we continued to have our meetings away from the Laboratory.

As I observed earlier, it is most important to publish the proceedings of these organized symposia, and for this Cold Spring Harbor had developed its own method of publication. To establish a similar mechanism, I approached a number of publishers — Academic Press and a few others — and they said they didn't want to print

FOREWORD

symposia proceedings. As you know, the whole picture has changed in the last 10 to 15 years.

I talked to Detlev Bronk, who originally organized the Johnson Foundation for Biophysics of the University of Pennsylvania and who had some connection with Wistar Press. He suggested that the Journal of Cellular and Comparative Physiology might publish our proceedings. The symposium proceedings were first issued as supplements to this Journal. As our publications grew in time beyond the capability of Wistar, we switched to other means of publication. Several volumes were translated into the Russian language.

In retrospect, I realize the benefit of consistently bringing out all of our proceedings through one publishing house so that the work is more readily recognized as a distinct series that reflects the interests of and connection with the Biology Division and affords greater distinction than these proceedings now have. Looking over the various titles of the thirty or so symposia that have been organized by the ORNL Biology Division, one can see that they truly demonstrate the broad interest of those who organized them and cooperated toward their successful presentation. It also indicates, by sequence and diversity, the changes that took place over the years. Published from a single source, a kind of history of the Division would have been more clearly evident.

It is my tendency to discuss not only the background and history of our activities in research but also to look ahead to what lies before us. It is, of course, not possible to predict what might come up in the future. Who would have thought ten years ago that monoclonal antibodies would be on the horizon? However, even in the late 1930s it was obvious to some of us that the nucleic acids would become the central theme for the future development of genetics. We surmised that radiation biology might serve as a sort of model for an understanding of certain approaches to the detection of damage created by adverse chemical agents on living cells and their subsequent repair processes.

In regard to the process of repair, which is being discussed with its very latest developments at this symposium, it is well to point out that it had been recognized years ago that a number of the steps in biochemical and biophysical reactions could not proceed without following an established pattern. In the multitude of reactions that are the basis for the function and growth of living systems, some steps could go wrong by the nature of biological activity or through outside interference. If the living system did not have the capability of repair or activation of alternate pathways to accomplish the same thing, we might not be here on this earth today! Many of the above points were discussed as an advanced development some 40 years ago, and they were not very popular when expressed at that time.

I believe we are really at the threshold of an entirely new development of the interplay between genetics, biochemistry, and biophysics that will have a tremendous impact on medical sciences, as well as the industrial development in the area of chemical production. Our new generation of investigators may have different ideas, and I am anxious to see our young colleagues develop their own thinking. They have done a very good job thus far; and I am certain that by improving upon our earlier efforts in these areas, many important new developments in the basic approaches to biological problems will continue to unfold. Some have not yet reached the stage of application and, traditionally, there has always existed a certain lag time between experimental technique developments and their application.

I already mentioned the important developments in repair techniques that have become a most beautiful tool for the understanding of basic radiation effects and now also for the effects caused by chemicals on living cells. What I am especially referring to here are the new advances in genetic engineering, hybridization, plasmid transfer, chromosome exchanges, and the effects of shifting genes. There are probably many others that we haven't yet discovered or that have not yet been reported. I can see that with the development of these new tools, very significant progress will be made leading to a more comprehensive understanding of the basic mechanisms, how the cell functions, and what controls the different steps in development. I am looking forward to the new array of practical discoveries.

The topics of our present symposium include the most important active areas of research in fundamental mutagenesis to date and should be of general interest to all workers in genetics, mutagenesis, carcinogenesis, molecular biology, etc. It is, therefore, a great pleasure for me to see that, through this symposium, the ORNL Biology Division has once again taken the initiative in bringing together these various research areas in an attempt to stimulate more progress in molecular mutagenesis.

                              Alexander Hollaender
                              Associated Universities, Inc.
                              Washington, D.C. 20036

ACKNOWLEDGMENT

Dr. Hollaender's work is supported in part by U.S. Department of Energy Contract EY-76-C-02-0016 with Associated Universities, Inc., Brookhaven National Laboratory.

PREFACE

   Mutagenesis is a fundamental cellular process that contributes to genetic variation, one that is observed to occur both "spontaneously" and as a result of exposure to DNA-damaging agents. Over the past two decades tremendous progress has been made in understanding cellular mechanisms of mutagenesis. Most of these basic studies have been focused on two areas: one involves a molecular description of DNA base-sequence changes in mutants long after the original mutational event, and the other is concerned with genetically controlled cellular processes that convert damaged DNA to an altered damage-free base sequence. This latter approach, largely dependent on use of mutant strains exhibiting aberrant DNA metabolism (repair, replication, recombination, or mutagenesis), has been particularly fruitful in leading to useful working hypotheses and concepts such as error-free and error-prone repair. Unfortunately, precise molecular mechanisms of mutagenesis have remained elusive, partly because adequate tools have not been available to elucidate specific gene action or to determine the mutational consequences of specific DNA lesions.

   Now that recombinant DNA technology has become a powerful tool of molecular biology, we can look forward to a new era of progress when genes controlling mutagenesis and DNA repair can be isolated and used as probes for direct studies of their own expression and regulation. The ability to clone, sequence, and manipulate specific genes has made it possible to enlarge our view of how a cell mutates its genome. For example, transposable elements now appear to be responsible for a significant fraction of spontaneous mutation events involving large insertions, deletions, or other gross DNA rearrangements. Point mutations involving single-base changes seem more likely to arise from damaged DNA, e.g., alkylated bases, pyrimidine dimers, apurinic/apyrimidinic sites, or other lesions with ambiguous base-pairing properties. How cells convert such noncoding lesions into transition, transversion, or frameshift mutations is still a difficult and challenging central problem in mutagenesis.

PREFACE

There now exists a large body of evidence suggesting that induced mutations result from copying errors introduced into DNA during repair and/or replication. Elucidation of error-prone repair or replicative error mechanisms for any organism will depend on biochemical studies of the interaction between cloned genes and purified gene products.

This book is concerned with several of the issues discussed above and is based on the proceedings of the recent "Symposium on Molecular and Cellular Mechanisms of Mutagenesis," held April 5-9, 1981, in Gatlinburg, Tennessee. Organized by the Biology Division of Oak Ridge National Laboratory with support from the U.S. Department of Energy, National Institute of Environmental Health Sciences (with assistance from Drs. F. J. de Serres and M. D. Shelby), and the University of Tennessee - Oak Ridge Graduate School of Biomedical Sciences, the conference had two main objectives: (1) to present a more expanded view of those biological phenomena that contribute to or affect mutagenesis, and (2) to promote greater interaction between investigators working on different problems with different organisms. The symposium was largely successful and included presentations on mismatch repair, DNA damage-inducible repair, the "adaptive-response" repair, site-specific mutagenesis, DNA replication errors in vitro and in vivo, mutagenesis by transposition, repair and mutagenesis of yeast mitochonfrial DNA, and DNA repair in human chromatin.

J.F. Lemontt[*] and W.M. Generoso

Biology Division
Oak Ridge National Laboratory
Oak Ridge, Tennessee 37830

---

[*]J.F. Lemontt's current address is Integrated Genetics, Inc., 51 New York Avenue, Framingham, Nassachusetts 01701.

CONTENTS

SECTION I CELLULAR RESPONSES TO MUTAGENIC AGENTS

Chapter 1 Mutagenesis from a Chemical Perspective: Nucleic Acid Reactions, Repair, Translation, and Transcription 1
    B. Singer

Chapter 2 Regulation and Functions of Escherichia coli Genes Induced by DNA Damage 43
    Graham C. Walker, Cynthia J. Kenyon, Anne Bagg, Stephen J. Elledge, Karen L. Perry, and William G. Shanabruch

Chapter 3 Methylation-Instructed Mismatch Correction as a Postreplication Error Avoidance Mechanism in Escherichia coli 65
    B. W. Glickman

Chapter 4 Cellular Defense Mechanisms Against Alkylation of DNA 89
    Tomas Lindahl, Björn Rydberg, Thomas Hjelmgren, Monica Olsson, and Anita Jacobsson

Chapter 5 Cellular Responses to Mutagenic Agents: A Summary and Perspective 103
    Sohei Kondo

SECTION II MUTAGENESIS AT SPECIFIC SITES

Chapter 6 Mechanisms of UV Mutagenesis in Yeast 109
    Christopher W. Lawrence, Roshan Christensen, and Ann Schwartz

Chapter 7	Site-specific Mutagenesis: A New Approach for Studying the Molecular Mechanisms of Mutation by Carcinogens .... Robert W. Chambers	121
Chapter 8	Single-Stranded Gaps as Localized Targets for In Vitro Mutagenesis .... David Shortle and David Botstein	147
Chapter 9	Mutagenesis at Specific Sites: A Summary and Perspective .... Michael Smith	157

SECTION III MUTATORS, ANTIMUTATORS, AND DNA REPLICATION ERRORS

Chapter 10	Polymerase Infidelity and Frameshift Mutation .... Lynn S. Ripley and Nadja B. Shoemaker	161
Chapter 11	In Vitro Replication of Mutagen-Damaged DNA: Sites of Termination........................ Peter D. Moore, Samuel D. Rabkin, and Bernard S. Strauss	179
Chapter 12	Depurination of DNA as a Possible Mutagenic Pathway for Cells .... Roeland M. Schaaper, Thomas A. Kunkel, and Lawrence A. Loeb	199
Chapter 13	Passive Polymerase Control of DNA Replication Fidelity: Evidence Against Unfavored Tautomer Involvement in 2-Aminopurine-Induced Base-Transition Mutations .... Myron F. Goodman, Susan M. Watanabe, and Elbert W. Branscomb	213
Chapter 14	Mutators, Antimutators, and DNA Replication Errors: A Summary and Perspective .... Maurice J. Bessman	231

SECTION IV TRANSPOSABLE ELEMENTS AND SPONTANEOUS MUTATION

Chapter 15	Low Level and High Level DNA Rearrangements in Escherichia coli .... Ahmad I. Bukhari and Hajra Khatoon	235

CONTENTS

Chapter 16 Mutants of Escherichia coli K12 Which Affect
            Excision of Transposon TN10.................. 245
            Victoria Lundblad and Nancy Kleckner

Chapter 17 Gene Conversion: A Possible Mechanism for
            Eliminating Selfish DNA 259
            Robin Holliday

Chapter 18 Transposons and Illegitimate Recombination
            in Prokaryotes: A Summary and Perspective ... 265
            Nancy Kleckner

SECTION V CHROMOSOMAL AND NONCHROMOSOMAL DNA

Chapter 19 Mutagenesis and Repair in Yeast
            Mitochondrial DNA 273
            E. Moustacchi and M. Heude

Chapter 20 Alterations in Chromatin Structure During
            DNA Excision Repair 303
            Michael W. Lieberman

Chapter 21 New Approaches to DNA Damage and Repair:
            The Ultraviolet Light Example 315
            William A. Haseltine, Lynn K. Gordon,
            Christina Lindan, Judith Lippke,
            Douglas Brash, Kwok Ming Lo, and
            Brigitte Royer-Pokora

Chapter 22 Chromosomal and Nonchromosomal DNA:
            A Summary and Perspective 333
            Ada L. Olins

SECTION VI MUTAGENESIS: FUTURE DIRECTIONS

Chapter 23 Comparison of the Induction of Specific
            Locus Mutations in Wild-Type and Repair-
            Deficient Strains of Neurospora Crassa 335
            F. J. de Serres

Chapter 24 Mammalian Mutagenesis: Future Directions 345
            Liane B. Russell and E. G. Bernstine

Chapter 25 Perspectives in Molecular Mutagenesis 361
            John W. Drake

CONTRIBUTORS 379

INDEX 383

CHAPTER 1

MUTAGENESIS FROM A CHEMICAL PERSPECTIVE: NUCLEIC ACID

REACTIONS, REPAIR, TRANSLATION, AND TRANSCRIPTION

B. Singer

Department of Molecular Biology and Virus Laboratory
University of California, Berkeley, California 94720

SUMMARY

Simple directly acting alkylating agents can be classified in terms of their mutagenic efficiency and their chemical reactivity. The most mutagenic are the N-nitroso compounds and these have a preference for reacting with nucleic acid oxygens in vitro and in vivo. In contrast, the alkyl sulfates are generally poor mutagens and react almost exclusively with base nitrogens. Other classes of alkylating agents also show correlations between oxygen reaction and mutagenicity. Ethylating agents are more oxygen-specific than the analogous methylating agent and, in a substantial number of cases, also more mutagenic at lower levels of treatment.

Sites of substitution by ethyl nitroso compounds (e.g., N-ethyl-N-nitrosourea, N-ethyl-N'-nitro-N-nitrosoguanidine) in double-stranded nucleic acids are as follows: phosphate $\gg$ N7-G > O^2-T, O^6-G > N3-A > O^4-T, O^2-C > other N. In single-stranded nucleic acids the reactivity of the O^2 of C, N1 of A, and N3 of U, T, or C is considerably greater. Certain of these derivatives have been shown in in vitro transcription or ribosome binding studies to mispair; namely, O^2-alkyl T, O^4-alkyl T, O^6-alkyl G, O^2-alkyl C, 3-alkyl C, 3-alkyl U or T and 1-alkyl A. In all cases, nonspecific mispairing occurs with high frequency. During in vivo replication such errors are probably relatively rare but nevertheless postulated to occur.

There is evidence that various types of repair enzymes exist in bacteria and mammalian cells that can remove, to varying extents, N-3 and N-7 alkyl purines, O^6-alkyl G, O^2-alkyl T, O^4-alkyl T, and O^2-alkyl C. Phosphotriesters in DNA appear to be very stable.

When substitution occurs on a site necessary for basepairing or in a site causing steric hindrance or electronic shielding of Watson-Crick sites, the result is ambiguity in transcription rather than termination. It is hypothesized that any or all unrepaired promutagenic lesions can be expressed as errors during replication.

Other mutagens described in terms of their chemical reactions and repair include simple nonalkylating agents, most of which change basepairing due to deamination or tautomeric shifts and the metabolic products of aromatic amines and polyaromatic hydrocarbons.

INTRODUCTION

Mutagenesis is defined as a heritable change that can occur through indirect, as well as direct, change in the genetic message. This paper, however, will focus on the direct effects of chemical modification on DNA or viral RNAs that are both genome and messenger. Since man is the species we are most concerned about, the effect of modification on mammalian cells or whole animals will be stressed whenever possible.

Many of the types of chemicals which have been assessed in terms of their mutagenicity in mammalian systems (Bartsch et al., 1980; Maher and McCormick, 1978; Montesano and Bartsch, 1976; Pienta, 1980) and for which there are data on the mechanism of chemical modification of nucleic acids are shown in Figs. 1 and 6.

CHEMICAL REACTIONS OF SIMPLE ALKYLATING AGENTS

A large group of mutagens are simple alkylating agents (Fig. 1), although these differ greatly in their mutagenicity. For this reason, we have been working for a number of years to elucidate the chemical reactions of nucleic acids with the "good" mutagens (e.g., N-nitroso compounds), as compared to the "poor" mutagens (e.g., alkyl sulfates).

Single-stranded (ss) homopolyribonucleotides were first used as models for RNA in most of these studies. Significant differences were found between methylating agents and the analogous ethylating agents, suggesting that the greater mutagenicity of, for example, ethyl methanesulfonate (EtMS) compared to methyl methanesulfonate (MeMS) could be due to reaction with oxygens or exocyclic amino groups. However, in RNA the exocyclic amino groups were not measurably reactive, while all oxygens, including phosphodiesters and ribose, could be modified with ethylnitrosourea (EtNU) (Table 1) (Singer, 1976, 1977). Both EtNU and ethyl nitrosoguanidine (ENNG) ethylated oxygens predominately. Two new derivatives of cytosine (C)

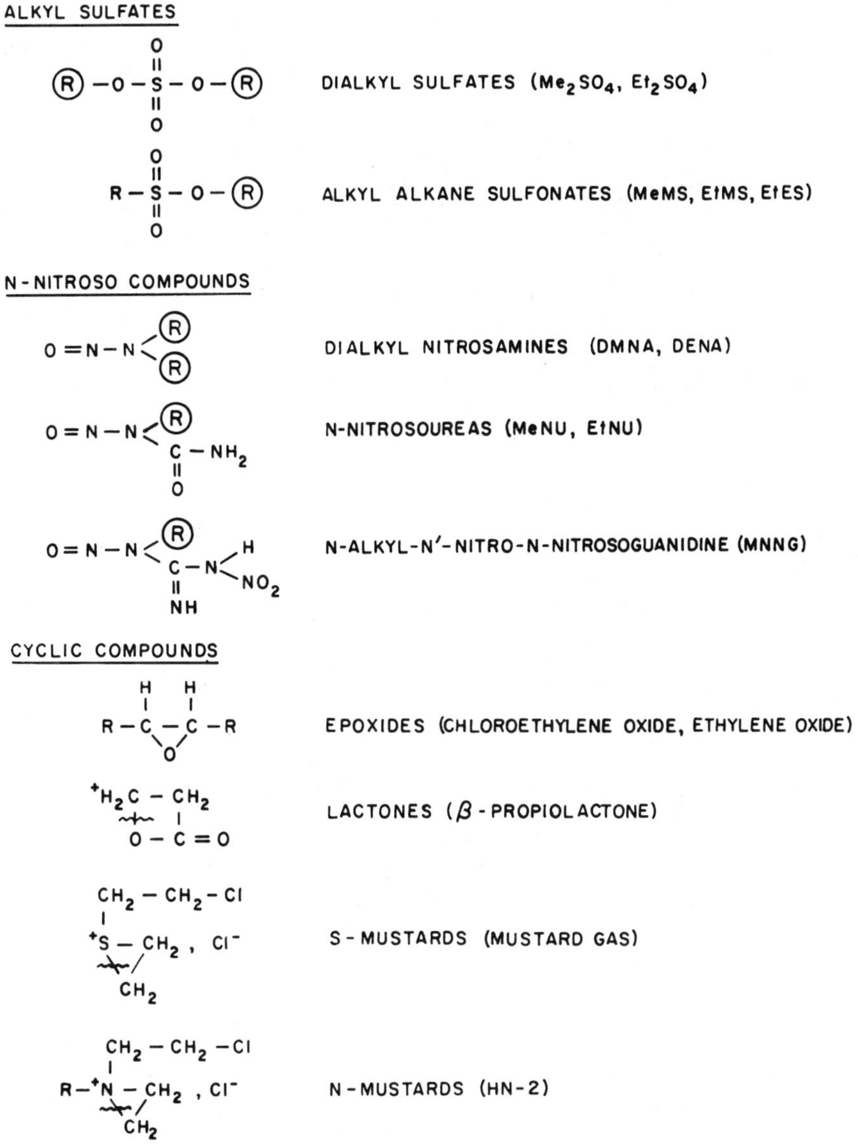

ALKYL SULFATES

$$\text{R}-\text{O}-\overset{\overset{\text{O}}{\|}}{\underset{\underset{\text{O}}{\|}}{\text{S}}}-\text{O}-\text{R}$$ DIALKYL SULFATES (Me_2SO_4, Et_2SO_4)

$$\text{R}-\overset{\overset{\text{O}}{\|}}{\underset{\underset{\text{O}}{\|}}{\text{S}}}-\text{O}-\text{R}$$ ALKYL ALKANE SULFONATES (MeMS, EtMS, EtES)

N-NITROSO COMPOUNDS

$O=N-N\big\langle\begin{smallmatrix}R\\R\end{smallmatrix}$ DIALKYL NITROSAMINES (DMNA, DENA)

$O=N-N\big\langle\begin{smallmatrix}R\\C-NH_2\\\|\\O\end{smallmatrix}$ N-NITROSOUREAS (MeNU, EtNU)

$O=N-N\big\langle\begin{smallmatrix}R\\C-N\langle\begin{smallmatrix}H\\NO_2\end{smallmatrix}\\\|\\NH\end{smallmatrix}$ N-ALKYL-N′-NITRO-N-NITROSOGUANIDINE (MNNG)

CYCLIC COMPOUNDS

$\text{R}-\overset{H}{\underset{}{C}}-\overset{H}{\underset{}{C}}-\text{R}$ with O bridge   EPOXIDES (CHLOROETHYLENE OXIDE, ETHYLENE OXIDE)

$^+H_2C-CH_2$ / $O-C=O$ LACTONES (β-PROPIOLACTONE)

^+S with CH_2-CH_2-Cl and CH_2-CH_2, Cl^- S-MUSTARDS (MUSTARD GAS)

$R-{}^+N$ with CH_2-CH_2-Cl and CH_2-CH_2, Cl^- N-MUSTARDS (HN-2)

Fig. 1. Structural formulas and abbreviations of some simple, direct-acting alkylating agents.

Table 1. Alkylation of Single-stranded Nucleic Acids In Vitro[a]

Alkylation Site	Percentage of Total Alkylation[b]					
	Me_2SO_4	MeMS	MeNU	Et_2SO_4	EtMS	EtNU
Adenine						
N-1	13.2	18	2.8	11	8	2
N-3	2.6	1.4	2.6	3	1	1.2
N-7	3.1	3.8	1.8	3	3	0.6
Guanine						
N-3	∼0.1	∼1	0.4	2	1	0.5
O^6	<0.2	nd	3	2	1	7
N-7	62	68	69	62	77	10
Uracil/thymine						
O^2				2		6
N-3				nd		
O^4				1		4
Cytosine						
O^2				2		5.5
N-3	9.5	10	2.3	11	5	1.7
Diester	<2	2	∼10	6	10	65
Ribose						12

[a] Analyses were from experiments using DNA from M13 phage and RNA from TMV, yeast, HeLa cells, animal ribosomes, and μ2 phage. Much of the data was compiled by Singer (1975). Other references are Singer (1976) and Singer and Fraenkel-Conrat (1975).

[b] The absolute amount of alkylation varied greatly but the proportion of derivatives was not noticeably affected; nd indicates that the derivative was not detected.

and uracil (U) were found, O^2-alkyl C and O^2-alkyl U, respectively, and one previously marginally detected, O^4-alkyl U, was also a significant product.

Turning to double-stranded (ds) nucleic acids, all the same reaction products could be quantitated, and again EtNU (and ENNG) reacted to a high extent with the O^6 of guanine (G), the O^2 of C and thymine (T), the O^4 of T and phosphodiesters (Singer, 1976, 1977) (Table 2). The specificity of eight reagents toward oxygens apparently correlates with the reported mutagenicities in various systems:

Table 2. Alkylation of Double-stranded Nucleic Acids In Vitro[a]

Alkylation Site	Percentage of Total Alkylation[b,c]					
	Me_2SO_4	MeMS	MeNU	Et_2SO_4	EtMS	EtNU
Adenine						
N-1	1.9	3.8	1.3	2	1.7	0.2
N-3	18	10.4	9	10	4.9	4.0
N-7	1.9	(1.8)	1.7	1.5	1.1	0.3
Guanine						
N-3	1.1	(0.6)	0.8	0.9	0.9	0.6
O^6	0.2	(0.3)	6.3	0.2	2	7.8
N-7	74	83	67	67	65	11.5
Thymine						
O^2			0.11	nd	nd	7.4
N-3			0.3	nd	nd	0.8
O^4			0.4	nd	nd	2.5
Cytosine						
O^2	(nd)	(nd)	0.1	nd	nd	3.5
N-3	(<2)	(<1)	0.6	0.7	0.6	0.2
Diester		0.8	17	16	13	57

[a] Analyses are from experiments using DNA from salmon sperm, calf thymus, salmon testes, rat liver and brain, human fibroblasts, and HeLa cells. Some of the data were compiled by Singer (1975). Other key references are: Beranek et al. (1980); Goth and Rajewsky (1974); Lawley et al. (1975); Newbold et al. (1980); Shackleton et al. (1979); Singer (1976, 1979); Sun and Singer (1975); Swenson and Lawley (1978).
[b] The absolute amount of alkylation varied greatly but the proportion of derivatives was not noticeably affected; nd indicates that the derivative was not detected.
[c] Parentheses indicate either that there was a single determination or that two very different results were averaged.

EtNU, ENNG > methylnitrosourea (MeNU), MNNG > EtMS > MeMS > Et_2SO_4. While the metabolically activated nitrosamines are not included in the Tables, data for their reaction in vivo indicates that they resemble the nitrosoureas in oxygen specificity.

There are several points in Tables 1 and 2 which should particularly be noted when discussing chemical reactivity. First the Tables deal with percentage of total alkylation, or proportion, rather than absolute amounts. Methylation is about 20 times more efficient in reacting with nucleic acids (in vitro or in vivo) than ethylation under the same reaction conditions. Thus, 6.3% O^6-MeG (MeNU) represents in moles about 15 times the amount of O^6-EtG (EtNU) (Table 2). It cannot be said, therefore, that quantitatively methylation of oxygens occurs to a lower extent than ethylation. Yet the biological effects of EtNU are greater than with MeNU at similar dose levels (Montesano and Bartsch, 1976).

Regarding phosphotriesters, which are clearly the major product of EtNU reaction, here too quantitatively there are as many or more methyltriesters than ethyltriesters, as is also the case when comparing triesters formed by MeMS and EtMS. All these comparisons, of course, are only true when comparing equimolar reactions under identical conditions of time, temperature, pH, etc., and in vivo there are the additional factors of repair and errors in replication that may differ for methyl and ethyl substituents.

Another point to be noted is that both the N-1 of adenine (A) and N-3 of C are hydrogen-bonded in ds nucleic acids and, in contrast to ss nucleic acids, are not expected to be reactive. However, a definite and unexpectedly high reactivity was observed for these sites (Table 2). Having convinced ourselves that the ds nucleic acids we were studying were completely ds, we then found that alkylation of hydrogen-bonded sites at, and even below, 37°C occurred as a result of thermal denaturation (Bodell and Singer, 1979). This finding, also described by others using different techniques (Lukashin et al., 1976; Mandal et al., 1979; McGhee and Von Hippel, 1977), opens up many possibilities and must be considered when looking for products of nucleic acid-mutagen interaction. In a later Section, the potential mutagenic effect of modification of the base-paired positions, the N-1 of A or N-3 of C or U/T, will be discussed. Among the other positions on bases that are involved in hydrogen-bonding in a Watson-Crick structure are the O^6 of G, O^4 of U/T, and O^2 of C. These, however, have a free electron pair as shown in Fig. 2, allowing alkylation to take place. This is well illustrated by the data for EtNU reaction with ss and ds nucleic acids. Here, in contrast to the hydrogen-bonded nitrogens, N-1 of A or N-3 of C, there is no significant difference in oxygen reactivity as a result of strandedness.

Both ss or ds polynucleotides have proven very useful in the identification of reaction products and in testing hypotheses, such as reactivity of hydrogen-bonded sites. Much of our technical expertise, such as separation methods, stability measurements, chemical parameters, etc., has come from the use of synthetic

MUTAGENESIS FROM A CHEMICAL PERSPECTIVE

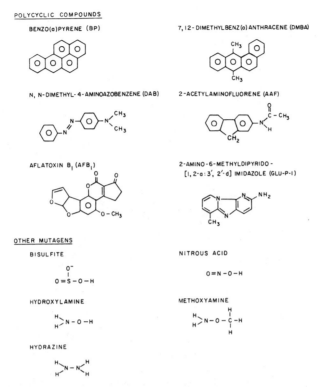

Fig. 2. Participation of electron pairs in basepairing. The ovals represent hydrogen bonds in a Watson-Crick structure. Solid pairs of dots represent electron pairs. Electron pairs inside ovals, when involved in hydrogen bonding, are unreactive. Others may react when in ds structures, if physically accessible. Note that the oxygens all have a free electron pair, reactive both in ss and ds polymers (Tables 1, 2). The three exocyclic amino groups may rotate, as indicated by arrows, so that the two hydrogens are equivalent. A substituent may, therefore, interfere with normal H-bonding if rotated into the base-pairing side (Singer and Spengler, 1981).

polynucleotides. Figure 3 shows some experiments that led to unequivocal identification of many alkyl derivatives. It must be noted, however, that a number of derivatives, clearly identified in polymer reactions with EtNU (Fig. 3), are not found in RNA and DNA reactions. These are shown on the right-hand side of Fig. 4 and include N^6-alkyl A, N^4-alkyl C, and N^2-alkyl G. The latter alkylation site has been found only as O^6, $-N^2$-diethyl G after EtNU reaction, or as N^2-benzylguanosine, reported for guanosine reaction with N-nitroso-N-benzylurea (Moschel et al., 1980). In no experiment could we identify a C-8 substituted purine.

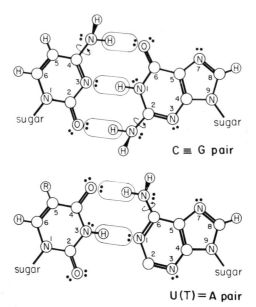

Fig. 3. Separation of products from neutral, aqueous reaction of EtNU with polynucleotides. A, HPLC fractionation of depurinated, thermally denatured poly(rA)·poly(rU) (Bodell and Singer, 1979). B,C, Chromatographic separation of products of poly(U) following hydrolysis with ribonuclease and phosphatases (B), and snake venom phosphodiesterase and phosphatases (C) (Kuśmierek and Singer, 1976). D, Chromatographic separation of products from ds poly(dG)·poly(dC). Depurination was followed by enzyme digestion for analysis (Singer et al., 1978). E, Chromatographic separation of enzyme-digested ethylated poly(rC) (Singer, 1979).

There thus appears little overlap regarding reactive sites between the products of many polycyclic aromatic hydrocarbon (PAH) compounds with nucleic acids and those formed by simple alkylating agents. This will be discussed in a later Section. An appropriate caution is that results with synthetic polymer data cannot always be extrapolated directly to nucleic acids. However, no reaction has been found to occur in nucleic acids that does not occur in model experiments.

If polymers are not the same as nucleic acids in terms of reactivity, then are test tube experiments valid as predictors of in vivo reactions? Data on reactions in animals and cells from our and other

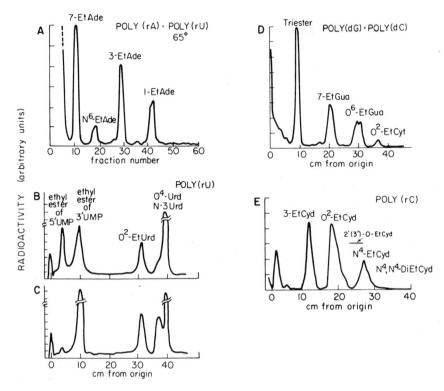

Fig. 4. Sites of reaction of N-nitroso compounds with nucleic acids or polynucleotides in neutral aqueous solution. All derivatives shown are formed with EtNU. The derivatives on the right side have been found only after reaction with synthetic polynucleotides. References are in Tables 1, 2, and Fig. 3.

laboratories are very similar to in vitro data in Table 2, although the extent of in vivo reaction is 2 to 3 orders of magnitude less than in many in vitro experiments. It has not yet been mentioned, but almost all the data in vivo (and usually in vitro) are obtained by using radioactive alkylating agents. The limitation of specific activity and dose when studying reactions in mammalian cells or whole animals makes the results more variable when trying to quantitate minor derivatives. The most complete data illustrating that nucleic acid reactions are independent of milieu or extent of reaction are shown in Table 3.

We would conclude that the reactive sites in EtNU-treated DNA are: phosphate diester $\gg$ N-7 G $> O^2$-T, O^6-G $>$ N-3 A $> O^2$C, O^4-T $>$ other N-.

CHEMICAL REACTIONS OF CYCLIC ALKYLATING AGENTS

Cyclic alkylating agents (Fig. 1) comprise a number of unrelated mutagens and carcinogens. Representative identified nucleic acid derivatives are shown in Fig. 5. The alkylating agents studied earliest were the N- and S-mustards, the former of which (HN-2) is also used therapeutically. Brookes and Lawley (1960, 1961, 1963), in a pioneering series of studies, found that mustard gas reacts with the N-7 of G and both a monoadduct and a cross-linked adduct are formed. Later, derivatives resulting from reaction of the N-1 of A and N-3 of C were identified so that it now appears that S-mustard resembles alkyl sulfates in its reactions except that, being bifunctional, inter- and intrastrand cross-linking also occurs. N-mustard also cross-links through the N-7 of G (Chun et al., 1969).

β-Propiolactone also reacts with the N-7 of G (Roberts and Warwick, 1963) and N-1 of A (Maté et al., 1977). There is less information on other reactions although it might be predicted that the N-3 of A could also be modified.

Aliphatic epoxides such as ethylene and propylene oxide, widely used as sterilants and in industry (Fishbein, 1969), are mutagenic, and ethylene oxide has been added recently to the small list of human carcinogens. The epoxides resemble typical alkylating agents and react with DNA and RNA at the N-7 of G and the N-1 and N-3 of A (Lawley and Jarman, 1972), forming the hydroxyethyl or hydroxypropyl derivatives. No O^6-alkyl derivative was detected using propylene oxide. N-7 alkyl G and N-3 alkyl A are also the principal products from dimethylsulfate-treated DNA (Table 2), which has also been termed a human carcinogen, but the known reactions of both dimethylsulfate and ethylene oxide with DNA are not those that would be likely to cause mutations. This point certainly deserves further investigation since the identified derivatives are the same as those formed by alkylating agents not now considered to be significantly mutagenic or carcinogenic.

Table 3. Products of Ethylnitrosourea Reaction with DNA

State of DNA	Percentage of Total Ethylation											
	Adenine			Guanine			Thymine			Cytosine		Diester
	1[a]	3	7	3	O6	7	O2	3[a]	O4	O2	3[a]	
In vitro[b]	0.2	4.0	0.3	0.6	7.8	11.5	7.4	0.8	2.5	3.5	0.22	57
Cell culture												
Human fibroblasts[c]		4.5			9.2	13.6	7.1		2.1	4.5		51
Fetal rat brain[c]		4.7	1.3	1.7	7.6	12.3	7.0		4.3	3.4		49
V-79[d]		5.0	1.3	1.5	9.1	12.3	8.5		1.4	1.8		66
HeLa[e]	0.1	4.6			7.5	17.0		0.4			0.3	56
In vivo												
BDIX rats												
Liver (10 days)[f]		2.7			5.7	14.0	7.2		2.4	1.4		61
Brain (10 days)[f]		4.6			5.9	12.3	7.8		2.0	0.7		59
Liver (adult)[g]		4.1			5.7	13.0						
C-58BL mice (8–10 weeks)												
Liver[h]		3.6	0.5	1.3	6.5	13.0				2.2		56
Brain[h]		3.8	0.5	1.0	7.4	13.0				1.7		48

[a] Position hydrogen bonded in double-stranded nucleic acid.
[b] Data from Table 2.
[c] Singer et al. (1978).
[d] Swenson et al. (1980).
[e] Sun and Singer (1975). Data for phosphotriesters have been recalculated since determining that the original value also represented the O-ethylpyrimidines.
[f] Singer et al. (1981).
[g] Goth and Rajewsky (1974). Data have been recalculated, using 13% for 7-EtG.
[h] Frei et al. (1978). Data have been recalculated, using 13% for 7-EtG.

	Nucleic acids *in vivo* and *in vitro*			Synthetic polynucleotides *in vitro*	
URIDINE OR THYMIDINE	N-3	O^2	O^4		
CYTIDINE				N^4	
GUANOSINE	O^6	N-7	N-3	N-1	$N^2 + O^6$
ADENOSINE	N-1			N^6	
BACKBONE	Phosphate	Ribose			

Chloroethylene oxide is a highly reactive and mutagenic metabolite of vinyl chloride (Barbin et al., 1975), also a human carcinogen. Like all epoxides, the ring opens easily to form chloroacetaldehyde. Both chloroethylene oxide and chloroacetaldehyde react with polynucleotides to form the cyclic etheno derivatives, εA, εC, and εG (Fig. 5). εA and εC have been identified as products in the liver of rats given vinyl chloride in drinking water (Green and Hathway, 1978) or by inhalation (Laib and Bolt, 1978). This work does not exclude that the third known derivative, $1,N^2$-ethanoguanine (Sattsangi et al., 1977), is also a product. The bromine analogue, bromoacetaldehyde, appears to be chemically identical but is evidently of very low mutagenicity (Kayasuga-Mikado et al., 1980).

Another bifunctional mutagen which adds a ring to a nucleoside is glycidaldehyde (Goldschmidt et al., 1968; Van Duuren and Loewengart, 1977). The structure of the modified G derivative from reaction with calf thymus DNA is shown in Fig. 5. This mutagen, like chloroacetaldehyde, should also form cyclic derivatives with other nucleosides.

The reactions of the widely used antineoplastic agent 1,3-bis-(2-chloroethyl)-1-nitrosourea (BCNU) have been studied in synthetic polynucleotides and three new derivatives identified: 3-(β-hydroxyethyl)C, $3,N^4$-ethano C, and 7-(β-hydroxyethyl)G (Ludlum et al., 1975). Cross-links have been measured indirectly (Bradley et al., 1980), and these could result from initial reaction at either the N^4 of C or the N-7 of G, which would make the entire molecule an alkylating agent. A diguanyl derivative, 1,2-(diguan-7-yl)ethane, has been found in model studies (Gombar et al., 1979), but this may not be the only cross-link. The fluorine analogue FCNU would be likely to react with nucleic acids in a similar manner but, as yet, only the N-3 derivative of C has been reported (Tong and Ludlum, 1978).

Fig. 5. Examples of mutagenic reaction resulting in cyclic or cross-linked derivatives. a, The glycidaldehyde product with guanosine (Goldschmidt et al., 1968; Van Duuren and Loewengart, 1977). b, $3,N^4$-Ethano C from BCNU reaction with cytidine (Ludlum et al., 1975). c, Etheno derivatives from reaction of cytidine, adenosine, and guanosine with chloroacetaldehyde (Barrio et al., 1972; Sattsangi et al., 1977). d, Cross-linked N-7 G resulting from reaction with nitrogen mustard (Brookes and Lawley, 1961). e, Two types of cross-links, G-G or G-A, resulting from nitrous acid reaction (Shapiro et al., 1977).

CHEMICAL REACTIONS OF AROMATIC AMINES AND POLYCYCLIC HYDROCARBONS

Aromatic amines and polycyclic hydrocarbons (Fig. 6) undergo complex enzymatic modifications to form some mutagenic metabolites which react with nucleic acids. [For reviews of specific metabolic pathways see Brookes (1977), Kriek and Westra (1979), Miller (1978), Sims and Grover (1974)]. Although there has been a great deal of progress in determining the ultimate mutagens and their products, particularly in the cases of benzo[a]pyrene (BP, acetylaminofluorene (AAF), and aflatoxin B_1 (AFB_1), much remains to be learned. Some of the nucleic acid derivatives identified after activation are shown in Figs. 7 and 8. There is a striking difference between these adducts and those formed by simple alkylating agents or bifunctional reagents, as illustrated in Table 4. This is not simply that the substituent is large and aromatic, but the favored sites of reaction are the exocyclic amino groups, particularly the N^2 of G and the C-8 of G. AFB_1 is an exception. Here, although the ultimate metabolite is bulky, reaction takes place on the N-7 of G and, to a lesser extent, the O^6 of G. Phosphotriesters appeared to be a product of 7,12-dimethylbenz[a]anthracene (DMBA) reaction with DNA. However, this evidence was indirect and more recently it was reported that triesters did not occur in either DNA or RNA treated with BP diol-epoxide (BPDE) (Gamper et al., 1980).

A caution: the derivatives most important in mutation may not yet be known and, particularly, species differences in metabolism may be of the greatest importance. The derivatives in Figs. 7 and 8 are not the only ones formed but are chosen to illustrate the various types of modifications found and identified.

SIMPLE, NONALKYLATING MUTAGENS

Simple, nonalkylating mutagens (Fig. 6) might be turned "classical" mutagens since their predominant effect is to change base-pairing by deamination (nitrous acid, bisulfite), or addition to the N^4 of C (hydroxylamine, methoxyamine, hydrazine), causing a shift in the tautomeric equilibrium (chemistry reviewed by Brown, 1974). The major products of each of these mutagens are shown in Fig. 9. In addition, bisulfite, hydroxylamine, methoxyamine, and hydrazine can add to the 5,6 double-bond of C or (except for methoxyamine) to the 5,6 double-bond of U. Hydroxylamine and methoxyamine also react with the N^6 of A (Brown and Osborne, 1971; Budowsky et al., 1975), which is a potentially mutagenic reaction. While nitrous acid primarily deaminates C→U and A→hypoxanthine (HX), deamination of G→xanthine (X) and evidence for cross-links have also been reported (Shapiro et al., 1977). Both these latter reactions are inactivating rather than mutagenic. The two types of cross-links identified in nitrous acid-treated DNA are shown in Fig. 9.

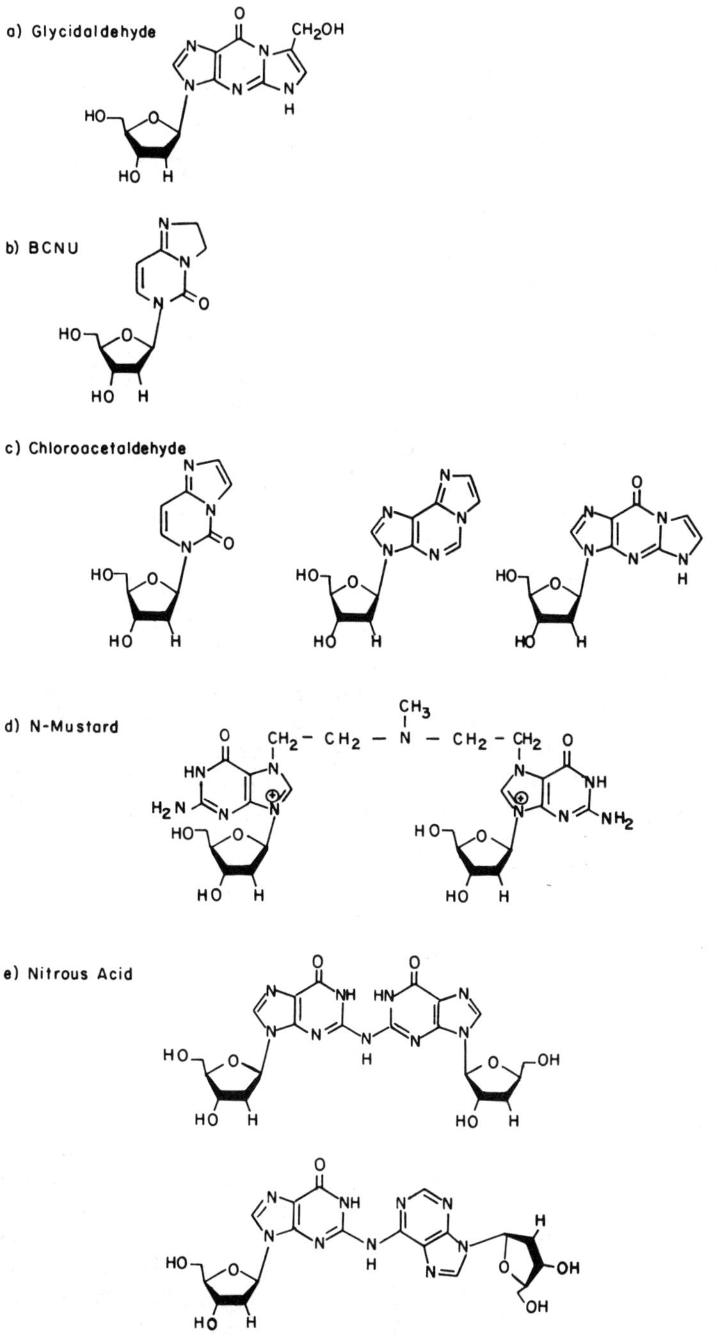

MUTAGENESIS FROM A CHEMICAL PERSPECTIVE 15

Fig. 6. Structural formulas and abbreviations of some metabolically activated polycyclic aromatic hydrocarbons (top) and of simple nonalkylating mutagens (bottom).

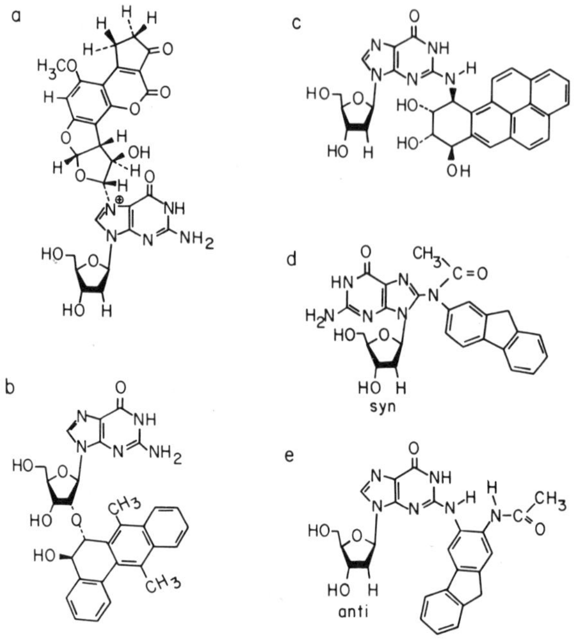

Fig. 7. Examples of reaction products of metabolically activated PAH with nucleic acids. Stereochemistry, where indicated, is drawn according to the specific authors. a, AFB$_1$ N-7 of G (Autrup et al., 1979; Essigmann et al., 1977; Lin et al., 1977; Martin and Garner, 1977;). b, DMBA 2'0 of ribose (Kasai et al., 1977). c, BP N^2 of G (Jeffrey et al., 1976b; Koreeda et al., 1976; Nakanishi et al., 1977; Weinstein et al., 1976). Derivatives of the N^6 of A (not shown) have also been identified (Jeffrey et al., 1979). d, AAF C-8 of G (Kriek et al., 1967). e, AAF N^2 of G (Westra et al., 1976).

EFFECT OF NUCLEIC ACID STRUCTURE ON REACTIVITY

Conformation and secondary structure of nucleic acids play a major role in chemical reactivity (Brown, 1974; Singer and Fraenkel-Conrat, 1969). That is, in general mononucleotides are more reactive than ss homopolynucleotides, which are more reactive than ss nucleic acids, which in turn are more reactive than ds nucleic acids. In particular, it has been found that hydroxylamine, methoxyamine, and bisulfite are poor reagents for ds nucleic acids even though their preferred sites of modification are not hydrogen-bonded. Bisulfite and formaldehyde are unreactive toward the N^4 of C or the N^6 of A, respectively, in ds polymers, and this selectivity has been used as a test for the extent of ds structure (Fraenkel-

Conrat, 1954; McGhee and von Hippel, 1977; Shapiro et al., 1972). Chloroacetaldehyde is much less reactive in a ds DNA compared to denatured DNA but is nevertheless a mutagen (Kimura et al., 1977).

All DNA when being transcribed or replicating is ss, and ss regions are more reactive than ds regions. Therefore, it follows that reaction on ss segments can be more extensive and on different sites (i.e., N-1 of A, N-3 of C) than on ds segments. In fact, DNA treated with alkylating agents in vivo is found to react at what are normally base-paired sites (Table 3).

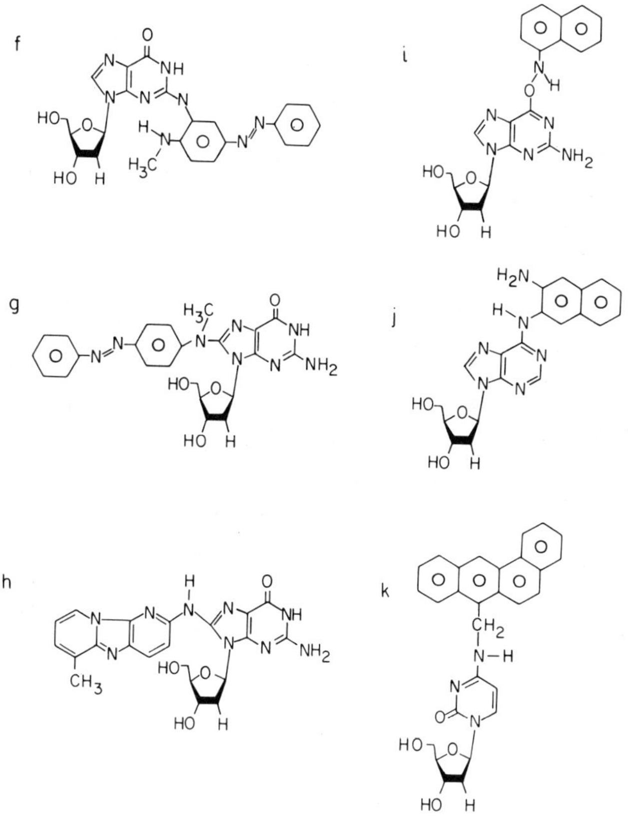

Fig. 8. Examples of reaction products of additional metabolically activated PAH with nucleic acids. f, DAB N^2 of G (Beland et al., 1980; Tarpley et al., 1980). g, DAB C-8 of G (Beland et al., 1980; Tarpley et al., 1980). h, Glu-P1 C-8 of G (Hashimoto et al., 1980). i, 1-NA O^6 of G (Kadlubar et al., 1978). j, 2-NA N^6 of A (Kadlubar et al., 1980). k, 7 BMBA N^4 of C (Dipple et al., 1971; Rayman and Dipple, 1973).

Table 4. Summary of Sites of Reaction of Nucleic Acids with Various Mutagens or Their Metabolites[a]

Proximate Mutagens[b]	Sites of Modification[c]													
	Adenine				Guanine						Cytosine			PO$_4$	Ribose
	1	3	N^6	7	1	N^2	3	O^6	C-8	7	O^2	3	N^4		
EtNU	✓	✓		✓			✓	✓		✓	✓	✓		✓	✓
S-Mustard	✓	✓		✓			✓			✓		✓			
β-propiolactone	✓									✓					
Ethylene oxide	✓	✓						–		✓	✓				
Propylene oxide	✓	✓						–		✓	✓				
BCNU	✓	✓	✓		✓			✓		✓		✓	✓		
Chloroacetaldehyde	✓		✓		✓	✓						✓	✓		
Glycidaldehyde	✓				✓	✓									
Glu-P-1						✓			✓						
AAF						✓			✓	[✓]					
BP			✓			✓								–	
DAB									✓	✓					
DMBA			✓			✓									
7BMBA			✓			✓				✓					✓
2-Naphthylamine									✓				✓		
1-Naphthylamine								✓	✓						
AFB$_1$										✓					

[a] Data are compiled from references in Figs. 5–8 and in the text. Other relevant references are Beland (1978), Carrell et al. (1981), Jeffrey et al. (1976a), Nakanishi et al. (1980), and Tong et al. (1981).
[b] Some ultimate mutagens have been identified and some are reactive without metabolic activation. Abbreviations are given in Figs. 1, 6, and 8 and in the text.
[c] Thymine and uracil are not included since they have been found reactive only with simple alkylating agents (Tables 1,2). Symbols: ✓, site found to be reactive without judgment as to the extent of reaction; [✓], lack of extensive evidence for reactivity; –, reactivity looked for but not seen.

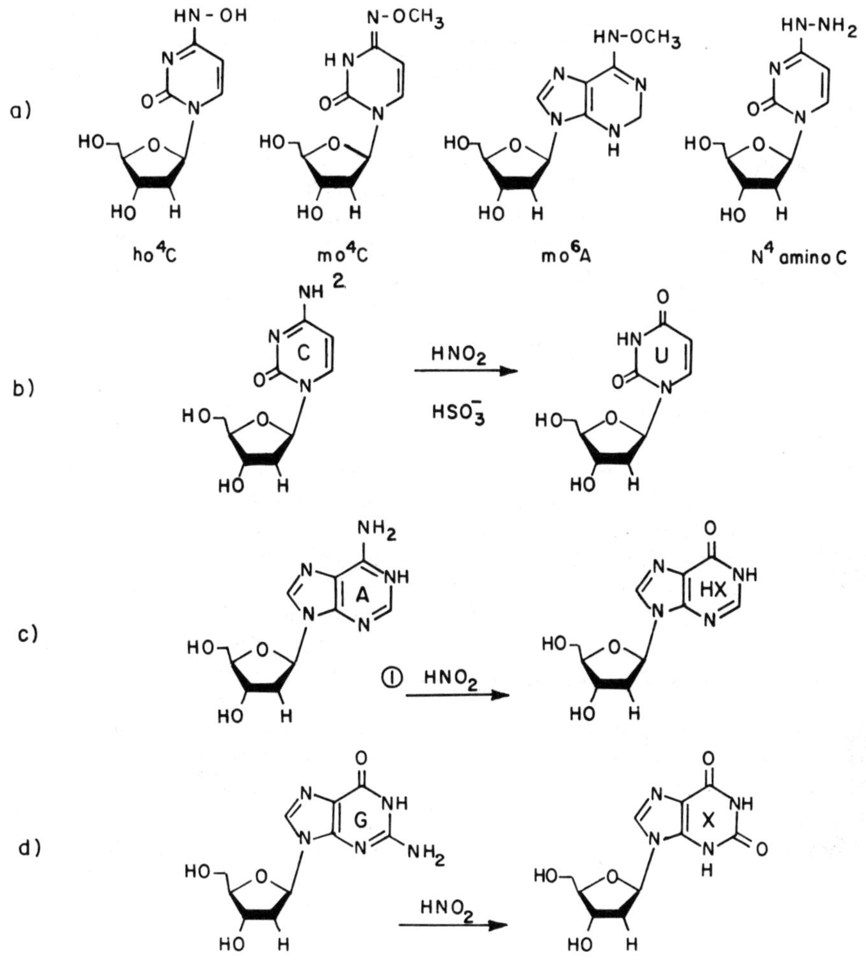

Fig. 9. Examples of reactions with mutagens producing shifts in the tautomeric equilibrium or deamination. a, Reactions of hydroxylamine (ho⁴C), methoxyamine (mo⁴C, mo⁶A) or hydrazine (N⁴-amino C) (Brown, 1974; Brown et al., 1966; Budowsky et al., 1969, 1972). b, Deamination of C to U by nitrous acid (Schuster and Schramm, 1958) or bisulfite (Hayatsu et al., 1970; Shapiro et al., 1970). c, Deamination of A to hypoxathine by nitrous acid (Schuster and Schramm, 1958). d, Deamination of G to xanthine by nitrous acid (Schuster and Schramm, 1958).

Simple mutagens have been used with RNA viruses and the effect on the genomic messenger is complex. This is due not only to secondary structure but to the protein coat, which can be selectively protective of specific base sites or can itself react causing inactivation, e.g., cross-links can be formed between nucleic acid and protein. A good discussion of this problem is given by Fraenkel-Conrat (1981).

Another class of mutagens are the bulky aromatic compounds which also exhibit a preference for reacting with ss nucleic acids or ss regions of base-paired nucleic acids. AAF has been most extensively studied (reviewed by Grunberger and Weinstein, 1979a). It reacts best with ss nucleic acids or polymers. BP, when metabolized, is more reactive in cell culture with RNA than DNA (Ivanovic et al., 1978).

QUANTITATION OF REPAIR OR REMOVAL OF MODIFIED NUCLEOSIDES

This Section is generally restricted to ds DNA since there are no known enzymes which recognize damage to RNA or ss DNA. Chemical modifications in an RNA virus are, therefore, either inactivating or mutagenic, while the effects in ds DNA are more variable. There are many excellent reviews on the general subject of DNA repair in eukaryotes (Hanawalt et al., 1979; Maher and McCormick, 1978; Roberts, 1978; Strauss et al., 1978; Verly, 1980) and only conclusions or data not yet in review will be discussed.

Most of the well-known and well-characterized repair enzymes have been isolated from prokaryotes, notably Escherichia coli (reviewed by Lindahl, 1979). These include enzymes repairing UV dimers and apurinic/apyrimidymic sites; glycosylases acting on 3-alkyl A, ring-opened 7-methyl G, deaminated A and C; and an enzyme dealkylating O^6-alkyl G by transfer of the alkyl group to an amino acid.

In vivo many nucleic acid derivatives discussed in this paper are found to be removed/repaired at rates varying according to the chemical stability of the derivative, amount of derivative, the cell or animal species, and the specific organ or cell studied. These factors affect the half-lives of derivatives and for this reason specific data are not given in this paper.

Chemical evidence for disappearance of mutagen products has been shown for all ethyl- and methyl-base alkylated derivatives identified in mammalian cells or animals treated with alkylating agents. (Examples are Bodell et al., 1979; Frei et al., 1978; Pegg et al., 1978; Singer et al., 1981.) Two laboratories have found that rat liver or human lymphoblasts contain a glycosylase for 7-MeG (Margison and Pegg, 1981; Singer and Brent, 1981). The lymphoblast extract also depurinates alkylated ds DNA containing 7-EtG and the other N-3 and N-7 methyl and ethyl purines, although the ethyl derivatives are poorer substrates than the methyl (Singer and Brent, 1981). Rat liver homogenates (Pegg, 1978) and a factor in rat liver chromatin (Renard and Verly, 1980) dealkylates O^6-alkyl G and it is now known that the mechanism is transfer of the alkyl group to the cysteine of an acceptor protein (Mehta et al., 1981) as previously reported for the E. coli enzyme (Foote et al., 1980; Karran et al., 1979; Olsson and Lindahl, 1980).

Three other O-alkyl nucleosides are also enzymatically lost, particularly in EtNU-treated rat liver (Singer et al., 1981) and human fibroblasts (Bodell et al., 1979). There are no data on the enzymes responsible other than that they probably differ from the O^6-alkyl G enzyme on the basis that O^6-alkyl G was found to be dealkylated in in vitro experiments, while O^2-EtT was not (Renard and Verly, 1980).

Repair of alkylphosphotriesters, which represent a high proportion of the reaction of N-nitroso alkylating agents with nucleic acids, is questionable. Shooter and Slade (1977), using an indirect method of quantitation (alkali strand breaks), conclude that methylphosphotriesters are decreased at a reasonable rate in vivo but ethylphosphodiesters apparently are not. Our in vivo data (Singer et al., 1981) suggest that ethylphosphotriesters persist for much longer times than do base-ethylated products. Whether loss of triesters is due to dealkylation, which re-forms normal diesters, or to an excision mechanism is still unclear.

The most labile derivative yet studied results from reaction of the N-7 of G with activated AFB_1. In vitro, 90% of the covalent AFB_1 derivative was released as the modified base in 12 h at pH 7.0 (Wang and Cerutti, 1980a). This can be contrasted with ~155 h to release 50% of 7-methylguanine from DNA (Singer, 1979). The AFB_1 residue bound to the N-7 of G is also cleaved rapidly at neutrality resulting in intact guanosine remaining in the DNA (Wang and Cerutti, 1980a). In 10T½ mouse embryo fibroblasts (Wang and Cerutti, 1980b) there was little difference between the half-lives of the AFB_1 guanosine derivative in vitro and in vivo, and it was concluded that neither the reversal reaction nor the glycosyl bond cleavage was enzymatic. However, there is a third product which results from opening of the imidazole ring of N-7 substituted guanosine, and this derivative may be stable in vivo. In addition, there are probably other products formed in varying amounts in specific organs of rats and mice (Wogan et al., 1980), but repair or removal has not been quantitated. The human cell line A549 did show apparent enzymatic removal of the total of all AFB_1 adducts (60% loss in 20 h) (Wang and Cerutti, 1979).

AAF forms three principal products in vivo (Miller, 1978). Two are C-8 derivatives of G and the other is an N^2G product. In rat liver and kidney, the N^2G derivative appears to persist while the C-8 products are enzymatically removed (Kriek, 1972). The C-8 product is also removed from human fibroblasts (Kaneko and Cerutti, 1980).

BP forms several adducts in vivo, but repair studies have focused on the covalent product of BPDE with the N^2 of G. Again, as with all studies of in vivo persistence or removal, the data vary according to the system studied. A good example of variability is reported by Shinohara and Cerutti (1977) who found greatly differing excision rates of BP-deoxyguanosine products in BHK and secondary MEF cells. There is evidence in all mammalian cells and organs studied that the N^2G adduct can be removed, but the rate may depend on the location of the adduct in the chromosome (Feldman et al., 1980; Kaneko and Cerutti, 1980).

7-Bromomethylbenz[a]anthracene (7-BMBA) has been extensively studied in a variety of mammalian cells. The three exocyclic amino derivatives formed in DNA are removed reasonably rapidly when the dose is low. The N^6A adduct is not excised as rapidly as the others (Dipple and Roberts, 1977; Lieberman and Dipple, 1972).

Methylaminobenzene (MAB) forms C-8 and N^2 derivatives of G. In mice the C-8 adduct is rapidly repaired while the N^2 adduct persists (Beland et al., 1980; Tarpley et al., 1980). This particular difference in enzyme recognition appears to be the same as reported for AAF adducts.

To summarize, most derivatives formed by mutagens reacting with DNA appear to be removed/repaired by mammalian enzymes, although such enzymes have not been purified. The extent or rate of loss of derivatives is strongly dependent on the initial dose of mutagen since there is much evidence that repair systems can be saturated. Regardless of how efficiently repair occurs, it is seldom complete, and certainly some potentially mutagenic derivatives are present during replication.

TRANSLATION AND TRANSCRIPTION OF MODIFIED NUCLEOTIDES IN NUCLEIC ACIDS

A modification of DNA which avoids repair may be mutagenic in at least three ways: it may directly mispair (Topal and Fresco, 1976a,b); it may cause frameshifts by stabilizing extrahelical forms as proposed by Streisinger et al. (1966); it may cause induction of a misrepair system: SOS repair (Witkin, 1976).

The mutagenicity of large aromatic derivatives appears to be partly a function of their conformation in the DNA. Termination of synthesis or termination with reinitiation (gaps) are most frequently found (Grunberger and Weinstein, 1979b; Hsu et al., 1977; Moore and Strauss, 1979; Moore et al., 1980). Secondary effects, such as the production of polymerases with decreased fidelity, have also been seen (Chan and Becker, 1979). In polymers modified with the simple alkylating agent β-propiolactone, polynucleotide replication reveals increased noncomplementary nucleotide incorporation with several polymerases (Sirover and Loeb, 1976).

A variety of modified RNAs have been used as messages in cell-free systems and generally there is a decrease in the level of protein synthesis. This may be due to an inhibition of formation of the initiation complex, as shown for BPDE-globin mRNA (Grunberger et al., 1980), or by termination of transcription when a modified nucleotide is encountered rather than miscoding (Braverman et al., 1975; Grunberger and Weinstein, 1971). In the case of alkylated brome mosaic virus RNA-4, also a monocistronic messenger, the coat protein made, though in lesser amounts, is identical in molecular weight to that made by untreated RNA, indicating that translation need not be terminated by ethyl or methyl groups (Fraenkel-Conrat and Singer, 1980).

When RNA viruses are treated with various mutagens, both mutagenicity and/or loss of infectivity are observed depending on the modification. Site-specific mutagenesis with N^4-hydroxy C incorporated in Qβ RNA gives a G→A transition with high frequency but also decreases the infectivity (Sabo et al., 1977). In tobacco mosaic virus (TMV)-RNA treated with AAF or BPDE and reconstituted shows a loss of infectivity but no mutation (Singer et al., 1980). Since both AAF and BPDE are mutagenic in bacteria, it is suggested that the failure to observe mutation reflects the lack of error-prone repair activity toward RNA (Singer et al., 1980). TMV-RNA is otherwise quite capable of expressing a variety of "simple" mutagenic lesions (Singer and Fraenkel-Conrat, 1974) in which the amino acid exchanges observed in the coat protein correlate with the expected C→U or A→G deamination reaction with nitrous acid and C→U tautomerism with hydroxylamine. In contrast, reaction of TMV-RNA with alkylating agents resulted in amino acid exchanges reflecting all possible base changes. This probably is due to the same ambiguity observed with alkyl derivatives in transcription experiments. The high mutability of influenza virus coat protein seen in vivo (and felt each year by the human population) is the result of an uncorrected replication error which produces the same kind of amino acid changes (G. Both, personal communication).

A variety of other mutagenic interactions studied in translation systems, such as codon-anticodon binding or modified polymer translations, are described by Singer and Kröger (1979) and Singer et al. (1979).

Transcription of polymers containing modified nucleotides has been used as an in vitro model of in vivo alterations which would be expected to cause mutation by directing the incorporation of one or more nucleotides. This direct mispairing is the simplest kind of mutational mechanism. It should be noted that such incorporations may be considered misincorporations only in a biological sense, in that they do not yield a faithful copy of the template. However, physico-chemically the interaction may be correct. For example, the substitutions found when C is in the imino form, or a purine glycosyl bond is syn, are considered part of the explanation for the background level of base changes observed which are both transitions and transversions (Drake and Baltz, 1976; Fresco et al., 1980; Topal and Fresco, 1976a,b). Modifications which change the proportion of a tautomer would presumably act this way, e.g. hydroxylamine. Figure 10, adapted from Fresco et al. (1980), shows the simplest tautomeric conversions.

When the presence of modified nucleotides in templates increases the level of misincorporation above the background — and it has been assumed that it is based on at least a transient hydrogen bonding which satisfies the polymerase (Singer, 1980; Singer and Kröger, 1979) — it is possible to conclude that the polymerase will insert any base which can form one or more hydrogen bonds into the transcript. However, when specific base-pairing sites are modified, the resulting ambiguity may not require any hydrogen bond formation. In living organisms, the chemistry of base pairs, including tautomeric equilibria and the equivalence of electron distributions, is tempered by the biochemistry of a variety of enzymes with specific proofreadings, alignment, and error levels.

Some types of modification have no measurable effect on fidelity in transcription (Table 5). Alkylations of ribose or substitutions at the C-5 of pyrimidines would not be considered mutagenic on this basis. The replacement of oxygen by sulfur also does not change the observed base pair, even though sulfur is larger. However, deamination of C to U or G to X does result in the expected transition incorporation.

The termination observed in replication experiments is not often seen in these in vitro transcription experiments, although it has been inferred when very large groups are the substituent. However, polymerases may pause at modified nucleotides as observed in transcription of 16S RNA containing N^2-methyl G (Youvan and Hearst, 1979). The pause is probably the time spent by the enzyme in choosing a "proper" triphosphate. In transcription of synthetic polymers containing N^2-methyl G, a level of ambiguity is found supporting the concept that a poor fit is better than none (Singer and Spengler, 1981).

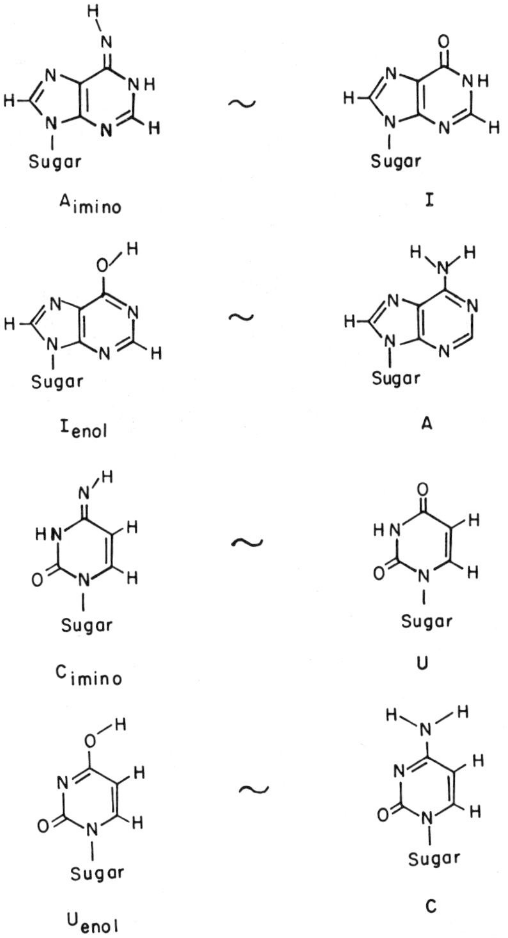

Fig. 10. Equivalence of the unfavored tautomeric forms (left side) to another base (right side) in the favored form. These shifts would lead to transitions. Adapted from Fresco et al. (1980).

It might be expected that alkylated bases substituted at essential Watson-Crick sites would terminate transcription but instead they are transcribed, introducing a complete ambiguity. Modifications on exocyclic groups not necessarily involved in base pairing produce a variety of mispairing patterns, even though they could all be nonmutagenic if the proton were oriented syn to the Watson-Crick side. N^4-methyl C is ambiguous, directing the incorporation of all four nucleotides in a transcript, while N^4-acetyl C, N^6-methyl A, N^6-hydroxyethyl A, and the large N^6-isopentenyl derivative are

Table 5. Template Activity of Modified Nucleosides Introduced into Polynucleotides[a]

Synthetic Polynucleotide			In Transcription Substitutes for[b,c]				References
Modified Nucleoside[d]	Percentage	In Copolymer with	A	G	U	C	
Adenosines							
1-Methyladenosine[+]	8	C	+	±	+		Kröger and Singer (1979b)
	9	U	±			+	Kröger and Singer (1979b)
3-Isoadenosine	38	C	+	–	–	–	Singer and Spengler (1981)
N⁶-Methyladenosine[+]	8	C	+	–	–	–	Kröger and Singer (1979b)
	11	U	+			–	Kröger and Singer (1979b)
	12	A				–	Singer and Spengler (unpublished data)
N⁶-Hydroxyethyl-adenoxine[+]	46, 65	C	+	–	–	–	Singer and Spengler (1981)
N⁶-Isopentenyladenosine[+]	12, 14	A	+	–	–	–	Singer and Spengler (1981)
	3–17	C	+	–	–	–*	Singer and Spengler (1981)
N⁶-Methoxyadenosine	21–33	C	+	+	–	–	Singer and Spengler (unpublished data)
1,N⁶-Ethenoadenosine[+]	6–25	C	+	±	+	–	Spengler and Singer (1981)
	~20	A					Spengler and Singer (1981)
2'-O-Methyladenosine	100	—	[–]	[–]			Gerard et al. (1972)
Inosine and Xanthosine							
Xanthosine	13	C	+	–	–	–	Kröger and Singer (1979b)
	~15	U					Kröger and Singer (1979b)
2'-O-Methylinosine	100	—	[–]		[–]		Kröger and Singer (1979b)

MUTAGENESIS FROM A CHEMICAL PERSPECTIVE

Compound	Value						Reference
Guanosines							
N²-Methylguanosine	7	C	±	±		+	±*	Singer and Spengler (1981)
7-Methylguanosine	3, 5	C		+	+			Ludlum (1970a)
	2, 4	U	+	-				Ludlum (1970a)
	5	C	+	+	+			Gerchman and Ludlum (1973)
O⁶-Methylguanosine	7-11	C		-	±			Singer and Spengler (unpublished data)
O⁶-Methyl-2'-deoxyguanosine	0.4	C	+	-	+			Mehta and Ludlum (1978)
Uridines							
3-Methyluridine†	7	C	+	±		+		Kröger and Singer (1979b)
	4	U		-		+	±	Kröger and Singer (1979b)
4-Thiouridine†	3-22	C	-	-		+		Singer and Spengler (1981)
5-Fluorouridine	33	A		-		+		Kröger and Singer (1979b)
5-Hydroxyuridine	11	C		-		±		Singer and Spengler (1981)
O²-Methyluridine	5-22	C	±*	-				Singer and Spengler (1981)
								(unpublished data)
	4, 10	U					±	Singer and Spengler (unpublished data)
O²-Ethyluridine	4-8	U		±			+	Singer et al. (1978)
	8	C	-		±*		+*	Singer et al. (1978)
	4-23	U	-				+	Singer and Spengler (unpublished data)
O⁴-Methyluridine	4-13	U					+	Singer et al. (1978)
	7-13	C		+				Singer et al. (1978)
	5-15	C	±	+	+			Singer and Spengler (unpublished data)
2'-O-Methyluridine	100	—			[+]			Gerard et al. (1972)

(continued on next page)

Table 5 (continued)

Cytidines	%	Base						Reference
2-Thiocytidine	100	—					—	Kröger and Singer (1979a)
	10	C	—	—	—	—	+	Kröger and Singer (1979a)
3-Methylcytidine†	6, 10	C	+	+	+			Ludlum (1970b)
	4	C	+	+	+			Fraenkel-Conrat and Singer (1971)
	13	C	+	+	+			Kröger and Singer (1979b)
	17	U	+	+	+		±	Kröger and Singer (1979b)
	7	C	+		+			Ludlum (1971)
3-Methyl-2'-deoxycytidine[e]	4, 6	C	+	—	+			Ludlum (1970b)
3-Ethylcytidine	9	C	—	±	+		±	Singer and Spengler (unpublished data)
N^4-Methylcytidine†	~20	C	+	+	+			Singer and Spengler (1981)
	~20	A	—	±	—			Singer and Spengler (1981)
N^4-Acetylcytidine†[e]	70	C	—	—	—		+	Singer and Spengler (1981)
N^4-Hydroxycytidine†	4	C	—	—	+		+	Fraenkel-Conrat and Singer (1971)
	8-16	C	+	+	+			Singer and Spengler (1981)
	22	A	—	—	—		±*	Singer and Spengler (1981)
N^4-Methoxycytidine†	3-16	C	+	—	+		—*	Singer and Spengler (1981)
	6	C	—		+		+	Fraenkel-Conrat and Singer (1971)
5-Fluorocytidine	20-67	C			—		+	Fraenkel-Conrat and Singer (1971)
5-Methyl-N^4-Methoxycytidine†	5, 10	C	±	±*	+			Singer and Spengler (unpublished data)
5-Methyl-N^4-Hydroxycytidine†	5, 10	C	±	±	+			Singer and Spengler (unpublished data)

Compound	%						Reference
5-Bromocytidine†e	100	—	—	—	—	+	Means and Fraenkel-Conrat (1971)
	97	U	—	—	—	+	Singer and Spengler (unpublished data)
5-Hydroxycytidine	15	C	—	—	—	+	Means and Fraenkel-Conrat (1971)
3,N^4-Ethenocytidine†	7-27	C	+	+	—	—	Spengler and Singer (1981)
	~20	A	—	+	—	—	Singer and Spengler (unpublished data)
2'-O-Methylcytidine†	100	—	—	—	—	[±]	Gerard et al. (1972)
	100	—	—	—	—	+	Singer and Spengler (1981)

[a] Experiments using chemically modified polynucleotides are not included, except as noted by the footnote (e), as it is difficult to eliminate the possibility of multiple reactions occurring.
[b] All experiments were performed with DNA-dependent RNA polymerase in the presence of Mn^{2+}. Brackets are used to designate data from experiments in which only two nucleoside triphosphates (NTP's) were used, one complementary to the carrier and one other. This type of noncompetitive experiment, in contrast to the use of all four NTP's, can give results that are not found under the more biological or competitive situation. Experiments from Ludlum's laboratory were generally performed with three nucleoside diphosphates.
[c] Asterisks are for data under competitive conditions but the results are borderline and not of the same significance as other data.
[d] The dagger symbol indicates experiments for which there are also data using combinations of two NTP's. When misincorporations occur using 4 NTP's, these results can be amplified under non-competitive conditions. In a few cases, notably polymers containing m^6A, misincorporation can be "forced" as a result of biasing the NTP pool.
[e] Chemically modified polymer. It is assumed that only a single modification occurs under the experimental conditions.

nonmutagenic derivatives since they act only like the unmodified base. We have proposed that these behaviors are a measure of the ability of the substituents on N to rotate <u>syn</u> or <u>anti</u> to the Watson-Crick side (Singer and Spengler, 1981). The misincorporations observed with O-alkyl derivatives would also require that the alkyl groups be <u>anti</u> to the base-pairing side (Singer, 1980).

The products of reaction with hydroxylamine (N^4-hydroxy C) and methoxyamine (N^4-methoxy C and N^6-methoxy A) illustrate the effect of rotation about the exocyclic amino group but specifically are used as the classic example of effect of a modification shifting the tautomeric equilibrium. N^4-Hydroxy C prefers the imino (U-like) form of C (10:1) but while in the amino form may have the hydroxyl group <u>syn</u> or <u>anti</u> to the base-pairing sites. As a result, in transcription experiments this derivative is completely ambiguous. N^4-Methoxy C differs in that it is completely in the tautomer and acts only as U. Methoxyamine modification of the N^6 of A also shifts the tautomeric equilibrium and in transcription N^6-methoxy A acts like both A and G with an estimated K_T of 10 for the amino form. On the basis of transcription results, N^4-methoxy C and N^6-methoxy A exist with the substituent <u>anti</u> to the Watson-Crick side.

The cyclic etheno derivatives 3,N^4-etheno C and 1,N^6-etheno A are products of chloroacetaldehyde modification. Because of its size and fluorescence, etheno C has been used as an analogue for A with several enzymes. It is not surprising, therefore, that it acts like A, predominantly, in transcription. However, it does not act like the parent nucleotide C. Likewise, etheno A is not transcribed as C, but as U predominantly, even though this is a bulky adduct in a fixed position (Spengler and Singer, 1981).

It should be recalled that in these transcription experiments the modification is in the polymer. The ability of the polymerase to use a modified triphosphate can also be measured (Budowsky et al. 1975; Engel and von Hippel, 1978). This is an interesting approach because triphosphates are more reactive than nucleic acids in cells. The failure of many polymerases to use these triphosphates may illustrate a partition of labor with polymerases only checking the incoming substrate while accepting the modified template.

The mutagenic effects of modified nucleotides in DNA and RNA studied in vitro in both translation and transcription systems can be summarized as follows. Translation of modified messages results in a decrease in the level of translation but mutant proteins have also been identified. Consideration of in vitro transcription studies has led us to propose that ambiguity will result whether hydrogen bonds of the appropriate number and length cannot be formed. The distortion may be due to steric hindrance or electron shielding of sites as well as loss of a receptor or donor by direct substitution. Stacking interactions and other neighborhood effects may determine the specific ambiguity observed when weak or no hydrogen bonds are formed. Thus, the mutagenic effects of unrepaired modifications of DNA can be elucidated in translation and transcription experiments in vitro, assuming that the process of base interaction seen there is the same as that in replication, exclusive of repair or proofreading.

ACKNOWLEDGMENT

This work was supported by Grant CA 12316 from the National Cancer Institute. The author is indebted to Dr. S. Spengler and Dr. H. Fraenkel-Conrat for their interest and criticism.

REFERENCES

Autrup, H., Essigmann, J. M., Croy, R. G., Trump, B. F., Wogan, G. N., and Harris, C. C., 1979, Metabolism of aflatoxin B_1 and identification of the major aflatoxin B_1-DNA adducts formed in cultured human bronchus and colon, Cancer Res., 39:694.

Barbin, A., Bresil, H., Croisy, A., Jacquignon, P., Malaveille, C., Montesano, R., and Bartsch, H., 1975, Liver microsome-mediated formation of alkylating agents from vinyl bromide and vinyl chloride, Biochem. Biophys. Res. Commun., 67:596.

Barrio, J. R., Secrist, III, J. A., and Leonard, N. J., 1972, Fluorescent adenosine and cytidine derivatives, Biochem. Biophys. Res. Commun., 46:597.

Bartsch, H., Malaveille, C., Camus, A.-M., Martel-Planche, G., Brun, G., Hautefeuille, A., Sabadie, N., Barbin, A., Kuroki, T., Drevon, C., Piccoli, C., and Montesano, R., 1980, Bacterial and mammalian tests: Validation and comparative studies on 180 chemicals, in: "Molecular and Cellular Aspects of Carcinogen Screening Tests," R. Montesano, H. Bartsch, and L. Tomatis, eds., IARC Scientific Publication No. 27.

Beland, F.A., 1978, Computer-generated graphic models of the N^{2-} substitued deoxyguanosine adducts of 2-acetylaminofluorene and benzo[a]pyrene and the O^6-substituted deoxyguanosine adduct of 1-naphthylamine in the DNA double-helix, Chem. Biol. Interact., 22:329.

Beland, F. A., Tullis, D. L., Kadlubar, F. F., Straub, K. M., and Evans, F. E., 1980, Characterization of DNA adducts of the carcinogen N-methyl-4-aminoazobenzene in vitro and in vivo, Chem. Biol. Interact., 31:1.

Beranek, D. T., Weis, C. C., and Swenson, D. H., 1980, A comprehensive quantitative analysis of methylated and ethylated DNA using high pressure liquid chromatography, Carcinogenesis, 1: 595.

Bodell, W. J., and Singer, B., 1979, Influence of hydrogen bonding in DNA and polynucleotides on reaction of nitrogens and oxygens toward ethylnitrosourea, Biochemistry, 18:2860.

Bodell, W. J., Singer, B., Thomas, G. H., and Cleaver, J. E., 1979, Evidence for removal at different rates of O-ethyl pyrimidines and ethyl-phosphotriesters in two human fibroblast cell lines, Nucleic Acids Res., 6:2819.

Bradley, M. O., Sharkey, N. A., Kohn, K. W., and Layard, M. W., 1980, Mutagenicity and cytotoxicity of various nitrosoureas in V-79 Chinese hamster cells, Cancer Res., 40:2719.

Braverman, B., Shapiro, R., and Szer, W., 1975, Modification of E. coli ribosomes and coliphage MS2 RNA by bisulfite: Effects on ribosomal binding and protein synthesis, Nucleic Acids Res., 2:501.

Brookes, P., 1977, Mutagenicity of polycyclic aromatic hydrocarbons, Mutat. Res., 39:257.

Brookes, P., and Lawley, P. D., 1960, The reaction of mustard gas with nucleic acids in vitro and in vivo, Biochem. J., 77:478.

Brookes, P., and Lawley, P. D., 1961, The reaction of mono- and difunctional alkylating agents with nucleic acids, Biochem. J., 80:496.

Brookes, P., and Lawley, P. D., 1963, Effects of alkylating agents on T2 and T4 bacteriophages, Biochem. J., 89:138.

Brown, D. M., 1974, Chemical reactions of polynucleotides and nucleic acids, in: "Basic Principles in Nucleic Acid Chemistry," Vol. II, P. O. P. Ts'o, ed., Academic Press, New York.

Brown, D. M., McNaught, A. D., and Schell, P., 1966, The chemical basis of hydrazine mutagenesis, Biochem. Biophys. Res. Commun., 24:967.

Brown, D. M., and Osborne, H. R., 1971, The reaction of adenosine with hydroxylamine, Biochem. Biophys. Acta, 247:514.

Budowsky, E. I., Sverdlov, E. D., and Monastyrskaya, G. S., 1969, Mechanism of mutagenic action of hydroxylamine. II. Reaction of hydroxylamine with the adenine nucleus, J. Mol. Biol., 44:205.

Budowsky, E. I., Sverdlov, E. D., and Spasokutotskaya, T. N., 1972, Mechanism of the mutagenic action of hydroxylamine. VII. Functional activity and specificity of cytidine triphosphate modified with hydroxylamine and O-methylhydroxylamine, Biochim. Biophys. Acta, 287:195.

Budowsky, E. I., Sverdlov, E. D., Spasokutotskaya, T. N., and Koudelka, J. A., 1975, Mechanism of mutagenic action of hydroxylamine. VIII. Functional properties of the modified adenosine residues, Biochim. Biophys. Acta, 300:1.

Carrell, H. L., Glusker, J. P., Moschel, R. C., Hudgins, W. R., and Dipple, A., 1981, Crystal structure of a carcinogen: nucleoside adduct, Cancer Res., 41:2230.

Chan, J. Y. H., and Becker, F. F., 1979, Decreased fidelity of DNA polymerase activity during N-2-fluorenylacetamide hepatocarcinogenesis, Proc. Natl. Acad. Sci. U.S.A., 76:814.

Chun, E. H. L., Gonzales, L., Lewis, F. S., Jones, J., and Rutman, R. J., 1969, Differences in the in vivo alkylation and crosslinking of nitrogen mustard-sensitive and -resistant lines of Lettŕe-Ehrlich ascites tumors, Cancer Res., 29:1184.

Dipple, A., Brookes, P., Mackintosh, D. S., and Rayman, M. P., 1971, Reaction of 7-bromomethylbenz[a]anthracene with nucleic acids, polynucleotides and nucleosides, Biochemistry, 10:4323.

Dipple, A., and Roberts, J. J., 1977, Excision of 7-bromomethylbenz[a]anthracene-DNA adducts in replicating mammalian cells, Biochemistry, 16:1499.

Drake, J. W., and Baltz, R. H., 1976, The biochemistry of mutagenesis, Annu. Rev. Biochem., 45:11.

Engel, J. D., and von Hippel, P. H., 1978, $d(m^6ATP)$ As a probe of the fidelity of base incorporation into polynucleotides by Escherichia coli DNA polymerase I, J. Biol. Chem., 253:935.

Essigmann, J. M., Croy, R. G., Nadzan, A. M., Busby, Jr., W. F., Reinhold, V. N., Buchi, G., and Wogan, G. N., 1977, Structural identification of the major DNA adduct formed by aflatoxin B1 in vitro, Proc. Natl. Acad. Sci. U.S.A., 74:1870.

Feldman, G., Remsen, J., Wang, T. V., and Cerutti, P., 1980, Formation and excision of covalent deoxyribonucleic acid adducts of benzo[a]pyrene 4,5-epoxide and benzo[a]pyrene epoxide I in human lung cells A549, Biochemistry, 19:1095.

Fishbein, L., 1969, Degradation and residues of alkylating agents, Ann. N. Y. Acad. Sci., 163:869.

Foote, R. S., Mitra, S., and Pal, B. C., 1980, Demethylation of O^6-methylguanine in a synthetic DNA polymer by an inducible activity in Escherichia coli, Biochem. Biophys. Res. Commun., 97:654.

Fraenkel-Conrat, H., 1954, Reaction of nucleic acid with formaldehyde, Biochim. Biophys. Acta, 15:308.

Fraenkel-Conrat, H., 1981, Chemical modification of viruses, in: "Comprehensive Virology," Vol. 17, H. Fraenkel-Conrat, and R. R. Wagner, eds., Plenum Press, New York.

Fraenkel-Conrat, H., and Singer, B., 1971, Template and messenger activities of mutagen-treated polynucleotides, in: "Biological Effects of Polynucleotides," R. F. Beers, Jr., and W. Braun, eds., Springer-Verlag, New York.

Fraenkel-Conrat, H., and Singer, B., 1980, Effect of introduction of small alkyl groups on mRNA function, Proc. Natl. Acad. Sci. U.S.A., 77:1983.

Frei, J. V., Swenson, D. H., Warren, W., and Lawley, P. D., 1978, Alkylation of deoxyribonucleic acid in vivo in various organs of C57BL mice by the carcinogens N-methyl-N-nitrosourea, N-ethyl-N-nitrosourea, and ethyl methanesulfonate in relation to induction of thymic lymphoma, Biochem. J., 174:1031.

Fresco, J. R., Broitman, S., and Lane, A.-E., 1980, Base mispairing and nearest neighbor effects in transition mutations, in:

"Mechanistic Studies of DNA Replication and Genetic Recombination," B. Alberts, and C. F. Fox, eds., Academic Press, New York.

Gamper, H. B., Bartholomew, J. C., and Calvin, M., 1980, Mechanism of benzo[a]pyrene diol epoxide induced deoxyribonucleic acid strand scission, Biochemistry, 19:3948.

Gerard, G. F., Rottman, F., and Boezi, J. A., 1972, Template activity of 2'-O-methylpolyribonucleotides with Pseudomonas putida DNA-dependent RNA polymerase, Biochem. Biophys. Res. Commun., 46: 1095.

Gerchman, L. L., and Ludlum, D. B., 1973, The properties of O^6-methylguanine in templates for RNA polymerase, Biochem. Biophys. Res. Commun., 308:310.

Goldschmidt, B. M., Blazej, T. P., and Van Duuren, B. L., 1968, The reaction of guanosine and deoxyguanosine with glycidaldehyde, Tetrahedron Lett., No. 13:1583.

Gombar, C. T., Tong, W. P., and Ludlum, D. B., 1979, Mechanism of action of the nitrosoureas: Formation of 1,2-(diguanosin-7-yl) ethane from the reaction of BCNU (1,3-bis-[2-chloroethyl]-1-nitrosourea) with guanosine, Biochem. Biophys. Res. Commun., 90:878.

Goth, R., and Rajewsky, M. F., 1974, Molecular and cellular mechanisms associated with pulse-carcinogenesis in the rat nervous system by ethylnitrosourea: Ethylation of nucleic acids and elimination rates of ethylated bases from the DNA of different tissues, Z. Krebsforsch., 82:37.

Green, T., and Hathway, D. E., 1978, Interactions of vinyl chloride with rat liver DNA in vivo, Chem. Biol. Interact., 22:211.

Grunberger, D., Pergolizzi, R. G., and Jones, R. E., 1980, Translation of globin messenger RNA modified by benzo[a]pyrene 7,8-dihydrodiol 9,10-oxide in a wheat germ cell-free system, J. Biol. Chem., 255:390.

Grunberger, D., and Weinstein, I. B., 1971, Modifications of ribonucleic acid by chemical carcinogens. III. Template activity of polynucleotides modified by N-acetoxy-2-acetylaminofluorene, J. Biol. Chem., 246:1123.

Grunberger, D., and Weinstein, I. B., 1979a, Conformational changes in nucleic acids modified by chemical carcinogens, in: "Chemical Carcinogenesis and DNA," Vol. I, P. L. Grover, ed., CRC Press, Boca Raton.

Grunberger, D., and Weinstein, I. B., 1979b, Biochemical effects of the modification of nucleic acids by certain polycyclic aromatic carcinogens, Prog. Nucleic Acid Res. Mol. Biol., 23:106.

Hanawalt, P. C., Cooper, P. K., Ganesan, A. K., and Smith, C. A., 1979, DNA repair in bacteria and mammalian cells, Annu. Rev. Biochem., 48:783.

Hashimoto, Y., Shudo, K., and Okamoto, T., 1980, Metabolic activation of a mutagen, 2-amino-6-methyldipyrido-[1,2-a:3',2'-d]imidazole. Identification of 2-hydroxyamino-6-methyldipyrido[1,2-a:3',2'-d]

imidazole and its reaction with DNA, Biochem. Biophys. Res. Commun., 92:971.
Hayatsu, H., Wataya, U., Kai, K., and Fida, S., 1970, Reaction of sodium bisulfite with uracil, cytosine, and their derivatives, Biochemistry, 9:2858.
Hsu, W.-T., Lin, E. J. S., Harvey, R. G., and Weiss, S. B., 1977, Mechanism of phage φX174 DNA inactivation by benzo[a]pyrene-7,8-dihydrodiol-9,10-epoxide, Proc. Natl. Acad. Sci. U.S.A., 74:3335.
Ivanovic, V., Geacintov, N. E., Yamasaki, H., and Weinstein, I. B., 1978, DNA and RNA adducts formed in hamster embryo cell cultures exposed to benzo[a]pyrene, Biochemistry, 17:1597.
Jeffrey, A. M., Blobstein, S. H., Weinstein, I. B., Beland, F. A., Harvey, R. G., Kasai, H., and Nakanishi, K., 1976a, Structure of 7,12-dimethylbenz[a]anthracene-guanosine adducts, Proc. Natl. Acad. Sci. U.S.A., 73:2311.
Jeffrey, A. M., Grzeskowiak, K., Weinstein, I. B., Nakanishi, K., Roller, P., and Harvey, R. G., 1979, Benzo[a]pyrene-7,8-dihydrodiol-9,10-oxide adenosine and deoxyadenosine adducts: Structure and stereochemistry, Science, 206:1309.
Jeffrey, A. M., Jennette, K. W., Blobstein, S. H., Weinstein, I. B., Beland, F. A., Harvey, R. G., Kasai, H., Miura, I., and Nakanishi, K., 1976b, Benzo[a]pyrene-nucleic acid derivative found in vivo: Structure of a benzo[a]pyrenetetrahydrodiol epoxide-guanosine adduct, J. Am. Chem. Soc., 98:5714.
Kadlubar, F. F., Miller, J. A., and Miller, E. C., 1978, Guanyl O^6-arylamination and O^6-arylation of DNA by the carcinogen N-hydroxy-1-naphthylamine, Cancer Res., 38:3628.
Kadlubar, F. F., Unruh, L. E., Beland, F. A., Straub, K. M., and Evans, F. E., 1980, In vitro reactions of the carcinogen N-hydroxy-2 naphthylamine with DNA at the C-8 and N-2 atoms of guanine and at the N^6 atom of adenine, Carcinogenesis, 1:139.
Kaneko, M., and Cerutti, P., 1980, Excision of N-acetoxy-2-acetylaminofluorene-induced DNA adducts from chromatin fractions of human fibroblasts, Cancer Res., 40:4313.
Karran, P., Lindahl, T., and Griffin, B., 1979, Adaptive response to alkylating agents involves alteration in situ of O^6-methylguanine residues in DNA, Nature, 280:76.
Kasai, H., Nakanishi, K., Frenkel, K., and Grunberger, D., 1977, Structures of 7,12-dimethylbenz[a]anthracene 5,6-oxide derivatives linked to the ribose moiety of guanosine, J. Am. Chem. Soc., 99:8500.
Kayasuga-Mikado, K., Hashimoto, T., Negishi, T., Negishi, K., and Hayatsu, H., 1980, Modification of adenine and cytosine derivatives with bromoacetaldehyde, Chem. Pharm. Bull., 28:932.
Kimura, K., Nakanishi, M., Yamamoto, T., and Tsuboi, M., 1977, A correlation between the secondary structure of DNA and the reactivity of adenine residues with chloroacetaldehyde, J. Biochem., 81:1699.

Koreeda, M., Moore, P. D., Yagi, H., Yeh, H. J. C., and Jerina, D. M., 1976, Alkylation of polyguanylic acid at the 2-amino group and phosphate by the potent mutagen (±)-7β, 8α-dihydroxy-9α, 10β-epoxy-7,8,9,10-tetrahydrobenzo[a]pyrene, J. Am. Chem. Soc., 98:6720.

Kriek, E., 1972, Persistent binding of a new reaction product with the carcinogen N-hydroxy-2-acetylaminofluorene with guanine in rat liver DNA in vivo, Cancer Res., 32:2042.

Kriek, E., Miller, J. A., Juhl, U., and Miller, E. C., 1967, 8-(N-2-Fluorenylacetamido) guanosine, an arylamidation reaction product of guanosine and the carcinogen N-acetyoxy-N-2-fluorenylacetamide in neutral solution, Biochemistry, 6:177.

Kriek, E., and Westra, J. G., 1979, Metabolic activation of aromatic amines and amides and interactions with nucleic acids, in: "Chemical Carcinogens and DNA," Vol. II, P. L. Grover, ed., CRC Press, Boca Raton.

Kröger, M., and Singer, B., 1979a, Influence of different levels of 2-thiocytidine on physical and template properties of cytidine-2-thiocytidine copolymers, Biochemistry, 18:91.

Kröger, M., and Singer, B., 1979b, Ambiguity and transcriptional errors as a result of methylation of N-1 of purines and N-3 of pyrimidines, Biochemistry, 18:3493.

Kuśmierek, J. T., and Singer, B., 1976, Sites of alkylation of poly (U) by agents of varying carcinogenicity and stability of products, Biochim. Biophys. Acta, 442:420.

Laib, R. J., and Bolt, H. M., 1978, Formation of 3,N^4-ethenocytidine moieties in RNA by vinyl chloride metabolites in vitro and in vivo, Arch. Toxicol., 39:235.

Lawley, P. D., and Jarman, M., 1972, Alkylation by propylene oxide of deoxyribonucleic acid, adenine, guanosine and deoxyguanylic acid, Biochem. J., 126:893.

Lawley, P. D., Orr, D. J., and Jarman, M., 1975, Isolation and identification of products from alkylation of nucleic acids: Ethyl- and isopropyl-purines, Biochem. J., 145:73.

Lieberman, M. W., and Dipple, A., 1972, Removal of bound carcinogen during DNA repair in nondividing human lymphocytes, Cancer Res., 32:1855.

Lin, J.-K., Miller, J. A., and Miller, E. C., 1977, 2,3-Dihydro-2-(guan-7-yl)-3-hydroxyaflatoxin B1, a major acid hydrolysis product of aflatoxin B1-DNA or ribosomal RNA adducts formed in hepatic microsome-mediated reactions and in rat liver in vivo, Cancer Res., 37:4430.

Lindahl, T., 1979, DNA glycosylases, endonucleases for apurinic/apyrimidinic sites, and base-excision repair, Prog. Nucleic Acids Mol. Biol., 22:135.

Ludlum, D. B., 1970a, The properties of 7-methylguanine-containing templates for ribonucleic acid polymerase, J. Biol. Chem., 245:477.

Ludlum, D. B., 1970b, Alkylated polycytidylic acid templates for RNA polymerase, Biochim. Biophys. Acta, 213:142.

Ludlum, D. B., 1971, Methylated polydeoxyribocytidylic acid templates for RNA polymerase, Biochim. Biophys. Acta, 247:412.

Ludlum, D. B., Kramer, B. S., Wang, J., and Fenselau, C., 1975, Reaction of 1,3-bis(2-chloroethyl)-1-nitrosourea with synthetic polynucleotides, Biochemistry, 14:5480.

Lukashin, A. V., Vologodskii, A. V., Frank-Kamenetskii, M. D., and Lyubchenko, Y. L., 1976, Fluctuational opening of the double helix as revealed by theoretical and experimental study of DNA interaction with formaldehyde, J. Mol. Biol., 108:665.

Maher, V. M., and McCormick, J. J., 1978, Mammalian cell mutagenesis by polycyclic aromatic hydrocarbons and their derivatives, in: "Polycyclic Hydrocarbons and Cancer," Vol. II, H. V. Gelboin, ed., Academic Press, New York.

Mandal, C., Kallenbach, N. R., and Englander, S. W., 1979, Base-pair opening and closing reactions in the double helix, J. Mol. Biol., 135:391.

Margison, G. P., and Pegg, A. E., 1981, Enzymatic release of 7-methylguanine from methylated DNA by rodent liver extracts, Proc. Natl. Acad. Sci. U.S.A., 78:861.

Martin, C. N., and Garner, R. C., 1977, Aflatoxin B-oxide generated by chemical or enzymic oxidation of aflatoxin B_1 causes guanine substitution in nucleic acids, Nature, 267:863.

Maté, U., Solomon, J. J., and Segal, A., 1977, In vitro binding of β-propiolactone to calf thymus DNA and mouse liver DNA to form 1-(2-carboxyethyl)adenine, Chem. Biol. Interact., 18:327.

McGhee, J. D., and von Hippel, P. H., 1977, Formaldehyde as a probe of DNA structure. 4. Mechanism of the initial reaction of formaldehyde with DNA, Biochemistry, 16:3276.

Means, G. E., and Fraenkel-Conrat, H., 1971, Effect of bromine on the template and messenger specificities of polynucleotides, Biochim. Biophys. Acta, 247:441.

Mehta, J. R., and Ludlum, D. B., 1978, Synthesis and properties of O^6-methyldeoxyguanylic acid and its copolymers with deoxycytidylic acid, Biochim. Biophys. Acta, 521:770.

Mehta, J. R., Ludlum, D. B., Renard, A., and Verly, W. G., 1981, Repair of O^6-ethylguanine in DNA by a chromatin fraction from rat liver: Transfer of the ethyl group to an acceptor protein. Proc. Natl. Acad. Sci. U.S.A., in press.

Miller, E. C., 1978, Some current perspectives on chemical carcinogenesis in humans and experimental animals: Presidential address, Cancer Res., 38:1479.

Montesano, R., and Bartsch, H., 1976, Mutagenic and carcinogenic N-nitroso compounds: Possible environmental hazards, Mutat. Res., 32:179.

Moore, P. D., Rabkin, S. D., and Strauss, B. S., 1980, Termination of in vitro DNA synthesis at AAF adducts in DNA, Nucleic Acids Res., 8:4473.

Moore, P., and Strauss, B. S., 1979, Sites of inhibition of in vitro DNA synthesis in carcinogen- and UV-treated φX174DNA, Nature, 278:664.

Moschel, R. C., Hudgins, W. R., and Dipple, A., 1980, Aralkylation of guanosine by the carcinogen N-nitroso-N-benzylurea, J. Org. Chem., 45:533.

Nakanishi, K., Kasai, H., Cho, H., Harvey, R. G., Jeffrey, A. M., Jennette, K. W., and Weinstein, I. B., 1977, Absolute configuration of a ribonucleic acid adduct formed in vivo by metabolism of benzo[a]pyrene, J. Am. Chem. Soc., 99:258.

Nakanishi, K., Komura, H., Miura, I., Kasai, H., Frenkel, K., and Grunberger, D., 1980, Structure of a 7,12-dimethylbenz[a]anthracene 5,6-oxide derivative bound to C-8 of guanosine, J. Chem. Soc. Chem. Comm., 82.

Newbold, R. F., Warren, W., Medcalf, A. S. C., and Amos, J., 1980, Mutagenicity of carcinogenic methylating agents is associated with a specific DNA modification, Nature, 282:596.

Olsson, M., and Lindahl, T., 1980, Repair of alkylated DNA in Escherichia coli. Methyl group transfer from O^6-methylguanine to a protein cysteine residue, J. Biol. Chem., 255:10569.

Pegg, A. E., 1978, Enzymatic removal of O^6-methylguanine from DNA by mammalian cell extracts, Biochem. Biophys. Res. Commun., 84:163.

Pegg, A. E., Hui, G., and Rogers, K. J., 1978, Effect of hypophysectomy on persistence of methylated purines in rat liver deoxyribonucleic acid after administration of dimethylnitrosamine, Biochem. Biophys. Acta, 520:571.

Pienta, R. J., 1980, Transformation of Syrian hamster embryo by diverse chemicals and correlation with their reported carcinogenic and mutagenic activities, in: "Chemical Mutagens. Principles and Methods for their Detection," Vol. 6, F. de Serres, and A. Hollaender, eds., Plenum Press, New York.

Rayman, M. P., and Dipple, A., 1973, Structure and activity in chemical carcinogenesis. Comparison of the reactions of 7-bromomethylbenz[a]anthracene and 7-bromomethyl-12-methylbenz[a]-anthracene with mouse skin deoxyribonucleic acid in vivo, Biochemistry, 12:1538.

Renard, A., and Verly, W. G., 1980, A chromatin factor in rat liver which destroys O^6-ethylguanine in DNA, FEBS Lett., 114:98.

Roberts, J. J., 1978, The repair of DNA modified by cytotoxic, mutagenic, and carcinogenic chemicals, Adv. Radiat. Biol., 7:211.

Roberts, J. J., and Warwick, G. P., 1963, The reaction of β-propiolactone with guanosine, deoxyguanosine and RNA, Biochem. Pharmacol., 12:1441.

Sabo, D. L., Domingo, E., Bandle, E. F., Flavell, R. A., and Weissman, C., 1977, A guanosine to adenosine transition in the 3' terminal extracistronic region of bacteriophage Oβ RNA leading to loss of infectivity, J. Mol. Biol., 112:235.

Sattsangi, P. D., Leonard, N. J., and Frihart, C. R., 1977, $1,N^2$-Ethenoguanine and N^2,3-ethenoguanine. Synthesis and comparison of the electronic spectral properties of these linear and angular triheterocycles related to the Y bases, J. Org. Chem., 42:3292.

Schuster, H., and Schramm, G., 1958, Bestimmung der biologisch wirksamen einheit in der ribosenucleinsaure des tabakmosaikvirus auf chemischen wege, Z. Naturforsch., 136:698.

Shackleton, J., Warren, W., and Roberts, J. J., 1979, The excision of N-methyl-N-nitrosourea-induced lesions from the DNA of Chinese hamster cells as measured by the loss of sites sensitive to an enzyme extract that excises 3-methylpurines but not O^6-methylguanine, Eur. J. Biochem., 97:425.

Shapiro, R., Dubelman, S., Freinberg, A., Crain, P., and McCloskey, J., 1977, Isolation and identification of cross-links of nucleoside from nitrous acid treated deoxyribonucleic acid, J. Am. Chem. Soc., 99:302.

Shapiro, R., Law, D. C. F., and Weisgras, J. M., 1972, A chemical probe for single-stranded RNA, Biochem. Biophys. Res. Commun., 49:358.

Shapiro, R., Serris, R. E., and Welcher, M., 1970, Reactions of uracil and cytosine derivatives with sodium bisulfite. A specific deamination method, J. Am. Chem. Soc., 92:422.

Shinohara, K., and Cerutti, P. A., 1977, Excision repair of benzo[a]pyrene-deoxyguanosine adducts in baby hamster kidney 21/C13 cells and in secondary mouse embryo fibroblasts C57BL/6J, Proc. Natl. Acad. Sci. U.S.A., 74:979.

Shooter, K. V., and Slade, T. A., 1977, The stability of methyl and ethyl phosphotriesters in DNA in vivo, Chem. Biol. Interact., 19:353.

Sims, P., and Grover, P. L., 1974, Epoxides in polycyclic aromatic hydrocarbon metabolism and carcinogenesis, Adv. Cancer Res., 20:165.

Singer, B., 1975, The chemical effects of nucleic acid alkylation and their relationship to mutagenesis and carcinogenesis, Prog. Nucleic Acid Res. Mol. Biol., 15:219.

Singer, B., 1976, All oxygens in nucleic acids react with carcinogenic ethylating agents, Nature, 264:333.

Singer, B., 1977, Sites in nucleic acids reacting with alkylating agents of differing carcinogenicity or mutagenicity, J. Toxicol. Environ. Health, 2:1279.

Singer, B., 1979, Guest Editorial. N-nitroso alkylating agents: Formation and persistence of alkyl derivatives as contributing factors in carcinogenesis, J. Natl. Cancer Inst., 61:1329.

Singer, B., 1980, Effect of base modification on fidelity in transscription, in: "Carcinogenesis: Fundamental Mechanisms and Environmental Effects," B. Pullman, P. O. P. Ts'o, and H. Gelboin, eds., D. Reidel, Dordrecht.

Singer, B., Bodell, W. J., Cleaver, J. E., Thomas, G. H., Rajewsky, M. F., and Thon, W., 1978, Oxygens in DNA are main targets for ethylnitrosourea in normal and xeroderma pigmentosum fibroblasts and fetal rat brain cells, Nature, 276:85.

Singer, B., and Brent, T. P., 1981, Human lymphoblasts contain DNA glycosylase activity excising N-3 and N-7 methyl and ethyl purines but not O^6-alkylguanine or 1-alkyladenine, Proc. Natl. Acad. Sci. U.S.A., 78:856.

Singer, B., and Fraenkel-Conrat, H., 1969, The role of conformation in chemical mutagenesis, Prog. Nucleic Acid Res. Mol. Biol., 9:1.

Singer, B., and Fraenkel-Conrat, H., 1974, Correlation between amino acid exchanges in coat protein of TMV mutants and the nature of the mutagens, Virology, 60:485.

Singer, B., and Fraenkel-Conrat, H., 1975, The specificity of different classes of ethylating agents toward various sites in RNA, Biochemistry, 14:772.

Singer, B., Fraenkel-Conrat, H., and Kuśmierek, J. T., 1978, Preparation and template activities of polynucleotides containing O^2- and O^4-alkyluridine, Proc. Natl. Acad. Sci. U.S.A., 75:1722.

Singer, B., and Kröger, B., 1979, Participation of modified nucleosides in translation and transcription, Prog. Nucleic Acid Res. Mol. Biol., 23:151.

Singer, B., Pergolizzi, R. G., and Grunberger, D., 1979, Synthesis and coding properties of dinucleoside diphosphates containing alkyl pyrimidines which are formed by the action of carcinogens on nucleic acids, Nucleic Acids Res., 6:1709.

Singer, B., Pulkrabek, P., Weinstein, I. B., and Grunberger, D., 1980, Infectivity and reconstitution of TMV RNA modified with N-acetoxy-2-acetylamino-fluorene or benzo[a]pyrene 7,8-dihydrol 9,10 oxide, Nucleic Acids Res., 8:2067.

Singer, B., and Spengler, S., 1981, Ambiguity and transcriptional errors as a result of modification of exocyclic amino groups of cytidine, guanosine and adenosine, Biochemistry, 20:1127.

Singer, B., Spengler, S., and Bodell, W. J., 1981, Tissue-dependent enzyme-mediated repair or removal of O-ethyl pyrimidines and ethyl purines in carcinogen-treated rats, Carcinogenesis, in press.

Sirover, M. A., and Loeb, L. A., 1976, Restriction of carcinogen-induced error incorporation during in vitro DNA synthesis, Cancer Res., 36:516.

Spengler, S., and Singer, B., 1981, Transcriptional errors and ambiguity resulting from the presence of $1,N^6$-ethenoadenosine or $3,N^4$-ethenocytidine in polyribonucleotides, Nucleic Acids Res., 9:365.

Strauss, B., Tatsumi, K., Karran, P., Higgens, N. P., Ben-Asher, E., Altamirano-Dimas, M., Rosenblatt, L., and Bose, K., 1978, Mechanisms of DNA excision repair in human cells, in: "Polycyclic Hydrocarbons and Cancer," Vol. 2, V. Gelboin, ed., Academic Press, New York.

Streisinger, G., Okada, Y., Emrich, J., Newton, J., Tsugita, A., Terzaghi, E., and Inouye, M., 1966, Frameshift mutations and the genetic code, Cold Spring Harbor Symp. Quant. Biol., XXXI: 77.

Sun, L., and Singer, B., 1975, The specificity of different classes of ethylating agents toward various sites of HeLa cell DNA in vitro and in vivo, Biochemistry, 14:1795.

Swenson, D. H., Harbach, P. R., and Trzos, R. J., 1980, The relationship between alkylation of specific DNA bases and induction of sister chromatid exchange, Carcinogenesis, 1:931.

Swenson, D. H., and Lawley, P. D., 1978, Alkylation of deoxyribonucleic acid by carcinogens dimethyl sulfate, ethyl methanesulfonate, N-ethyl-N-nitrosourea and N-methyl-N-nitrosourea. Relative reactivity of the phosphodiester site thymidylyl (3'-5') thymidine, Biochem. J., 171:575.

Tarpley, W. G., Miller, J. A., and Miller, E. C., 1980, Adducts from the reaction of N-benzoyloxy-N-methyl-4-aminoazobenzene with deoxyguanosine or DNA in vitro and from hepatic DNA of mice treated with N,N-dimethyl-4-aminobenzene, Cancer Res., 40:2493.

Tong, W. P., Kirk, M. C., and Ludlum, D. B., 1981, Molecular pharmacology of the haloethyl nitrosoureas: Formation of 6-hydroxyethylguanine in DNA treated with BCNU (N,N'-bis[2-chloroethyl]-N-nitrosourea, Biochem. Biophys. Res. Commun., 100:351.

Tong, W. P., and Ludlum, D. B., 1978, Mechanism of action of the nitrosoureas. I. Role of fluoroethylcytidine in the reaction of bis-fluoroethyl nitrosourea with nucleic acids, Biochem. Pharmacol., 27:77.

Topal, M. D., and Fresco, J. R., 1976a, Complementary base pairing and the origin of substitution mutations, Nature, 263:285.

Topal, M. D., and Fresco, J. R., 1976b, Base pairing and fidelity in codon-anticodon interaction, Nature, 263:289.

Van Duuren, B. L., and Loewengart, G., 1977, Reaction of DNA with glycidaldehyde. Isolation and identification of a deoxyguanosine reaction product, J. Biol. Chem., 252:5370.

Verly, W. G., 1980, Prereplicative error-free DNA repair, Biochem. Pharmacol., 29:977.

Wang, T. V., and Cerutti, P., 1979, Formation and removal of aflatoxin B_1-induced DNA lesions in epithelioid human lung cells, Cancer Res., 39:5165.

Wang, T. V., and Cerutti, P., 1980a, Spontaneous reactions of aflatoxin B_1 modified deoxyribonucleic acid in vitro, Biochemistry, 19:1692.

Wang, T. V., and Cerutti, P., 1980b, Effect of formation and removal of aflatoxin B_1: DNA adducts in 10T½ mouse embryo fibroblasts on cell viabilities, Cancer Res., 40:2904.

Weinstein, I. B., Jeffrey, A. M., Jennette, K. W., Blobstein, S. H., Harvey, R. G., Harris, C., Autrup, H., Kasai, H., and Nakanishi, K., 1976, Benzo[a]pyrene diol-epoxides as intermediates in nucleic acid binding in vitro and in vivo, Science, 193:592.

Westra, J. G., Kriek, E., and Hittenhausen, H., 1976, Identification of the persistently bound form of the carcinogen N-acetyl-2-aminofluorene to rat liver DNA in vivo, Chem. Biol. Interact., 15:149.

Witkin, E. M., 1976, Ultraviolet mutagenesis and inducible DNA repair in Escherichia coli, Bacteriol. Rev., 40:869

Wogan, G. N., Croy, R. G., Essigmann, J. M., Bennett, R. W., 1980, Aflatoxin-DNA interactions: Qualitative, quantitative and kinetic features in relation to carcinogenesis, in: "Carcinogenesis: Fundamental Mechanisms and Environmental Effects," B. Pullman, P. O. P. Ts'o, and H. Gelboin, eds., D. Riedel, Dordrecht.

Youvan, D. C., and Hearst, J. E., 1979, Reverse transcriptase pauses at N^2-methylguanine during in vitro transcription of Escherichia coli 16S ribosomal RNA, Proc. Natl. Acad. Sci. U.S.A., 76:3751.

CHAPTER 2

REGULATION AND FUNCTIONS OF Escherichia coli GENES

INDUCED BY DNA DAMAGE

Graham C. Walker, Cynthia J. Kenyon, Anne Bagg,
Stephen J. Elledge, Karen L. Perry, and
William G. Shanabruch

Biology Department, Massachusetts Institute of
Technology, Cambridge, Massachusetts 02139

SUMMARY

We have used the operon fusion vector Mud(Ap, lac) to generate a set of fusions within the Escherichia coli chromosome in which β-galactosidase (the product of the lacZ gene carried by the phage) is induced in response to DNA damaging agents such as UV, mitomycin C, 4-nitroquinoline-1-oxide, and N-methyl-N'-nitro-N-nitrosoguanidine. This induction is not seen in $recA^-$ or $lexA^-$ cells. We have identified two members of this set of inducible genes as being uvrA and uvrB; the products of these two genes are required for excision repair of pyrimidine dimers and other lesions. In addition we have recently isolated a Mud(Ap, lac) insertion in the umuC gene. The product of this gene is specifically required for most chemical mutagenesis in E. coli and for inducible (Weigle) reactivation of UV-irradiated bacteriophage λ. Expression of β-galactosidase in the umuC::Mud(Ap, lac) fusion strain was induced by UV and a variety of agents in a $recA^+ lexA^+$-dependent fashion. In all, the expression of ten E. coli genes is now known to be induced by DNA damage. Genetic analysis and biochemical experiments indicate that the lexA gene product is the direct repressor of these genes and that proteolytic cleavage of the lexA protein by the recA protease is required for their induction.

We have also been attempting to determine the cellular processes involved in chemical mutagenesis. We have recently cloned the umuC gene. In addition we have been studying pKM101, which is a 35.4 kb plasmid derived from the clinically isolated plasmid R46. pKM101 increases the susceptibility of cells to mutagenesis by a variety of agents and also increases their resistance to killing by UV. Because of its ability to increase chemical mutagenesis, pKM101 was introduced into the Ames Salmonella tester strains, and it has played a major role in the success of the system. The effects of the plasmid on mutagenesis and resistance are not seen in recA⁻ or lexA⁻ cells. Moreover, pKM101 suppresses the nonmutability of a umuC mutant. The simplest interpretation of these results is that pKM101 carries an analog of the chromosomal umuC gene. By a combination of insertion mutagenesis, subcloning, and construction of deletion derivatives, we have identified an approximately 2000 bp region of the pKM101 DNA (the muc gene), which is responsible for these effects. Interestingly this region of pKM101 DNA is flanked by inverted repeats raising the possibility that the gene(s) required for mutagenesis are, or once were, transposable. By use of in vitro techniques, we have constructed a muc-lac fusion, and we have used this to demonstrate that the pKM101 muc function is inducible by UV.

INTRODUCTION

In Escherichia coli, DNA damage elicits a diverse set of biological responses, termed the SOS responses (Radman, 1975; Witkin, 1976). For example, certain prophage, such as λ, are induced, the key event being proteolytic cleavage of the phage repressor (Roberts et al., 1978). A cellular system, often referred to as error-prone repair, is induced that processes damaged DNA in such a way that mutations result (Witkin, 1974, 1976). The damaged cells induce a capacity to reactivate and mutate UV-irradiated bacteriophage (Weigle-reaction and Weigle-mutagenesis) (Defais et al., 1971). The recA protein is synthesized at high levels following DNA damage (McEntee et al., 1976). DNA damage also induces filamentous growth, apparently resulting from inhibition of septum formation (Castellazzi et al., 1972). These and other responses are induced by treatment of cells by agents such as UV; mitomycin C; 4-nitroquinoline-1-oxide, or a variety of chemical carcinogens; or nalidixic acid, an inhibitor of DNA gyrase (Witkin, 1976). Moreover, treatments that interfere with DNA replication, for example, shifting a temperature sensitive dnaB mutant to the restrictive temperature, also elicit the same set of responses (Shuster et al., 1973; Witkin, 1976).

Genetic studies have indicated that the SOS responses are coordinately regulated by a process involving the products of the recA and lexA genes. One type of recA allele, termed recA⁻, prevents the induction of the set of SOS responses, whereas another type of recA allele, termed tif, (Castellazzi et al., 1972) causes the expression of SOS functions after a shift to 42°C. In addition, one type of allele of lexA, termed lexA⁻ (Mount et al., 1972), prevents induction of the set of SOS responses, whereas another type of lexA allele, termed spr (Mount, 1977), results in the constitutive expression of SOS functions. In contrast, a few mutations have been identified that specifically prevent the expression of individual SOS responses.

For example, a sfiA mutation prevents filamentous growth but not responses such as λ induction (George et al., 1975). Similarly a umuC⁻ mutant lacks the "error-prone repair" responses and Weigle-reactivation/mutagenesis (Kato and Shinoura, 1977) but still can induce λ, the recA protein, and undergo filamentous growth (Kato and Shinoura, 1977; Walker and Dobson, 1979).

The biochemistry and regulation of recA is understood in considerable detail. The recA gene product is a 38,000 dalton (Sancar et al., 1980) protein with two distinct classes of biochemical activities. It is a specific protease that can cleave both λ repressor (Roberts et al., 1978) and the lexA protein (Little and Harper, 1979). It can catalyze a number of reactions that look as though they could be involved in homologous recombination, a process that has an absolute requirement for recA function in vivo. For example, recA protein will catalyze a "strand assimilation" reaction, an invasion of a homologous single strand into a duplex DNA to form a D loop (McEntee et al., 1979; Shibata et al., 1979).

The basis of the regulation of the recA gene is outlined schematically in Fig. 1. The lexA protein is the direct repressor of the recA gene (R. Brent and M. Ptashne; J. Little and D. Mount, personal communications). In uninduced cells, the recA protein is present at a relatively low level; however, the amount present is evidently sufficient for normal homologous recombination. The recA protease does not seem to be active in uninduced cells. Exposure of the cells to UV or other SOS-inducing treatments initiates a series of events within the cell that leads to activation of the recA protease activity by a mechanism that has not yet been conclusively established. The recA protein then proteolytically cleaves the lexA repressor leading to increased synthesis of recA protein whose protease activity is then similarly activated. The activated recA protein can also cleave λ repressor leading to λ induction. A lexA⁻ mutation makes the lexA protein much less susceptible to cleavage by the recA protease and consequently prevents the induction of the recA gene (Little and Harper, 1979).

In this paper we will describe our use of an operon fusion technique to study the molecular basis of the regulation of SOS phenomena. In addition, we will summarize our recent work concerning the cellular processes involved in UV and chemical mutagenesis.

GENERAL PROCEDURES

The bacterial strains used have been described previously (Kenyon and Walker, 1980, 1981; Langer et al., 1981; Shanabruch and Walker, 1980), as have the relevant procedures for phage growth, isolation of Mud(Ap, lac) fusions, plasmid transfers, isolation of

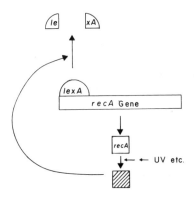

Fig. 1. Regulation of the recA gene of E. coli.

insertion mutants, and genetic mapping. The media used have been described (Kenyon and Walker, 1980; Shanabruch and Walker, 1980). β-galactosidase assays were described by Kenyon and Walker (1980). Restriction endonuclease mapping and gel electrophoresis followed the procedures of Shanabruch and Walker (1980).

THE Mud(Ap, lac) OPERON FUSION VECTOR

In order to study the molecular basis of SOS regulation, we have taken advantage of the Mud(Ap, lac) bacteriophage, a very powerful tool constructed by Casadaban and Cohen (1979) that greatly facilitates the isolation of operon fusions. Mud(Ap, lac) is a derivative of bacteriophage Mu, which, like λ, can integrate into the E. coli chromosome and then elaborate a repressor that shuts off most of the phage genes, thereby allowing it to remain stably integrated. However, unlike λ, which prefers to integrate at one unique site, Mu can integrate into the chromosome essentially at random. As shown in Fig. 2, Mud(Ap, lac) is missing some of the original phage genes and now carries a β-lactamase gene conferring ampicillin resistance. Thus Mud(Ap, lac) has similarities to transposable elements such as Tn5 in that it can insert randomly into the chromosome, and its insertion can easily be detected by the acquisition of antibiotic resistance by the organism. In addition, Mud(Ap, lac) also carries the structural genes of the lac operon including the lacZ gene whose product is β-galactosidase. However, the phage does not carry any promoter for the lac genes (Casadaban and Cohen, 1979).

Thus the lac genes are not expressed when Mud(Ap, lac) integrates into the chromosome unless it inserts, in the correct orientation, into a chromosomal gene that is being constitutively

REGULATION AND FUNCTIONS OF GENES

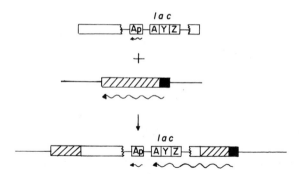

Fig. 2. Creation of an operon fusion by the insertion of Mud(Ap, lac) into a chromosomal gene. In such a fusion the expression of β-galactosidase is now regulated by the control region of the chromosomal gene.

expressed (Fig. 2). In that case, RNA polymerase initiates transcription at the promoter for the chromosomal gene and continues on into the lac genes on the phage. The net result of such insertions is that the synthesis of β-galactosidase is now regulated by the control region of the chromosomal genes. Two points concerning Mud(Ap, lac) fusions are of particular importance. (i) These are operon fusions not gene fusions. The lacZ gene has a translational start signal so that normal β-galactosidase is synthesized no matter where the mRNA synthesis was initiated. (ii) At the same time that an operon fusion is formed, the chromosomal gene is usually mutated because of the insertion of the phage into it.

SCREENING FOR DAMAGE-INDUCIBLE GENES

We first used the Mud(Ap, lac) phage to search directly for genes whose expression was increased by DNA damage (Kenyon and Walker, 1980). Random insertions of Mud(Ap, lac) into a chromosome of a Δlac E. coli strain were isolated by selecting for ampicillin-resistant colonies. The colonies were then replica-plated onto indicator plates for β-galactosidase activity that either contained, or did not contain, the DNA damaging agent mitomycin C. We then screened for Mud(Ap, lac) fusions that expressed β-galactosidase at higher levels in the presence of mitomycin C than in its absence. From 40,000 random insertions of Mud(Ap, lac), C Kenyon obtained ten independent fusions to din (damage inducible) genes; these subsequently were mapped to five different genetic loci. It is probably worth noting that this search we carried out is a specific application of a general approach, that of searching for a set of genes that are members of a common regulatory network without having to know the function of the genes in advance.

Before characterizing the din::Mud(Ap, lac) fusions, we constructed a Mud (Ap, lac)-generated fusion of lac to a prophage λ promoter because λ genes were known to be expressed in response to DNA damage. This fusion served as the standard of comparison in our subsequent induction studies.

INDUCTION OF din::Mud(Ap, lac) FUSIONS

Expression of β-galactosidase in the din::Mud(Ap, lac) and λ::Mud(Ap, lac) fusions is induced not only by mitomycin C but by a variety of DNA-damaging agents, including UV (Fig. 3), 4-nitroquinoline-1-oxide, and N-methyl-N'-nitro-N-nitrosoguanidine. Moreover, other agents or treatments that induce SOS functions, such as nalidixic acid and thymine starvation, also induce β-galactosidase synthesis in these strains.

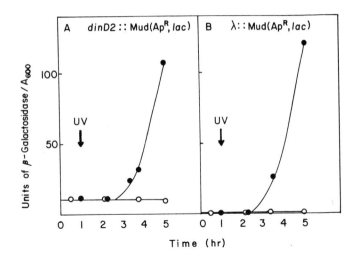

Fig. 3. Kinetics of UV induction of β-galactosidase in dinD2::Mud(Ap, lac) (A) and in λ::Mud(Ap, lac) (B) fusion strains (Kenyon and Walker, 1980). Exponential cultures were exposed to UV light (60 J/m^2) at the times indicated. Aliquots were removed periodically and the total activity of the β-galactosidase in the extract was determined. Cell density was determined by measuring the absorbance at 600 nm. ○, Unirradiated cultures; ●, irradiated cultures.

REGULATION AND FUNCTIONS OF GENES

INDUCTION OF uvrA AND uvrB

One of our initially isolated din::Mud(Ap, lac) fusions mapped at the uvrA locus and made the cells very sensitive to killing by UV (Fig. 4). In this fusion strain, β-galactosidase synthesis was induced by DNA damaging agents and, as we will discuss later, this induction was under $recA^+lexA^+$ control. We were surprised by this result because $uvrA^+$-dependent excision repair is observed both in cells in which protein synthesis has been inhibited (Boyle and Setlow, 1970) and in $recA^-$ and $lexA^-$ cells (Clark and Volkert, 1978), findings that had led to the conclusion that the uvrA gene is constitutively expressed and not under the control of the $recA^+lexA^+$ regulatory circuitry (Radman, 1975).

At first we were initially concerned that some secondary event, such as a Mu-mediated deletion (Faelen et al., 1978), could have resulted in the fusion of the Mud(Ap, lac) insertion at the uvrA locus to the regulatory region of some gene other than uvrA. We therefore decided to isolate Mud(Ap, lac) insertions in the uvrA gene without imposing the requirement that the resulting fusions possess specific regulatory characteristics. We conducted a search for uvrA::Mud(Ap, lac) mutants based on the criterion of UV sensitivity. After screening about 110,000 random Mud(Ap, lac) insertions, Kenyon identified 13 insertions in the uvrA gene (Fig. 4) (Kenyon and Walker, 1981). Eight of these were shown by chromosome mobilization experiments to have the same orientation in the chromosome as the original uvrA::Mud(Ap, lac) fusion and to be inducible to comparable extents. The remaining uvrA::Mud(Ap, lac) insertions had the opposite orientation in the chromosome and showed no β-galactosidase expression under any conditions. Our conclusion that the uvrA gene is induced by DNA damage has recently been confirmed by B. M. Kacinski, D. W. Rupp, and E. Seeberg (personal communication) who have found that the levels of the uvrA gene product are increased by treatments or conditions that result in the expression of SOS functions.

In this same experiment, 5 Mud(Ap, lac) insertions were obtained that mapped at the uvrB locus. Another uvrB::Mud(Ap, lac) fusion was independently isolated by Fogliano and Schendel (1981). Although, curiously, these fusions show little mitomycin C inducibility, they were induced by UV and nalidixic acid to the same extent as the uvrA fusions (Fogliano and Schendel, 1981; Kenyon and Walker, 1981). Thus at least two of the genes induced in response to DNA damage are ones whose products have known roles (Seeberg, 1978) in DNA repair processes.

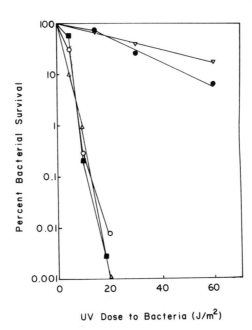

Fig. 4. Mud(Ap, lac) insertions in uvrA. Cells were grown in supplemented minimal medium to 1×10^8 cells/ml. Aliquots were irradiated as indicated. Survival was determined by plating appropriate dilutions on supplemented M9 plates. ▽, uvrA$^+$; △, uvrA6; ○, uvrA215::Mud(Ap, lac); ●, uvrA215::Mud(Ap, lac)/pDR2000 (pDR2000 is a small plasmid carrying the uvrA$^+$ gene); ■, uvrA215::Mud(Ap, lac)/pDR2026 (pDR2026 is a derivative of pDR2000 carrying a uvrA::γδ mutation) (Sancar et al., 1981).

INDUCTION OF umuC

Since there appears to be an absolute requirement for umuC function for much chemical mutagenesis (Kato and Shinoura, 1977; Steinborn, 1978) and since this type of mutagenesis seems to be an inducible process (Witkin, 1974), it seemed likely that the umuC gene might be inducible. When we did not obtain an insertion of Mud(Ap, lac) in umuC during our first screen for din genes, we then decided to screen for insertions within umuC by looking directly for loss of umuC function. By screening 17,000 random Mud(Ap, lac) insertions looking for nonmutable phenotypes, A. Bagg isolated an insertion that mapped at the position of umuC and had the phenotypic characteristics of known umuC mutants: (i) nonmutability with UV and other agents, (ii) deficiency in Weigle reactivation, (iii) modest sensitivity to UV killing, (iv) unimpaired ability to carry out other SOS functions, such as λ induction (Bagg et al., 1981). As shown in Fig. 5, β-galactosidase expression in this Mud(Ap, lac) fusion was strongly induced by DNA damage. Thus, in order to be mutagens in E. coli many chemicals have to do two things: (i) introduce a lesion into the DNA that can be processed in such a way that a mutation results, and (ii) induce at least one function that is required for that processing.

ANALYSIS OF din GENE REGULATION

To date, a total of ten E. coli genes have been shown to be inducible by DNA-damaging agents (Fig. 6). Mud (Ap, lac) fusions have been obtained to uvrA (Kenyon and Walker, 1981), uvrB (Fogliano and Schendel, 1981), umuC (Bagg et al., 1981), sfiA (Huisman and D'Ari, 1981), dinA, dinB, dinD, and dinF (Kenyon and Walker, 1980). Both recA (McEntee et al., 1976) and lexA (Brent and Ptashne, 1980) have been shown to be inducible by other means.

Obtaining lac fusions to din genes greatly facilitated the analysis of their regulation. Studies of the regulation of the SOS responses themselves are difficult for a couple of reasons. First, the individual SOS responses, such as chemical mutagenesis or filamentous growth, are often experimentally unwieldly. Second, some proteins may play dual regulatory and mechanistic roles. For example, the protease activity of the recA protein could be involved in the regulation of a din gene, yet one of the recA protein's "recombinational" activities could be required in the pathway in which the product of that din gene subsequently acts.

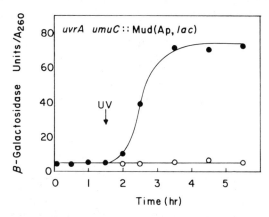

Fig. 5. UV induction of β-galactosidase synthesis in a uvrA umuC:: Mud(Ap, lac) fusion strain. Experimental procedure was similar to that described in the legend to Fig. 3 except that a dose of 1.5 J/m^2 was used.

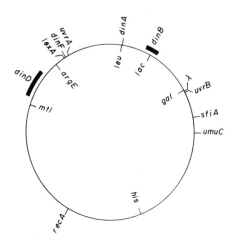

Fig. 6. Map position of E. coli genes whose expression is stimulated by DNA damage.

Our genetic analyses of the regulation of uvrA, uvrB, umuC, dinA, dinB, dinD, and dinF have indicated that all of these din genes are regulated in a similar fashion by a process requiring the products of the recA and lexA genes. Huisman and D'Ari (1981) have obtained similar results for sfiA. A classical recA mutation, which causes loss of both protease and "recombinational" functions of the recA protein, prevents din gene induction. The effect of a recA⁻ mutation in the induction of uvrA::Mud(Ap, lac) is shown in Fig. 7. Moreover, a second type of recA mutation [previously referred as lexB (Morand et al., 1977)] that causes a deficiency in the protease activity of recA protein (Roberts and Roberts, 1981) but not in its "recombination" functions caused a considerable reduction in the induction of the din::Mud(Ap, lac) fusions. These data indicate that there normally is a requirement for recA function(s) for din gene induction and that the protease function of recA is important in this process.

A lexA⁻ mutant produces a lexA protein that is much less susceptible to cleavage by the recA protease (Little and Harper, 1979). The introduction of a lexA⁻ mutation into the din-lac fusion strain prevented induction of the din genes. Figure 7 shows the effect of a lexA⁻ mutation on the induction of a uvrA::Mud(Ap, lac) fusion. In contrast, another allele of lexA termed spr has been identified that seems to result in loss of lexA activity (Mount, 1977). The introduction of a spr mutation into the din-lac fusions strains resulted in high level constitutive expression of β-galactosidase in these strains. An example of this for the case of uvrA is

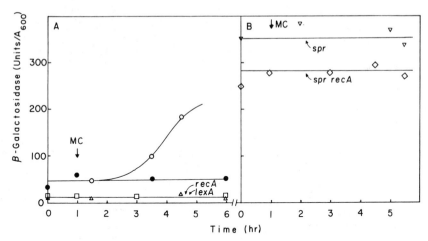

Fig. 7. Kinetics of induction of β-galactosidase in uvrA::Mud(Ap, lac) fusion strains (Kenyon and Walker, 1981). Exponential cultures growing in supplemented M9 media were treated with 1 μg/ml mitomycin C (MC) as shown and the specific activity of β-galactosidase was determined. (A) ●, uvrA215::Mud(Ap, lac), no MC; ○, uvrA215::Mud(Ap, lac); □, recA uvrA215::Mud(Ap, lac); △, lexA uvrA215::Mud(Ap, lac). (B) ▽, spr uvrA215::Mud(Ap, lac); ◇, spr recA uvrA215::Mud(Ap, lac).

shown in Fig. 7. These data indicate that lexA is playing a negative regulatory role in the control of the din genes.

There are two simple models for din gene regulation consistent with the observations that lexA seems to be acting as a repressor and that recA protease activity is required for din gene induction in a normal cell; these are shown in Fig. 8. In the first model, lexA functions as a direct repressor of din genes, such as the uvrA gene. (This type of model is a direct analogy to the regulation of the recA gene). A spr mutation, which eliminates lexA gene function, directly allows the expression of the din genes. In a second model, a repressor other than the lexA protein regulates the expression of the din genes. (This type of model is a direct analogy to the regulation of λ). In this case, a spr mutation would affect din gene in an indirect fashion by leading to high levels of synthesis of the recA protein (since lexA is the direct repressor of the recA gene) and the recA protein would then inactivate the repressor(s) of the din genes.

These models can be distinguished by the fact that the first model predicts that constitutive high level din gene expression in a spr background would not require recA function, whereas the second model predicts that it would require recA function. As shown in Fig. 7 for the case of the uvrA gene, we found that all the din:: Mud(Ap, lac) fusions we have examined to date expressed β-galactosidase at high levels even in a spr recA background. These data are consistent with the first model.

CLONING THE umuC GENE

Because of the key role of umuC in UV and chemical mutagenesis, we were interested not only in studying the regulation of this gene but in determining the biochemical role of its gene product. As a first step in this endeavor we have cloned the umuC gene. We initially attempted to clone umuC by preparing a bank of random E. coli chromosome fragments in the multicopy vector pBR322 and then screening these hybrid plasmids for their ability to complement a umuC strain. We did not identify the umuC gene by this procedure, perhaps because overproduction of umuC or some linked gene is deleterious to cell survival or results in an alternative phenotype. As an alternative approach, S. Elledge obtained a Tn5 insertion in the umuC gene and then cloned out a restriction fragment containing part of Tn5 (including the gene for neomycin phosphotransferase) as well as a small piece of the adjacent chromosomal DNA. This restriction fragment was then used to screen λ banks containing random fragments of E. coli DNA. A λ phage was identified whose DNA hybridized to DNA from the umuC region of the chromosome. Since this particular λ phage also complemented the nonmutability of a umuC mutant (Table 1), we conclude that it carries the umuC gene. We anticipate that

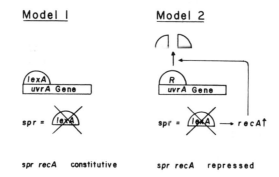

Fig. 8. Two simple models for din gene regulation.

Table 1. Complementation of a umuC⁻ Mutant by a λ-umuC⁺ Bacteriophage[a]

Strain	Bacteriophage	Arg⁺ Revertants/10^7 Survivors	
		(0 J/m^2)	(20 J/m^2)
AB1157(pKB280)	—	1.5	50
AB1157(pKB280)	λumuC⁺	1.3	100
GW2000(pKB280)	—	1.1	1
GW2000(pKB280)	λumuC⁺	1.0	75

[a]The genotype of AB1157 has been described (Walker, 1977). GW2000 is AB1157 umuC::Tn5. pKB280 is a derivative of pMB9 carrying the λ cI gene (Backman and Ptashne, 1978). Cells were grown to a density of 4×10^8, the medium was adjusted to 10 mM MgSO$_4$, and the cells were infected or mock-infected with λ umuC⁺ at a m.o.i. of 3. Cells plus phage were incubated together at 37°C with gentle shaking for 15 min. The cells were then UV irradiated at a fluence of 0.25 J/m^2. Mutagenesis and survival were measured as described previously (Shanabruch and Walker, 1980). Experiments using wild-type λ or Charon 28 derivatives, which lack umuC sequence homology, fail to enhance reversion rates of arg or his alleles.

analysis of this phage will be of considerable help in attempting to identify the role of the umuC gene product in mutagenesis.

ANALYSIS OF THE MUTAGENESIS-ENHANCING PLASMID pKM101

An additional approach we have been taking towards determining the cellular basis of chemical mutagenesis has been to study the mutagenesis-enhancing plasmid pKM101. pKM101 is a 35.4 kb plasmid that was derived from its clinically isolated parent plasmid R46 by the in vivo deletion of a single 14 kb region (Langer and Walker, 1981; Langer et al., 1981; Mortelmans and Stocker, 1979). Both pKM101 and R46 are members of a subset of naturally occurring plasmids that have both the ability to make cells more resistant to killing by UV and to make them more susceptible to mutagenesis by a variety of chemical agents (Drabble and Stocker, 1968; Mortelmans and Stocker, 1976). Because of its ability to enhance chemical mutagenesis, pKM101 was incorporated into the Ames Salmonella tester strains for the detection of carcinogens as mutagens (McCann et al., 1975), and it has played a major role in the success of the system

(McCann and Ames, 1976; Walker, 1981). The plasmid not only increases the susceptibility of the cells to base substitution but also to frameshift mutagenesis with the appropriate compounds.

The effects of pKM101 on mutagenesis and resistance to UV killing are dependent on the recA$^+$lexA$^+$ genotype, as are the SOS functions (Walker, 1977). However pKM101 does not seem to exert its effects by causing constitutive expression of SOS functions since the presence of pKM101 does not cause recA induction (Walker and Dobson, 1979) or λ-induction (Goze and Devoret, 1979). Rather it appears possible that pKM101 may possibly carry an analog of the chromosomally encoded umuC gene since pKM101 suppresses the nonmutability of E. coli umuC mutants (Fig. 9), including the umuC::Mud(Ap, lac) mutant (Bagg et al., 1981; Walker and Dobson, 1979). Moreover, the introduction of pKM101 into a cell restores the capacity of the cell to carry out inducible phage reactivation (Weigle reactivation) (Walker, 1978a; Walker and Dobson, 1979) suggesting that some plasmid-encoded function might be induced by DNA damage.

We were able to obtain both N-methyl-N'-nitro-N-nitrosoguanidine-derived (Walker, 1978b) and insertion mutants (Shanabruch and Walker, 1980) of pKM101 that were no longer able to increase the susceptibility of cells to base substitution mutagenesis (Fig. 9). These same plasmid mutants were also defective in their ability to increase frameshift mutagenesis and make cells resistant to UV killing, suggesting that all three processes are closely related. Shanabruch mapped the position of 20 independent Tn5 insertions in pKM101 that eliminated the abilities of pKM101 to enhance mutagenesis or UV resistance. All of these mapped within an approximately 2000 bp region on pKM101 that we have termed the muc (mutagenesis: UV and chemical) genes (Fig. 10) (Langer et al., 1981; Shanabruch and Walker, 1980). As can be seen in Fig. 9, muc$^-$ mutants of pKM101 are no longer capable of suppressing the nonmutability of umuC mutants. Thus, if pKM101 does, in fact, act by providing an analog of the umuC gene product, then the muc region codes for that umuC analog. Interestingly, the muc region is surrounded by inverted repeats, raising the possibility that the mutagenesis/repair functions are, or once were, on a transposable element.

Recently, Perry has cloned the muc region from pKM101 onto the vector pBR322 and found that the hybrid plasmid, which contains a 2.1 kb fragment of pKM101 DNA, is capable of increasing both the susceptibility of cells to chemical mutagenesis and their resistance to UV killing. Thus, the muc gene(s) of pKM101 is not only necessary but apparently sufficient for plasmid-mediated mutagenesis and UV resistance.

In addition, Elledge recently used in vitro technology (Casadaban et al., 1980) to construct a plasmid carrying a fusion

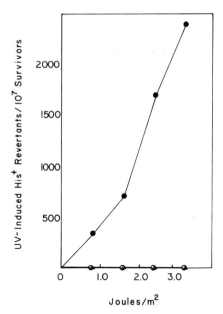

Fig. 9. Effect of pKM101 and pKM101 muc⁻ mutants on the number of UV-induced Arg⁺ revertants/survivor in a umuC⁻ background. ○, umuC⁻; ●, umuC⁻ (pKM101); △, umuC⁻ (pKM101 muc::Tn5).

of the amino terminus of the muc gene to the carboxy terminus of the β-galactosidase gene. The fusion protein has β-galactosidase activity and its synthesis is stimulated by DNA-damaging agents. It seems likely that the pKM101 muc function(s) may be regulated in a similar fashion to the chromosomally encoded din functions with lexA being the direct repressor of the plasmid-encoded muc gene(s).

CONCLUSIONS

Damage to the DNA of E. coli results in the increased expression of a set of genes — recA, lexA, uvrA, uvrB, umuC, sfiA, dinA, dinB, dinD and dinF — whose products play biological roles in processes as diverse as chemical mutagenesis and filamentous growth. It is clear that at least some of the SOS responses result from increased expression of these genes. Moreover, there may well be additional genes that are members of this regulatory network that have not yet been identified.

Our current understanding of the regulation of din genes is outlined schematically in Fig. 11. The lexA protein appears to

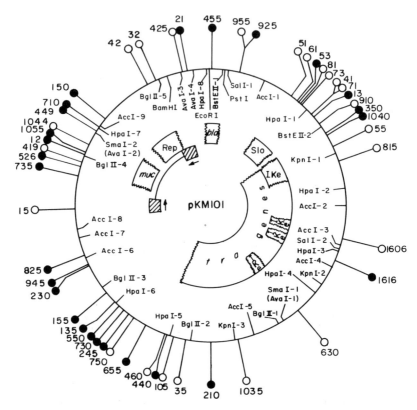

Fig. 10. pKM101 restriction map showing position of Tn5 insertions and regions of pKM101 DNA essential for pKM101 phenotypes (Langer et al., 1981). Restriction enzyme cleavage sites are numbered according to their clockwise order from the EcoRI site, which has been given the position of "0." Positions of Tn5 insertions are indicated by lines with open circles (Tn5 DNA is in the "-" orientation) or with filled circles (Tn5 DNA is in the "+" orientation) (Shanabruch and Walker, 1980). The names of insertions are numbers next to the circles. The regions of DNA essential for pKM101 phenotypes are indicated: bla-ampicillin resistance, Slo-slow growth, Ike-sensitivity to phage Ike, tra-conjugal transfer, muc-enhancement of UV and chemical mutagenesis, Rep-region essential for plasmid replication. The pKM101 inverted repeat and loop region is indicated with two shaded boxes connected with a line. Arrows indicate inverted repeats. Only 6 of the 20 Tn5 muc insertions characterized by Shanabruch and Walker (1980) are shown on this map.

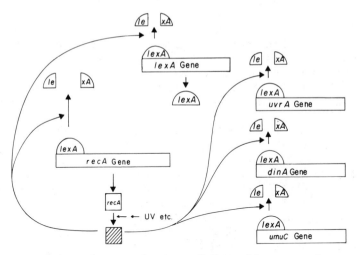

Fig. 11. Model for the regulation of E. coli genes that are induced by DNA damage.

function as the direct repressor of all of the above genes. The recA protein is made in moderately low levels in uninduced cells. UV irradiation and other inducing treatments initiate a series of events, at this stage still incompletely understood, that result in activation of the recA protease. Biochemical experiments have suggested single-stranded DNA might be the ultimate inducing signal. The recA protease then cleaves the lexA protein resulting in increased synthesis of the recA protein. Since the pool of lexA product is being depleted as the result of cleavage by recA protein, the various din genes, such as uvrA and umuC, begin to be expressed at high levels. The point at which a particular din gene would begin to be expressed at increased levels would depend on the Kd of binding of lexA protein to the regulatory region of that din gene. Direct biochemical evidence has been obtained that the lexA protein is the direct repressor of the recA and lexA genes (R. Brent and M. Ptashne; J. Little and D. Mount, personal communications). Recently C. Kenyon cloned the promoters of the dinA, dinB, and dinD genes, and in collaboration with R. Brent and M. Ptashne we have preliminary evidence that purified lexA protein will inhibit transcription from the promoters of these genes. The binding of the lexA protein to the operator of the lexA gene probably plays a role in regulating the level of lexA protein in uninduced cells. The system is probably shut off by a reduction in the amount of an inducing signal such as single-stranded DNA. Simply having large amounts of recA protein within cells is not enough to cause expression of the SOS functions; apparently the protease function of recA must be activated. Our recent experiments with pKM101 suggest that some plasmid-coded genes may be regulated as part of this same network.

It is conceivable that the whole regulatory picture may be even more complex than this. For example, the possibility exists that some din genes may be regulated by a repressor that is not the lexA product (as in Fig. 8, Model 2), but we have simply not yet identified an example of this type of din gene. Moreover, we cannot rule out the possibility that the recA protease has some additional roles. For example, it could proteolytically modify constitutively produced cellular proteins or even the gene products of certain din genes.

Finally, one of the most interesting of the genes induced by DNA damage is the umuC gene. Its product plays a key and, as yet undefined, role in UV and chemical mutagenesis. An analogous function may well be encoded by the pKM101 muc gene. We are optimistic that our studies of the umuC and muc genes will help to elucidate the biochemical basis of chemical mutagenesis.

ACKNOWLEDGMENTS

We thank J. Geiger, J. Krueger, B. Mitchell, P. Pang, and S. Winans for their many useful discussions and L. Withers for help in preparing the manuscript.

This work was supported in part by a Rita Allen Scholar Award to G. C. W. and Public Health Service Grant 5-R01-CA21615-04 from the National Cancer Institute. C. J. K. was a National Science Foundation Predoctoral Fellow and now is supported by a Johnson and Johnson Associated Industries Fund Fellowship. W. G. S., K. L. P., and S. J. E. were supported by National Institutes of Health Predoctoral Training Grant GM07287. A. B. was supported in part by the Undergraduate Opportunities Program at the Massachusetts Institute of Technology.

REFERENCES

Backman, K., and Ptashne, M., 1978, Maximizing gene expression on a plasmid using recombination in vitro, Cell, 13:65.
Bagg, A., Kenyon, C.J., and Walker, G.C., 1981, Inducibility of a gene product required for UV and chemical mutagenesis in Escherichia coli, Proc. Natl. Acad. Sci. U.S.A., in press.
Boyle, J. M., and Setlow, R. B., 1970, Correlations between host-cell reactivation, ultraviolet reactivation and pyrimidine dimer excision in the DNA of bacteriophage λ, J. Mol. Biol., 51:131.

Brent, R., and Ptashne, M., 1980, The lexA gene product represses its own promoter, Proc. Natl. Acad. Sci. U.S.A., 77:1932.

Casadaban, M. J., Chou, J., and Cohen, S. N., 1980, In vitro gene fusions that join an enzymatically active β-galactosidase segment to amino-terminal fragments of exogenous proteins: Escherichia coli plasmid vectors for the detection and cloning of translational initiation signals, J. Bacteriol. 143:971.

Casadaban, M. J., and Cohen, S. N., 1979, Lactose genes fused to exogenous promoters in one step using a Mu-lac bacteriophage: In vivo probe for transcriptional control, Proc. Natl. Acad. Sci. U.S.A., 76:4530.

Castellazzi, M., George, J., and Buttin, G., 1972, Prophage induction and cell division of Escherichia coli. Further characterization of the thermosensitive mutation tif-1 whose expression mimics the effect of UV irradiation, Mol. Gen. Genet., 139.

Clark, A. J., and Volkert, M. R., 1978, A new classification of pathways repairing pyrimidine dimer damage in DNA, in: "DNA Repair Mechanisms," P. C. Hanawalt, E. C. Friedbert, and C. F. Fox, eds., Academic Press, New York.

Defais, M., Fauquet, P., Radman, M., and Errera, M., 1971, Ultraviolet reactivation and ultraviolet mutagenesis of λ in different genetic systems, Virology, 43:495.

Drabble, W. T., and Stocker, B. A. D., 1968, R (transmissible drug-resistance) factors in S. typhimurium: Pattern of transduction by phage P22 and ultraviolet protection effect, J. Gen. Microbiol., 53:109.

Faelen, M., Huisman, O., and Toussaint, A., 1978, Involvement of phage Mu-1 early functions in Mu-mediated chromosomal rearrangements, Nature, 271:580.

Fogliano, M., and Schendel, P. F., 1981, Evidence for the inducibility of the uvrB operon, Nature, 289:196.

George, J., Castellazzi, M., and Buttin, G., 1975, Prophage induction and cell division in E. coli. III. Mutations sfiA and sfiB restore division in tif and lon strains and permit the expression of mutator properties of tif, Mol. Gen. Genet., 140:309.

Goze, A., and Devoret, R., 1979, Repair promoted by plasmid pKM101 is different from SOS repair, Mutat. Res., 61:163.

Huisman, O., and D'Ari, R., 1981, An inducible DNA replication-cell division coupling mechanism in Escherichia coli, Nature, 290:797.

Kato, T., and Shinoura, Y., 1977, Isolation and characterization of mutants of Escherichia coli deficient in induction of mutations by ultraviolet light, Mol. Gen. Genet., 156:121.

Kenyon, C. J., and Walker, G. C., 1980, DNA-damaging agents stimulate gene expression at specific loci in Escherichia coli, Proc. Natl. Acad. Sci. U.S.A., 77:2819.

Kenyon, C. J., and Walker, G. C., 1981, Expression of the uvrA gene of Escherichia coli is inducible, Nature, 289:810.

Langer, P. J., Shanabruch, W. G., and Walker, G. C., 1981, Functional organization of the plasmid pKM101, J. Bacteriol., 145:1310.

Langer, P. J., and Walker, G. C., 1981, Restriction endonuclease cleavage map of pKM101: Relationship to parental plasmid R46, Mol. Gen. Genet., in press.

Little, J. W., and Harper, J. E., 1979, Identification of the lexA gene product of Escherichia coli K-12, Proc. Natl. Acad. Sci. U.S.A., 76:6147.

McCann, J., and Ames, B. N., 1976, Detection of carcinogens as mutagens in the Salmonella/microsome test: Assay of 300 chemicals: Discussion, Proc. Natl. Acad. Sci. U.S.A., 73:950.

McCann, J., Spingarn, N. E., Kobori, J., and Ames, B. N., 1975, Detection of carcinogens as mutagens: Bacterial tester strains with R factor plasmids, Proc. Natl. Acad. Sci. U.S.A., 72:979.

McEntee, K., Hesse, J. E., and Epstein, W., 1976, Identification and radiochemical purification of the recA protein of Escherichia coli, Proc. Natl. Acad. Sci. U.S.A., 73:3979.

McEntee, K., Weinstock, G. M., and Lehman, I. R., 1979, Initiation of general recombination catalyzed in vitro by the recA protein of Escherichia coli, Proc. Natl. Acad. Sci. U.S.A., 76:2515.

Morand, P., Blanco, M., and Devoret, R., 1977, Characterization of lexB mutations in Escherichia coli K-12, J. Bacteriol., 131:572.

Mortelmans, K., and Stocker, B. A. D., 1979, Segregation of the mutator property of plasmid R46 from its ultraviolet-protecting property, Mol. Gen. Genet., 167:317.

Mortelmans, K. E., and Stocker, B. A. D., 1976, Ultraviolet light protection, enhancement of ultraviolet light mutagenesis, and mutator effect of plasmid R46 in Salmonella typhimurium, J. Bacteriol., 128:271.

Mount, D. W., 1977, A mutant of Escherichia coli showing constitutive expression of the lysogenic induction and error-prone DNA repair pathways, Proc. Natl. Acad. Sci. U.S.A., 74:300.

Mount, D. W., Low, K. B., and Edmiston, S. J., 1972, Dominant mutations (lex) in Escherichia coli K-12 which affect radiation sensitivity and frequency of ultraviolet light-induced mutations, J. Bacteriol., 112:886.

Radman, M., 1975, SOS repair hypothesis phenomenology of an inducible DNA repair which is accompanied by mutagenesis, in: "Molecular Mechanisms for Repair of DNA," Part A, P.C. Hanawalt, and R.B. Setlow, eds., Plenum Press, New York.

Roberts, J. W., and Roberts, C. W., 1981, Two mutations that alter the regulatory activity of E. coli recA activity, Nature, 290:422.

Roberts, J. W., Roberts, C. W., and Craig, N. L., 1978, Escherichia coli recA gene product inactivates phage λ repressor, Proc. Natl. Acad. Sci. U.S.A., 75:4717.

Sancar, A., Stachelek, C., Konigsberg, W., and Rupp, W. D., 1980, Sequences of the recA gene and protein, Proc. Natl. Acad. Sci. U.S.A., 77:2611.

Sancar, A., Wharton, R. P., Seltzer, S., Kacinski, B. M., Clarke, N. D., and Rupp, W. D., 1981, Identification of the uvrA gene product, J. Mol. Biol., 148:45.

Schuster, H., Beyersmann, D., Mikolajczk, M., and Schlicht, M., 1973, Prophage induction by high temperature in thermosensitive dna mutants lysogenic for bacteriophage lambda, J. Virol., 11:879.

Seeberg, E., 1978, Reconstitution of an Escherichia coli repair endonuclease activity from the separated $uvrA^+$ and $uvrB^+/uvrC^+$, Proc. Natl. Acad. Sci. U.S.A., 75:2569.

Shanabruch, W. G., and Walker, G. C., 1980, Localization of the plasmid (pKM101) gene(s) involved in $recA^+lexA^+$-dependent mutagenesis, Mol. Gen. Genet., 179:289.

Shibata, T., DasGupta, C., Cunningham, R. P., and Radding, C. M., 1979, Purified Escherichia coli recA protein catalyzes homologous pairing of superhelical DNA and single-stranded fragments, Proc. Natl. Acad. Sci. U.S.A., 76:1638.

Steinborn, G., 1978, Uvm mutants of Escherichia coli K12 deficient in UV mutagenesis, Mol. Gen. Genet., 165:87.

Walker, G. C., 1977, Plasmid(pKM101)-mediated enhancement of repair and mutagenesis: Dependence on chromosomal genes in Escherichia coli K-12, Mol. Gen. Genet., 152:93.

Walker, G. C., 1978a, Inducible reactivation and mutagenesis of UV-irradiated bacteriophage P22 in Salmonella typhimurium LT2 containing the plasmid pKM101, J. Bacteriol., 135:415.

Walker, G. C., 1978b, Isolation and characterization of mutants of the plasmid pKM101 deficient in their ability to enhance mutagenesis and repair, J. Bacteriol., 133:1203.

Walker, G. C., 1981, Molecular principles underlying the Ames Salmonella/microsome test: Elements and design of short term mutagenesis test, in: "In Vitro Toxicity Testing of Environmental Agents," A. Kolber, ed., Plenum, New York, in press.

Walker, G. C., and Dobson, P. P., 1979, Mutagenesis and repair deficiencies of Escherichia coli umuC mutants are suppressed by the plasmid pKM101, Mol. Gen. Genet., 172:17.

Witkin, E. M., 1974, Thermal enhancement of ultraviolet mutability in a tif-1 uvrA derivative of Escherichia coli B/r: Evidence that ultraviolet mutagenesis depends upon an inducible function, Proc. Natl. Acad. Sci. U.S.A., 71:1930.

Witkin, E. M., 1976, Ultraviolet mutagenesis and inducible DNA repair in Escherichia coli, Bacteriol. Rev., 40:869.

CHAPTER 3

METHYLATION-INSTRUCTED MISMATCH CORRECTION AS A POSTREPLICATION
ERROR AVOIDANCE MECHANISM IN Escherichia Coli

B. W. Glickman

National Institute for Environmental Health Sciences
P. O. Box 12233, Research Triangle Park
North Carolina   27709

SUMMARY

The error rate for DNA replication in Escherichia coli has been estimated as one error per 10^8-10^{11} base pairs replicated. This level of accuracy is considerably higher than that predicted on the basis of free-energy calculations or that achieved by DNA polymerases in vitro. The existence of an excision repair mechanism acting upon mismatched base pairs could account, at least in part, for the accuracy achieved in vivo. Such a repair mechanism has been postulated to account for a number of genetic observations related to gene conversion and high negative interference phenomena. Furthermore, the increased spontaneous mutability of Pnemococcus hex⁻ and E. coli uvrE⁻ mutants that appear to be deficient in the repair of some mismatched base pairs indicates the involvement of mismatch repair in the suppression of spontaneous mutation rates. The existence of a mismatch repair system capable of efficiently correcting mispaired bases implies that a strand discrimination system must exist which allows the discrimination between the "incorrect" newly synthesized daughter strand and the "correct original parental strand.

As DNA methylation is a postreplicational process (i.e., newly synthesized strands are undermethylated). DNA methylation is one possible mechanism allowing the discrimination between parental and daughter DNA strands. In accordance with this hypothesis is the observation that dam⁻ mutants that are deficient in 6-methyladenine residues occurring at the 5'G-A-T-C3' sequence are spontaneous mutators. These strains are also hypermutable by base analogs, e.g. 2-aminopurine.

Further mutants of E. coli have been isolated that are defective in mismatch repair. These mutants, which include mutH, mutL, mutS, and uvrE, increase spontaneous base transition and frameshift rates by as much as 10^3. These mutations also affect the level of negative interference seen in phage crosses for closely linked markers as well as cellular sensitivity to alkylating agents. More direct evidence in favor of the above hypothesis has been obtained from phage λ heteroduplex assays where the DNA strands are specifically methylated or nonmethylated. These experiments show the preferential loss of the genetic marker carried on the nonmethylated DNA strand consistent with this hypothesis.

INTRODUCTION

Although DNA replication is an extraordinarily accurate process, the basis of this accuracy has not yet been fully resolved. Estimates of the error rate in E. coli wild-type strains are between 10^{-9} and 10^{-11} per base pair replicated (Cox, 1976; Drake, 1969). Properties of the DNA template and substrates and of DNA polymerase can be expected to determine an error rate of about 10^{-5} to 10^{-7} (Loeb, 1974; Topal and Fresco, 1976). This level of accuracy is in the range attained with purified DNA polymerases and associated proteins (Argwal et al., 1979; Clayton et al., 1979; Hibner and Alberts, 1980). In this chapter, a postreplication error-avoidance mechanism is described which may account for the discrepancy between replicational accuracy due to the polymerizing proteins and that actually observed in vitro.

THE EXISTENCE OF MISMATCH CORRECTION AND A ROLE FOR DNA METHYLATION

This error-avoidance mechanism involves the excision of incorrectly inserted (i.e., mismatched) bases from newly synthesized DNA. The existence of an excision repair system acting upon mismatched bases in DNA has been postulated to account for gene conversion (Holliday, 1964; Whitehouse and Hastings, 1965), high negative interference (White and Fox, 1974), and map expansion phenomena (Fincham and Holliday, 1970). Evidence for mismatch repair has been obtained from studies using transfection assays with bacteriophage λ heteroduplex DNA (Nevers and Spatz, 1975; Wildenberg and Meselson, 1975), φX174 DNA (Baas and Jansz, 1972) in E. coli, and in mammalian cells using heteroduplexed SV40 DNA (Wilson, 1977). Heteroduplex repair has also been invoked to explain the predominance of pure mutant clones following UV irradiation in yeast (Eckardt and Haynes, 1977; Hannan et al., 1976; James and Kilbey, 1977; Nasim and Auerbach, 1967). Similar explanations have been offered to explain the predominance of pure mutant clones following 5-bromouracil (5BU) treatment in E. coli (Witkin and Sicurella, 1964). Consistent with the involvement of mismatch repair in base-analog mutagenesis are the observations made by Rydberg (1977) showing the reversibility of 5BU mutagenesis when the mutagenic treatment was followed by a period of amino-acid starvation, and of the hypermutability of uvrD and uvrE derivatives which are thought to be defective in mismatch repair as determined by the loss of mismatch correction of heteroduplexed λ DNA (Nevers and Spatz, 1975).

The possible importance of mismatch repair to spontaneous mutation rates is indicated by the observation that Pneumococcus hex⁻ and E. coli uvrE mutants, both of which are probably deficient in the repair of some mismatched base pairs (Nevers and Spatz, 1975; Tiraby and Fox, 1970) are also spontaneous mutators. Such observations imply that mismatch repair acts in a directed manner to recognize and correct spontaneous errors of DNA metabolism. Such an error-avoidance pathway requires a mechanism permitting discrimination between the "correct" (parental) strand and the "error-containing" (daughter) strand. The suggestion that the transient undermethylation of newly synthesized DNA might provide a basis for such discrimination was originally proposed by Dr. M. Meselson (Harvard University) at the 1975 EMBO Recombination Workshop in Nethybridge, Scotland, and subsequently published by Wagner and Meselson (1976).

Among the DNA methylases is one which methylates adenine residues yielding 6-methyladenine in the sequence 5'-GATC-3'/3'-CTAG-5' (Gomez-Eichelmann, 1979; Lacks and Greenberg, 1977). Moreover, Okazaki fragments are greatly undermethylated at this DNA sequence (Marinus, 1976). Thus, as predicted, newly synthesized DNA is indeed undermethylated and might provide a basis for discriminating between parental and daughter strands.

THE dam GENE OF E. coli

Adenine methylation in E. coli is primarily under the control of the dam gene (Marinus, 1976; Marinus and Morris, 1975). If this methylation is involved in strand discrimination during mismatch correction, the loss of DNA methylation should increase the spontaneous error rate. Consistent with this model, dam⁻ mutants of E. coli are spontaneous mutators (Glickman, 1979; Glickman et al., 1978; Marinus and Morris, 1975). dam⁻ mutants also exhibit a complex phenotype including hyper-recombination, frequent spontaneous induction of phage λ in lysogens, lethality in combination with recA, recB, lexA, and polA mutations, and a high sensitivity to some base analogs, including 2-aminopurine (2AP) (Babcock-Harms and Glickman, unpublished data; Bale et al., 1979; Glickman et al., 1978; Marinus, 1976; Marinus and Konrad, 1976; Marinus and Morris, 1975).

A MODEL ATTEMPTING TO ACCOUNT FOR THE PLEIOTROPIC EFFECTS OF THE dam MUTATION

Figure 1 outlines a unifying hypothesis which attempts to account for the pleiotropic effects of dam mutants. The details are described in the legend but the salient features are listed below.

(1) Mismatch correction is a postreplicational process in which strand selection is instructed by dam-regulated adenine methylation.

(2) The loss of DNA methylation in dam⁻ mutants allows mismatch correction to occur in both the parental and daughter strands producing secondary lesions requiring $recA^+$, $lexA^+$, $recB^+$, and $polA^+$ functions for their repair.

(3) The sensitivity of dam⁻ mutants to 2AP is the result of multiple, and possibly overlapping, excision repair tracts due to the attempted correction of mismatch sites in both the parental and daughter strands. This latter point may account for the induction of prophage λ by 2AP in dam⁻ strains.

THE ISOLATION OF MUTATOR MUTANTS DEFICIENT IN MISMATCH REPAIR

Implicit in the model presented in Fig. 1 is the idea that the sensitivity of dam⁻ strains to 2AP is due to the enzymatic steps postulated to occur at mismatched sites. If this assumption is correct, other mutants also defective in mismatch repair could result in the suppression of the 2AP sensitivity of the Dam⁻ strain. 2AP-Resistant revertants would include not only Dam⁺ revertants but also second-site mutants which disturbed any of the steps of mismatch repair involved in mismatch recognition, incision, and/or excision of the mismatched base. We have undertaken such a search for second-site revertants and have isolated several having properties consistent with what might be predicted for mismatch repair mutants (Glickman and Radman, 1980). Among 300 independent revertants examined, none were true dam⁺ revertants, and all showed spontaneous mutation rates for streptomycin and nalidixic acid resistance which were higher than in the original dam⁻ strain. Thirty-two independent 2AP-resistant second-site revertants have been mapped: 8 near mutH, 13 near mutL and 11 near mutS. Although the identities of these mutators were not confirmed by complementation tests, we have concluded that they are alleles of the named mutators for two reasons. First, the mutants, like the known mutators, exhibit increased mutation frequencies which are independent of $recA^+$ for their expression and show similar mutational spectra (causing frameshifts and transitions) (Cox, 1976; Glickman and Moessen, manuscript in preparation). Second, the introduction of the known mutators into the dam⁻ strain results in 2AP resistance.

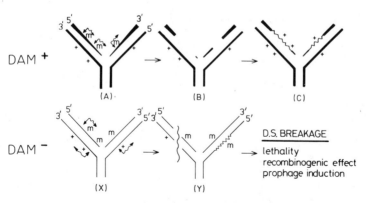

Fig. 1. The hypothesis of methylation-instructed mismatch repair and an interpretation of the sensitivity of a dam mutant to the base analog 2AP. The thickness of the DNA strands indicates their degree of methylation. A replicational error or a base-analog-induced mismatch is symbolized by an "m." Arrows perpendicular to the DNA strands symbolize endonucleolytic attack at mismatched sites, while wavy lines parallel to DNA strands indicate the 5' → 3' direction of excision repair tracts. The wavy portion of the single-stranded DNA indicates an excision repair tract. The wild-type (dam$^+$) situation: (A) Endonuclease/exonuclease-mediated mismatch removal occurs specifically from the undermethylated, newly synthesized strands. (B) Exonucleolytic degradation and resynthesis leads to the intact, fully conserved copy of the parental strand shown in (C). The dam$^-$ situation: (X) Endonuclease/exonuclease can act on both the newly synthesized and the parental DNA strands due to the absence of methylation. (Y) The left arm depicts the production of a double-strand break due to overlapping excision tracts extending over several thousand DNA bases (Wildenberg and Meselson, 1975). The right arm illustrates a situation where mismatch repair results in error fixation rather than error avoidance, thus illustrating the mutator phenotype of dam mutants. The creation of double-strand breaks can account for several pleiotropic effects of the dam$^-$ mutations: sensitivity to base analogs, and the lethality of the dam-recA and dam-recB combinations (see text). A consequence of this scheme is that a deficiency in any early step in mismatch correction would reverse many of the dam$^-$ effects, except for the mutator effect (Glickman and Radman, 1980; Radman et al., 1980).

THE GENETIC CHARACTERIZATION OF THE MUTATOR MUTANTS

Mutation Rates in Multiple Mutants

In order to determine whether the mutator mutants belong to the same pathway, we examined the spontaneous rates of mutH, mutL, and mutS in a series of otherwise isogenic strains. mutH101 and mutL101 are stronger mutators than either mutS101 or dam-3. The introduction of dam-3 into the mutator strains did not increase their mutation rates. All the possible mutator combinations were constructed, and, although additive effects cannot be eliminated, the multiple mutants show mutation rates very close to that of the strongest mutator in a given combination. The precise increase in mutation rates for any given mutator is dependent upon the mutational system. In the case of valine resistance, mutH and mutL increase the spontaneous rate 500 to 1000-fold; whereas in the dam and mutS mutants, this increase is 50 to 100-fold over the wild type (Glickman and Radman, 1980).

Restoration of Viability of dam recA Double Mutants by the Mismatch Mutators

As hypothesized in Fig. 1, the lethality of the recA dam⁻ double mutant may reflect a need for the recA product in repairing double-strand breaks occurring as a consequence of mismatch repair in parental rather than daughter strands. Hence, if the mutator mutants eliminate mismatch repair, the mutators should restore the viability of dam recA double mutants. A dam⁻ mutant carrying the recA200, a temperature-sensitive recA allele, was constructed. As predicted, the introduction of mutH, mutL, or mutS resulted in the restoration of colony formation at the restrictive temperature. mutL and mutS second-site revertants have also been isolated by their ability to relieve the lethality of dam recA double mutants (McGraw and Marinus, 1980).

The Effect of the uvrE Mutation on the Dam Phenotype

Since uvrE is suspected to result in deficient mismatch repair, we examined the effects of this mutator mutation on certain Dam-related properties. Its introduction into the dam⁻ strain did not result in resistance to 2AP nor were any of the dam mut combinations sensitized to 2AP by the introduction of the uvrE mutation. The uvrE mutation, itself a mutator, did not enhance the mutation rates of the mutH, mutL, or mutS strains or any of their combinations. We favor the interpretation that uvrE plays a role in the same repair pathway as the dam and mut mutations.

The Hyper-rec Phenotype of the Mismatch Repair Mutants

As previously noted, one of the properties of the dam mutation is its hyper-recombinogenic ("hyper-rec") phenotype (Marinus and Konrad, 1976). The possibility that the mutator mutations might also suppress this Dam trait was examined in crosses between a wild-type λ residing as a prophage and a phage carrying a Ram2 mutation. As a measure of recombination, the rescue of the Ram$^+$ allele was determined (Glickman and Radman, 1980). The hyper-rec effect of the dam$^-$ mutation was confirmed, but the presence of mutH, mutL, and mutS did not reverse this phenotype. On the contrary, the mutator mutants are themselves strongly hyper-rec.

One possible interpretation of these recombination results is that the dam and mutator mutants exert their hyper-rec effect via different mechanisms. In the case of the dam$^-$ mutants, the increased incidence of single-strand breaks in the parental strands may supply an impetus for recombination. In the case of the mutator strains, the incision step at mismatch sites may not occur, the net consequence of which would be fewer breaks of the kind suggested to be responsible for the hyper-rec effect seen in the dam$^-$ strain. The reduction in strand breakage in these strains may alter the superhelicity of the DNA, facilitating strand aggression. Alternatively, it is not unlikely that mutH, mutL, mutS, dam, and recL (a suspected allele of uvrE) are in some way directly involved in resolving some recombinational intermediate.

Effect of Mismatch Repair Mutants on Recombination Between Close and Distant Markers

Recombination between closely linked markers is often not produced by reciprocal crossing over but rather involves the formation of hybrid DNA followed by mismatch correction. This phenomenon, known variously as gene conversion, map expansion, and negative interference, is likely to depend upon the same system of mismatch repair as does error avoidance. In crosses using two tightly linked markers, a lack of mismatch repair would be expected to reduce the number of apparent recombinants (while widely separated markers should show no such dependence). Crosses were carried out in phage λ using two closely linked markers, Pam3 and Pam80, and two well-separated "outside" markers, cI and Rts2. The mutL and mutS mutations decreased intragenic recombination without affecting intergenic recombination, consistent with a role for these genes in mismatch repair. However, mutH demonstrated very high levels of intragenic recombination, increasing its frequency from 1.3% in the wild type to almost 17%. While this high frequency would seem to conflict with predictions based upon the requirement of mismatch repair for recombination between very close markers, these results might be explained by the extreme hyper-rec properties of the mutant (Radman, unpublished data) though other explanations will occur to the thoughtful reader.

TESTING THE HYPOTHESIS: TRANSFECTION WITH METHYLATED AND UNDERMETHYLATED HETERODUPLEX DNA

One way to test the hypothesis of methylation-instructed mismatch correction is to prepare heteroduplex DNA having each possible combination of strand methylation. In the initial experiments (Radman et al., 1980), heteroduplexes were prepared from methylated and undermethylated DNA without prior strand separation; the two parental DNAs were mixed in the ratio 1:1, denatured and reannealed. If the mismatch repair process outlined in Fig. 1 was highly efficient, 25% of the DNA duplexes would be subject to correction, thus altering the yield of the two types of phage. The results of these experiments were supportive of the working hypothesis and experiments with separated strands have now been performed in the laboratories of M. Radman (Dohet et al., 1981) and M. Meselson (Pukkila et al., 1981). The conclusions from these kinds of experiments are that (1) not all mismatches are corrected with equal enthusiasm, but (2) when correction occurs, it is indeed unequivocally in the direction predicted by the model of methylation-instructed error avoidance. These results thus lend considerable support to the mismatch correction model. Moreover, Dohet et al., (1981) and Pukkila et al. (1981) shown heteroduplex repair to be greatly reduced in the mutH, mutL, and mutS strains; transfection of these strains with heteroduplexed DNA leads primarily to mixed bursts. This is of course fully consistent with the methylation-instructed mismatch model and the predicted properties of the mutators mutH, mutL, and mutS.

CURIOUS CLUES: HOW DOES MISMATCH REPAIR ACTUALLY WORK?

In the absence of an established enzymology, it is difficult to determine the precise details of a molecular process. In the case of Ustilago a likely candidate for mismatch repair enzyme has been identified (Ahmad et al., 1975; Pukkila, 1978) but still remains largely uncharacterized. In the case of E. coli we sought to develop an in vitro assay making use of heteroduplex bacteriophage M13 DNA, but the approach was unsuccessful and we are presently considering other approaches. Thus, even though we seem to have interesting mutants of the mismatch repair system, we are still ignorant of the simplest enzymology concerning this process.

Fortunately, some insights into the mechanics of mismatch repair have been obtained indirectly (Pukkila et al., 1981). Although mismatch repair of heteroduplex DNA was always observed to occur preferentially in the direction predicted, a considerable amount of repair occurred in the opposite direction. The possibility that this was due to the in vivo methylation of the hemi-methylated λ heteroduplexes prior to replication was tested by carrying out the transfection experiments in a dam⁻ strain; however, the dam⁻ host did not improve the directionality of repair. Alternatively, directionality may be impaired because the methylated strand is not sufficiently methylated. When the methylated strands were further methylated in vitro using the purified product of the dam⁺ gene adenine methylase, strand selectivity was indeed enhanced. Furthermore, during these experiments it was observed that when both strands became fully methylated, the percentage of bursts that were "mixed" rose from about 1% to greater than 30%. From these results it would appear that fully methylated DNA is a poor substrate for this enzyme or enzyme-complex.

Interestingly, similar results were obtained when the methylated DNA was extracted from an adenine-methylase-overproducing strain. This overproducing strain is also a spontaneous mutator (Herman and Modrich, 1981), as might be predicted if an excess of the methylase completed the methylation of the newly synthesized strand too rapidly to allow correction.

MISMATCH REPAIR AND INDUCED MUTAGENESIS

There appear to be two major pathways responsible for induced mutagenesis in E. coli: (1) "indirect mutagenesis" provoked by nonpairing DNA lesions which inhibit DNA synthesis, and (2) "direct mutagenesis" provoked by subtle modifications of the DNA template or incorporated precursors capable of efficiently mispairing.

In the case of indirect mutagenesis, the nonpairing lesion (e.g., possibly a pyrimidine dimer in the case of UV-induced mutagenesis) obstructs DNA synthesis and results in the induction of a recA⁺-dependent error-prone repair system (also called induced repair, SOS repair, etc.; see Witkin, 1976 for a review) which has been proposed to enable the cell to bypass a "blocking" lesion at the cost of replicational fidelity (Radman, 1975; Radman et al., 1977).

Direct mutagenesis, on the other hand, can result when tautomers and isomers of the normal bases cause mispairing during replication (Topal and Fresco, 1976; Watson and Crick, 1953). Direct mutagenesis can also be the result of mispairing caused by base analogs or the altered properties of the template or substrate following treatment by agents such as hydroxylamine, bisulphite, and certain alkylating agents. By our definition, direct mutagenesis would also result from agents reducing the fidelity of DNA polymerase, possibly as in the case of some metals. The hallmark of direct mutagenesis, however, is its recA$^+$ independence.

Direct Mutagenesis and Mismatch Repair

The methylation-instructed error-avoidance system might be expected to reduce mutation frequencies related to direct mutagenesis. Consistent with this suggestion is the hypermutability of dam$^-$ mutants to the base analogs 2AP and 5BU (Glickman et al., 1978) and the mismatch repair-deficient mutators mutH, mutL, mutS, uvrD, and uvrE, to 5BU (Rydberg, 1977, 1978). Such a role for mismatch repair during base analog-induced mutagenesis may also help to explain the observation of Witkin and Sicurella (1964) that pure mutant clones arise following treatment with 5BU.

In the case of the alkylating agents, the picture is complicated due to the diversity of chemical products following mutagenic treatment and the large number of often overlapping repair pathways available to handle these lesions (Ripley, 1981; Todd et al., 1981). The dam$^-$ mutants are slightly sensitive to killing by ethyl methanesulfonate (EMS) and mutate at approximately twice the frequency of wild-type strains (Glickman et al., 1978; Mohn et al., 1980). The specificity of mutagenesis is changed quite dramatically, however. Using the lacI system (Coulondre and Miller, 1977a,b), recent studies in our laboratory show that while treatment of the wild-type strain with EMS results almost exclusively in G·C → A·T transitions, treatment of the dam$^-$ strain produced a significant component (15%) of mutants arising via transversion pathways (Brouwer and Glickman, manuscript in preparation).

There a number of possible explanations for the altered EMS mutational spectrum in the dam$^-$ strain. The appearance of transversions may indicate that the error-avoidance system, when fully operational as in the wild type, is capable of efficiently recognizing and correcting particular mismatch lesion(s) which would have led to transversions. Alternatively, since dam$^-$ mutants lack strand discrimination, it is also possible that the transversions arise as errors in strand selection, e.g., that the alkylated lesion resides in the substrate rather than the template strand. Interestingly, it has recently been demonstrated that the nucleotide triphosphate pool is effectively alkylated and may, therefore, be of importance in alkylation mutagenesis (Baker and Topal, 1981). Alternatively, the increased mutagenesis and altered spectrum in the dam$^-$ strain may reflect the efficient induction of the recA$^+$-dependent error-prone repair system. Experiments questioning the role of indirect mutagenesis in the dam$^-$ strain using a umuC derivative defective in this process (Kato and Shinoura, 1977) are in progress.

In the case of methylating agents, the situation appears to be somewhat different. We originally observed (Glickman and Guijt, unpublished data) that while dam⁻ mutants demonstrate an adaptive response to methyl methanesulfonate (MMS) (see Samson and Cairns, 1979), the Arg^+ reversion frequency of an AB1157 dam⁻ derivative was considerably lower than that of the dam⁺ strain. In later and more sensitive experiments, we failed to detect any Arg^+ revertants with either MMS or N-methyl-N'-nitro-N-nitrosoguanidine (MNNG) in dam-4 derivatives of E. coli 343/133, even though the wild-type strain mutated efficiently (Mohn et al., 1980). Our recent studies with dam⁻ strains using the lacI system have also failed to detect any induction of mutants (in terms of either absolute numbers or increased proportions of amber and ochre mutants) following treatment with MMS, MNNG or methyl nitrosourea although dam⁻ cells are more sensitive to killing by these agents than is the wild type (Brouwer and Glickman, manuscript in preparation).

Furthermore, when we asked whether, as in the case of 2AP sensitivity, the introduction of mutH into the dam⁻ strain resulted in the suppression of killing by these agents, we found that dam⁻ sensitivity to methylating agents was indeed suppressed. Unlike the 2AP situation, however, mutH strains are far more resistant to killing by MMS than even the wild type. Possibly this resistance might relate to the hyper-rec properties of mutH strains. We have also observed that mutH derivatives are more mutable by MNNG and MMS than is the wild-type strain.

The mechanism of mutagenesis by methylating agents is not well understood (see Ripley, 1981), and the observation that dam⁻ mutants are very poorly mutated by a series of methylating agents having widely different properties does not fit well into any current model. Many interpretations are possible, including the conjecture that mutagenesis induced by methylating agents is the result of inaccurate mismatch repair. This possibility seems rather slight, however, considering the efficient mutagenesis of T4 by such methylating agents even though most studies on T4 suggest that mismatch repair is at most weak in that organism (see Drake, 1981). Instead, we presently favor the view that the lesions which result in mutagenesis in dam⁺ strains result in lethality in dam⁻ strains. The finding that mutH strains are resistant to killing and demonstrate increased levels of mutagenesis after treatment with these agents in no way proves the model but is certainly consistent with it.

Indirect Mutagenesis and Mismatch Repair

As stated in the Introduction, heteroduplex repair has been invoked to account for the production of pure mutant clones following UV irradiation. Lethal sectoring may also contribute to the formation of such clones, but it is not solely responsible since many of these studies were carried out by pedigree analysis (e.g., James and Kilbey, 1977). Evidence that pure mutant clones can also arise in E. coli as a consequence of UV mutagenesis has also been obtained both by pedigree analysis (Haefner, 1968) and chemostat experiments (Kubitschek, 1964). The recovery of pure mutant clones following UV irradiation is most simply understood on the assumption that the molecular events leading to mutation fixation occur prior to DNA replication. Indeed, it seems most likely that mutations arising in excision-proficient strains occur as rare errors during the excision repair process itself (Hanawalt et al., 1979; Witkin, 1976) and that they occur prior to DNA replication (e.g., Bridges and Mottershead, 1978).

There is some genetical evidence for common steps in the pathways of mismatch repair and the excision repair of UV-induced lesions. The UV-sensitive mutants uvrE and uvrD are defective in the repair of heteroduplex DNA (Nevers and Spatz, 1975) but the function of these genes in DNA excision repair is not known. We therefore wondered if the defect leading to the high lethality in these mutants might not be related to the enzymatic production of a lethal lesion as a result of an abortive attempt at mismatch repair. If this were the case, then, with reasoning parallel to that used to explain the suppression of 2AP sensitivity in the dam⁻ strain, perhaps mutH would suppress the UV sensitivity of the uvrE and/or uvrD strains. Indeed, the introduction of mutH into the uvrE strain did result in considerable suppression of UV sensitivity (Fig. 2). The extent of the protection was equally great in the wild-type strain as well as the dam⁻ mutant (Fig. 3). In Fig. 2 it can be seen that the effects of the various mismatch repair deficiencies on the survival of the uvrE mutant are complex. One important point to note is that mutH, mutL, and mutS behave quite differently in their abilities to suppress the UV sensitivity of the uvrE strain.

We examined the dependence of mutH suppression of UV sensitivity on the functional states of other repair pathways (Glickman, Dunn, and Guijt, manuscript in preparation). These results are summarized in Table 1. Of the mutants examined, only uvrB was not effectively protected by mutH. It was also noted that the addition of the dam mutation to the uvrB strain reduced its survival. The uvr¹-dependence of the mutH protective effect may be consistent with the observation that only uvr⁺ strains of yeast produce pure mutant clones following UV irradiation; it is under these conditions that excision repair tracts occur prior to DNA replication.

A number of hypotheses can be launched to explain the complicated phenomenology observed here. There is, however, no solid information concerning the chemical identity of a specific lethal lesion. Furthermore, the position of that lesion may also be

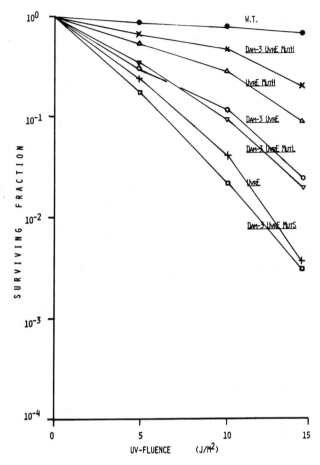

Fig. 2. The effect of various combinations of mismatch repair defects upon the UV sensitivity of the uvrE strain. The strains were isogenic derivatives of NR3752 differing for the genotypes indicated. Irradiation was carried out as described by Glickman et al. (1978).

important in determining its lethal potential. The protection of the recA strain by mutH would argue against the hyper-rec property of this strain being responsible for UV protection, although we have yet to confirm that the hyper-rec effect is $recA^+$-dependent. At present, we favor the idea that attempts at mismatch repair in the sensitive area of the replication fork can occasionally result in the production of lethal secondary lesions. The mutH mutation reduces mismatch repair and, as a consequence, even though the numbers of lesions are small, affords an appreciable level of UV protection.

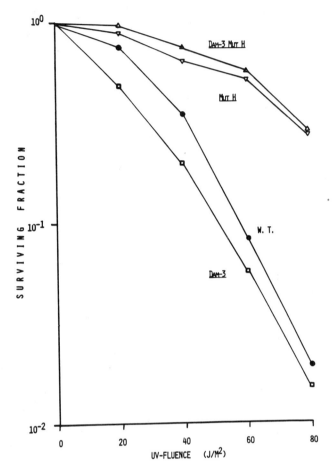

Fig. 3. UV-survival curves showing the protective effect of mutH in wild-type and dam⁻ strains. Description as in Fig. 2.

Attempts to examine the effects of mutH on UV mutagenesis have proven difficult due to the high background levels of mutagenesis already observed in these strains.

The increased resistance of mutH mutants is reminiscent of the protection afforded by the mutator plasmid pKM101, a derivative of R46 (Mortelmans and Stocker, 1976). Because of the possibility that the mechanism of UV protection might function by the same pathway, we determined the effect of PKM101 on the survival of the mutators strains. All three mutators, mutH, mutL, and mutS, showed increases in the level of protection similar to that seen in the wild-type strain. This can be interpreted as an indication that the modes of protection differ. This interpretation is supported by the

Table 1. Effect of dam and mut Mutations on UV Sensitivity

Genetic Background	Dose Modification Factor of Additional Marker[a]						
	dam⁻	mutH	dam⁻mutH	mutL	dam⁻mutL	mutS	dam⁻mutS
Wild type	0.9	2.0	2.7	1.0	1.2	1.2	0.9
uvrE	1.2	2.0	2.7			0.9	
recL	1.0	1.7	1.7				
uvrD	1.0	1.5	1.0				
recF	1.0	2.3					
recB recC	[b]	2.2					
recA	[b]	4.0					
uvrB	0.6	1.2–1.3[c]	0.6–0.7[c]				

[a]Dose modification factor calculated as ratio of dose required for 1% survival. (Larger factors indicate greater resistance.)
[b]Inviable combination.
[c]Range of values for three experiments.

observation that the spontaneous mutator effect of pKM101 is $recA^+$-dependent while the mutator effect of mutH is $recA^+$-independent (Glickman and Radman, 1980), and, finally, the spontaneous mutational spectra differ between the dam⁻- and pKM101-bearing strains (Mohn et al., 1980).

SIMILARITIES AND DIFFERENCES AMONG mutH, mutL, AND mutS

These three mutators were isolated because each had the ability to suppress the 2AP sensitivity of dam⁻ strains. Such mutants were envisioned to be defective in either the incision or excision steps of mismatch repair. In fact, the mutations could affect either these processes or other as yet unidentified component(s) of the repair system. The mutation rates of multiple mutants did not show synergistic effects indicative of separate pathways (Glickman and Radman, 1980), but additive effects, which would be expected if overlapping pathways or pathways having common intermediates were involved, could not be eliminated on this basis.

Strains carrying mutH, mutL, or mutS do have distinctly different properties. However, mutH and mutL lead to greater mutation rates than mutS. These mutations also result in different protection patterns in a uvrE⁻ strain (see Fig. 2 and Table 1). In the λ recombination crosses where close markers (Pam3 X Pam80) were examined, mutH showed increased conversion frequencies to P^+ while mutL and mutS showed reduced rates.

A possible explanation for some differences in properties might be that the mutators are defective for certain kinds of mismatches. Support for this suggestion can be found in the work of Feinstein and Brooks-Low (personal communication), who found the hyper-rec effect of each of these mutators to be specific for certain combinations of lacZ crosses, as might be predicted if the different mutators recognize different mismatches with different efficiencies. Further support for this notion comes from our own determination of the spontaneous mutational spectra of mutH, mutL, and mutS in the Miller lacI system. These mutations each have their own characteristic effects on mutational specificity (Table 2). The strains carrying dam-4 and mutL show similar levels of amber mutants among the lacI⁻ spontaneous mutants (circa 5%), while mutH and uvrE strains produce only about a third of that level. Evidence is accumulating which suggests different roles for these mutator mutations in mismatch repair. However, our final understanding of the roles of mutH, mutL, and mutS is likely to depend upon their biochemical characterization.

Table 2. A Comparison of the Distribution of Spontaneous lacI⁻ Mutants Between Amber and Ochre Sites.

Genotype	Number of Mutants Screened	Ambers		Ochres	
		Number	Percentage	Number	Percentage
Wild type	3280	27	0.8	11	0.4
dam-4	4000	163	4.1	195	4.9
mutH	10038	146	1.5	386	3.9
mutL	7307	398	5.5	292	4.0
mutS	1500	42	2.8	78	5.2
uvrE	13282	190	1.4	347	2.6

PRACTICAL APPLICATIONS OF THE dam MUTATION

Strains carrying dam⁻ have been used to provide unmodified DNA for recombinant DNA and sequencing experiments. We have investigated the use of the dam mutation as a sensitizer for specific mutagens in short-term screening assays (Mohn et al., 1980). While the introduction of the dam⁻ mutation into the E. coli K-12 tester strain 343/113 increased the spontaneous background at certain loci sufficiently to hinder the detection of some mutagens, the dam⁻ derivatives showed extraordinary sensitivity to the frameshifting agent 9-aminoacridine. In fact, this mutagen could be detected in the dam⁻ derivatives at concentrations 2 to 3 orders of magnitude lower than in the original strain. The significance of this increased sensitivity in terms of detecting environmental mutagens is being examined by G. Mohn (Leiden) where the dam⁻ derivative of 343/113 is included in their standard mutagen screening assay. As yet, insufficient data have accumulated to determine the effectiveness of this strain in identifying potential environmental hazards.

THE ROLE OF DNA METHYLATION AS AN INSTRUCTIVE COMPONENT OF MISMATCH REPAIR IN EUKARYOTES

Methylation in eukaryotes appears to be restricted to 5-methylcytosine (see Taylor, 1979 for review) and, therefore, higher organisms lack the 6-methyladenine residues associated with mismatch repair instruction in E. coli. Numerous functions have been proposed for methylation in eukaryotes, including the control of DNA replication, genetic recombination, gene expression, and differentiation (cf. Razin and Riggs, 1980).

We have attempted to ask whether mutagenesis, in a mammalian system, might be influenced by DNA methylation (Knaap et al., 1981). Our approach was based upon the observation that the carcinogen ethionine (the ethyl analog of methionine) blocks DNA methylation (Boehm and Drakovsky, 1979; Cox and Irving, 1977). We aked whether treatment by this analog would produce a "Dam⁻" phenotype in V-79 Chinese hamster cells.

The cells were treated with subtoxic levels of ethionine for up to 12 days and challenged with the base analog 6-hydroxyaminopurine. While this anlog turned out to be an effective mutagen in V-79 cells, the effect of the ethionine pretreatment was to reduce its mutagenic potency. This is in contrast to the hypermutability of dam⁻ mutants of E. coli when exposed to a base analog. The interpretation of these experiments, however, must be limited. We do not know with any degree of certainty that the ethionine treatment actually blocked DNA methylation, nor do we know that 6-hydroxyaminopurine exerts its mutagenic effect as a "classical" base analog.

While there is some evidence that mammalian cells are able to carry out mismatch repair (Wilson, 1977), we know of no evidence suggestive of a role for DNA methylation in strand instruction. Nonetheless, a variety of potential mechanisms allowing discrimination between parental and daughter strands can be envisioned (e.g., the segregation of histones) which would allow directed mismatch repair to occur. The advantages of a postreplicative error-avoidance mechanism in preserving genetic integrity make it seem likely that this kind of DNA maintenance system will function in higher organisms, though the actual mechanism(s) may vary.

ACKNOWLEDGMENTS

The contributions of G. Moessen, R. Dunn, M. Babcock-Harms, N. Guijt and J. Brouwer in the laboratory are gratefully recognized. The cooperation of Drs. M. Meselson, P. Pukkila, and M. Radman in providing access to unpublished data is greatly appreciated.

REFERENCES

Ahmad, A., Holloman, W. K., and Holliday, R., 1975, Nuclease that preferentially inactivates DNA containing mismatched bases, Nature, 258:54.

Argwal, S. S., Duke, D. K., and Loeb, L. A., 1979, On the fidelity of DNA replication: Accuracy of Escherichia coli DNA polymerase I, J. Biol. Chem., 254:101.

Baker, M. S., and Topal, M. D., 1981, Cellular DNA precursor triphosphates are targets of MNU, J. Supramol. Struc. Suppl., 5 (Cell Biochem.), 170.

Bale, A., d'Alarcao, M., and Marinus, M. G., 1979, Characterization of DNA adenine methylation mutants of Escherichia coli K12, Mutat. Res., 59:157.

Baas, P. D., and Jansz, H. S., 1972, Asymmetric information transfer during φX174 DNA replication, J. Mol. Biol., 63:557.

Boehm, T. L., and Drahovsky, D., 1979, Effect of carcinogen ethionine on enzymatic methylation of DNA sequences with various degrees of repetitiveness, Eur. J. Cancer, 15:1167.

Bridges, B. A., and Mottershead, R. P., 1978, Mutagenic DNA repair in Escherichia coli. VII. Constitutive and inducible manifestations, Mutat. Res., 52:151.

Clayton, L. K., Goodman, M. F., Branscomb, E. W., and Galas, D. J., 1979, Error induction and correction by mutant and wild type T4 DNA polymerases, J. Biol. Chem., 254:1902.

Coulondre, C., and Miller, J. H., 1977a, Genetic studies of the lac repressor. III. Additional correlation of mutational sites with specific amino acid residues, J. Mol. Biol., 117:525.

Coulondre, C., and Miller, J. H., 1977b, Genetic studies of the lacI repressor. IV. Mutagenic specificity of the lacI gene of Escherichia coli, J. Mol. Biol., 117:577.

Cox, E. C., 1976, Bacterial mutator genes and the control of spontaneous mutation, Annu. Rev. Genet., 10:135.

Cox, R., and Irving, Ch. C., 1977, Inhibition of DNA methylation by S-adenosylethionine with the production of methyl-deficient DNA in regenerating rat liver, Cancer Res., 37:222.

Dohet, C., Bourguignon-Van Horen, M. F., and Radman, M., 1981, Mismatch correction mutants: Their effect on lambda recombination, Proc. Natl. Acad. Sci. U.S.A., submitted

Drake, J. W., 1969, "The Molecular Basis of Mutation," Holden-Day, San Francisco.

Drake, J. W., 1981, Perspectives in molecular mutagenesis, this volume.

Eckardt, F., and Haynes, R. H., 1977, Induction of pure and sectored mutant clones in excision-proficient and deficient strains of yeast, Mutat. Res., 43:327.

Fincham, J. R. S., and Holliday, R., 1970, An explanation of fine structure map expansion in terms of excision repair, Mol. Gen. Genet., 109:307.

Glickman, B. W., 1979, Spontaneous mutagenesis in Escherichia coli strains lacking 6-methyladenine residues in their DNA: An altered mutational spectrum in dam⁻ mutants, Mutat. Res., 61:153.

Glickman, B. W., and Radman, M., 1980, Escherichia coli mutator mutants deficient in methylation-instructed DNA mismatch correction, Proc. Natl. Acad. Sci. U.S.A., 77:1063.

Glickman, B. W., Van der Elsen, P., and Radman, M., 1978, Induced mutagenesis in dam⁻ mutants of Escherichia coli: A role for 6-methyladenine in mutation avoidance, Mol. Gen. Genet., 163:307.

Gomez-Eichelmann, M. C., 1979, DNA adenine and cytosine methylation in Salmonella typhimurium and Salmonella typhi, J. Bacteriol., 140:574.

Haefner, K., 1968, Concerning the mechanism of ultraviolet mutagenesis. A micromanipulatory pedigree analysis in Schizosaccharomyces pombe, Genetics, 57:169.

Hanawalt, P. C., Cooper, P. K., Ganesan, A. K., and Smith, C. A., 1979, DNA repair in bacteria and mammalian cells, Annu. Rev. Biochem., 48:783.

Hannan, M. A., Duck, P., and Nasim, A., 1976, UV-induced lethal sectoring and pure mutant clones in yeast, Mutat. Res., 36:171.

Herman, G. E., and Modrich, P., 1981, Escherichia coli K-12 colonies that overproduce dam methylase are hypermutable, J. Bacteriol., 145:644.

Hibner, U., and Alberts, B. M., 1980, Fidelity of DNA replication catalyzed in vitro on a natural DNA template by the T4 bacteriophage multienzyme complex, Nature, 285:300.

Holliday, R. A., 1964, A mechanism for gene conversion in fungi, Genet. Res., 5:282.
James, A. P., and Kilbey, B. J., 1977, The timing of UV mutagenesis in yeast: A pedigree analysis of induced recessive mutation, Genetics, 87:237.
Kato, T., and Shinoura, Y., 1977, Isolation and characterization of mutants of Escherichia coli deficient in induction of mutations by ultraviolet light, Mol. Gen. Genet., 156:121.
Knaap, Ada G. A. C., Glickman, B. W., and Simons, J. W. I. M., 1981, Effects of ethionine on the replicational fidelity in V-79 Chinese hamster cells, Mutat. Res., in press.
Kubitschek, H., 1964, Mutation without segregation, Proc. Natl. Acad. Sci. U.S.A., 52:1374.
Lacks, S., and Greenberg, B., 1977, Complementary specificity of restriction endonucleases of Diplococcus pneumoniae with respect to methylation, J. Mol. Biol., 114:153.
Loeb, L. A., 1974, DNA replication fidelity, in: "The Enzymes," H. D. Boyer, ed., Academic Press, New York.
Marinus, M. G., 1976, Adenine methylation of Okazaki fragments in Escherichia coli map, J. Bacteriol., 128:853.
Marinus, M. G., and Konrad, B., 1976, Hyper-recombination in dam mutants of Escherichia coli K-12, Mol. Gen. Genet., 149:273.
Marinus, M. G., and Morris, R. N., 1975, Pleiotropic effects of a DNA adenine methylation mutation (dam-3) in Escherichia coli K12, Mutat. Res., 28:15.
McGraw, B. R., and Marinus, M. G., 1980, Isolation and characterization of Dam$^+$ revertants and suppressor mutations that modify secondary phenotypes of dam-3 strains of E. coli K-12, Mol. Gen. Genet., 178:309.
Mohn, G. R., Guijt, N., and Glickman, B. W., 1980, Influence of DNA adenine methylation dam mutation and of plasmid PKM101 on the spontaneous and induced mutability of certain genes in Escherichia coli K-12, Mutat. Res., 74:255.
Mortelmans, K. E., and Stocker, B. A. D., 1976, Ultraviolet light protection, enhancement of ultraviolet mutagenesis and mutator effect of plasmid R46 in Salmonella typhimurium, J. Bacteriol., 128:271.
Nasim, A., and Auerbach, C., 1967, The origin of complete and mosaic mutants from mutagenic treatment of single cells, Mutat. Res., 4:1.
Nevers, P., and Spatz, H., 1975, Escherichia coli mutants uvrD and uvrE deficient in gene conversion of λ-heteroduplexes, Mol. Gen. Genet., 139:233.
Pukkila, P. J., 1978, The recognition of mismatched base pairs in DNA by DNase I from Ustilago maydis, Mol. Gen. Genet., 161:245.
Pukkila, P. J., Peterson, J., Herman, G., Modrich, P., and Meselson, M., 1981, Changes in the amount of DNA methylation influence mismatch correction, Proc. Natl. Acad. Sci. U.S.A., in press.

Radman, M., 1975, SOS repair hypothesis: Phenomenology of inducible repair which is accompanied by mutagenesis, in: "Molecular Mechanisms for the Repair of DNA," P. C. Hanawalt, and R. B. Setlow, eds., Plenum Press, New York.

Radman, M., Villani, G., Boiteux, S., Defais, M., Caillet-Fauquet, P., and Spidari, S., 1977, On the molecular mechanism of induced mutagenesis, in: "Origins of Human Cancer," H. Hiatt, J. D. Watson, and J. D. Winstein, eds., Cold Spring Harbor Laboratory, New York.

Radman, M., Wagner, R. E., Glickman, B. W., and Meselson, M., 1980, DNA methylation, mismatch correction and genetic stability, in: "Progress in Environmental Mutagenesis," M. Alacevic, ed., Elsevier-North Holland, Amsterdam.

Razin, A., and Riggs, A. D., 1980, DNA methylation and gene function, Science, 210:604.

Ripley, L. S., 1981, Mutagenic specificity of nitrosomethylurea in bacteriophage T4, Mutat. Res., in press.

Rydberg, B., 1977, Bromouracil mutagenesis in Escherichia coli: Evidence for involvement of mismatch repair, Mol. Gen. Genet., 152:19.

Rydberg, B., 1978, Bromouracil mutagenesis and mismatch repair in mutator strains of Escherichia coli, Mutat. Res., 52:11.

Samson, L., and Cairns, J., 1979, A new pathway for DNA repair in Escherichia coli, Nature, 267:281.

Taylor, J. H., 1979, "Molecular Genetics, Part 3, Chromosome Structure," Academic Press, New York.

Tiraby, J. G., and Fox, M. S., 1973, Marker discrimination in transformation and mutation of pneumococcus, Proc. Natl. Acad. Sci. U.S.A., 70:3541.

Todd, P. A., Brouwer, J., and Glickman, B. W., 1981, Influence of DNA repair deficiences on MMS and EMS induced mutagenesis in E. coli K-12, Mutat. Res., in press.

Topal, M. D., and Fresco, J. R., 1976, Complementary base pairing and the origin of substitution mutations, Nature (London), 263:285.

Wagner, Jr., R., and Meselson, M., 1976, Repair tracts in mismatches DNA heteroduplexes, Proc. Natl. Acad. Sci. U.S.A., 73:4135.

Walker, G. C., 1977, Plasmid (pKM101)-mediated enhancement of repair and mutagenesis. Dependence on chromosomal genes in Escherichia coli K-12, Mol. Gen. Genet., 152:93.

Watson, J. D., and Crick, F. H. C., 1953, A structure for deoxyribose nucleic acids, Nature (London), 171:737.

White, R., and Fox, M., 1974, Genetic consequences of transfection with heteroduplex bacteriophage λ DNA, Mol. Gen. Genet., 141:163.

Whitehouse, H. L. K., and Hastings, P. J., 1965, The analysis of genetic recombination on the poloron hybrid DNA model, Genet. Res., 6:27.

Wildenberg, J., and Meselson, M., 1975, Mismatch repair in heteroduplex DNA, Proc. Natl. Acad. Sci. U.S.A., 72:2202.

Wilson, J. H., 1977, Genetic analysis of host range mutant viruses suggests an uncoating defect in simian virus 40-resistant monkey cells, Proc. Natl. Acad. Sci. U.S.A., 74:3503.

Witkin, E. M., 1976, Ultraviolet mutagenesis and inducible DNA repair in Escherichia coli, Bacteriol. Rev., 40:869.

Witkin, E. M., and Sicurella, N. A., 1974, Pure clones of lactose-negative mutants obtained in Escherichia coli after treatment with 5-bromouracil, J. Mol. Biol., 8:610.

CHAPTER 4

CELLULAR DEFENSE MECHANISMS AGAINST ALKYLATION OF DNA

Tomas Lindahl, Björn Rydberg, Thomas Hjelmgren,
Monica Olsson, and Anita Jacobsson

Department of Medical Biochemistry, Gothenburg
University, Gothenburg, Sweden

SUMMARY

Several repair enzymes are induced in Escherichia coli as a consequence of exposure to methylating agents such as N-methyl-N'-nitro-N-nitrosoguanidine. These include at least two different glycosylase functions, which act specifically on alkylated DNA by catalyzing the release of 3-methyladenine and 7-methylguanine in free form. The apparently most important repair function induced during this adaptive response to alkylating agents, however, reduces the mutagenic activity of these agents by allowing improved repair of O^6-methylguanine (Cairns, 1980; Schendel and Robins, 1978). This reaction can be faithfully duplicated in vitro employing cell-free extracts from adapted E. coli. The active factor has been shown to be an induced protein of molecular weight about 17,000, which catalyzes the transfer of a methyl group from the O^6-position of an alkylated DNA guanine residue to one of its own cysteine residues. The protein acts in an analogous fashion on ethylated DNA by removing the ethyl group from O^6-ethylguanine. The reaction is associated with suicide inactivation of the protein, since it appears to lack the capacity to release the blocking alkyl group. There is no detectable change in molecular weight of the protein accompanying methylation, so the inactivation probably depends on the modification of a reactive sulfhydryl group. S-Methylcysteine has been identified in the methylated protein by amino acid analysis and by conversion of this residue to its sulfone derivative by hydrogen peroxide treatment. S-Methylcysteine has not been found previously in enzymatically methylated proteins. Moreover, the reaction represents the only known case of protein methylation that is not dependent on S-adenosylmethionine as methyl donor. E. coli mutants unable to respond to alkylating agents by adaptation (isolated by P. Jeggo) also lack the methyltransferase, while mutants that express the adaptive response in a constitutive fashion (isolated by B. Sedgwick) also produce the methyltransferase constitutively. Thus, the DNA methyltransferase acting on O^6-methylguanine appears to be primarily responsible for the adaptive response to the mutagenic effect of simple alkylating agents.

INTRODUCTION

N-nitroso compounds are powerful mutagens which also act as carcinogens in animals. In experimental work on the alkylation and repair of DNA, directly acting methylating and ethylating agents such as N-methyl-N'-nitro-N-nitrosoguanidine (MNNG), N-methyl-N-nitrosourea, and N-ethyl-N-nitrosourea have often been employed. Three major purine lesions, 7-methylguanine, 3-methyladenine, and O^6-methylguanine (Lawley and Thatcher, 1970), occur in double-stranded DNA after treatment with these methylating agents either in vivo or in vitro. In addition, a number of minor alkylating products are formed, that is, other methylated purines, methylated pyrimidines, and phosphotriesters. After treatment with ethylating agents, the analogous ethylated DNA base derivatives occur (Singer, 1976).

Escherichia coli contains several different DNA repair enzymes which act specifically on alkylated DNA. Here, we review briefly the properties of these enzyme activities. Recent studies in several laboratories have shown that mammalian cells contain repair enzymes very similar to those found in bacteria that act on alkylated DNA (Brent, 1979; Cathcart and Goldthwait, 1981; Ishiwata and Oikawa, 1979; Margison and Pegg, 1981; Montesano et al., 1980; Renard et al., 1981; Singer and Brent, 1981). Therefore, it seems likely that the mechanisms of cellular defense against alkylation damage are basically the same in human cells as in E. coli.

DNA ALKYLATION WITH S-ADENOSYLMETHIONINE

It seems fairly obvious that living cells have developed effective repair mechanisms to deal with major DNA lesions caused by ultraviolet light, ionizing radiation, and heat-induced hydrolysis. The reason why cells contain an effective arsenal of repair enzymes specific for various types of alkylation damage is less clear. This is particularly so in the case of bacteria that lack the ability to metabolize many compounds which require activation to convert them into the alkylating agents which produce mutations. Even directly acting alkylating agents, such as MNNG, would be expected to have short half-lives under natural conditions, and living cells may only be exposed to agents of this kind under exceptional conditions. While the reaction of nitrite with various secondary amines has often

been mentioned as a possible cause for the occurrence of traces of N-nitroso compounds in the environment, an alternative explanation for recurrent alkylation damage to DNA might be that methylating agents are continuously generated intracellularly from small molecules which serve functions in normal metabolism. In this regard it is of interest that the major methyl donor in cells, S-adenosylmethionine, is a fairly unstable compound at neutral pH, so it could be converted to a shortlived alkylating derivative as a consequence of spontaneous, nonenzymatic hydrolysis. In fact, we have recently observed that when purified DNA is incubated in neutral solution with [^{3}H-methyl]S-adenosylmethionine of high specific radioactivity, small amounts of methylated purine residues are formed in the DNA, as determined by subsequent analysis of nucleic acid hydrolysates by high pressure liquid chromatography (B. Rydberg and T. Lindahl, manuscript in preparation). The reaction is time dependent and at a level of S-adenosylmethionine similar to that present intracellularly, that is, at $3 \cdot 10^{-5}$ M about one DNA base residue in 10^7 is methylated per 2 h at 37°C. The methylated residues detected are 7-methylguanine, and, at a tenfold lower amount, 3-methyladenine. Thus, the reaction products are the same as those obtained by treatment of DNA with chemical methylating agents such as methyl methanesulfonate and dimethyl sulfate, and they are entirely different from those (5-methylcytosine and N^6-methyladenine) generated by the action of DNA methylases. It seems reasonable to ascribe the reaction observed to a direct alkylation of DNA by S-adenosylmethionine rather than to some experimental artifact: (1) the methylated compounds remain associated with DNA on ethanol precipitation and CsCl density gradient centrifugation; (2) the reaction occurs in a similar fashion with different batches of radioactive S-adenosylmethionine (from different suppliers), as well as with S-adenosylmethionine that has been purified by ion exchange chromatography; and (3) Paik et al. (1975) have shown previously that carboxyl residues in certain proteins can be methylated with S-adenosylmethionine in a nonenzymatic reaction. While S-adenosylmethionine is far from an effective DNA alkylating agent, the level of spontaneous methylation would nevertheless be expected to be of physiological significance because it corresponds to the formation of about 3000 methylated purine residues per day in the genome of a mammalian cell. This frequency of DNA alteration is similar to that which produces, by spontaneous hydrolysis, the depurination of nonmodified DNA (Lindahl and Nyberg, 1972). The alkylation of DNA by S-adenosylmethionine may well have made it necessary to evolve specific DNA repair mechanisms in order to remove methylated base residues and avoid a high spontaneous mutation rate due to the presence of miscoding or noncoding bases in DNA.
It may be speculated that the price the cell must pay for having such an effective methyl donor as S-adenosylmethionine readily available for a large number of enzyme-catalyzed reactions in the nucleus and cytoplasm could be a low, but significant, level of premutagenic "background alkylation" of the cellular DNA.

EXCISION OF 3-METHYLADENINE

When E. coli cells growing in standard media are exposed to external alkylating agents, the 3-methyladenine formed in DNA is rapidly excised, while O^6-methylguanine is much less effectively repaired, and little or no active removal of 7-methylguanine is observed (Lawley and Orr, 1970). The efficient excision of 3-methyladenine is due to a DNA glycosylase, which specifically cleaves the base-sugar bond between 3-methyladenine and deoxyribose in alkylated DNA (Lindahl, 1976). The E. coli enzyme has been purified extensively but it is not presently available in homogeneous form (Riazuddin and Lindahl, 1978). A similar DNA glycosylase has been found in Micrococcus luteus (Laval, 1977). It initiates a repair event by catalyzing the release of the methylated purine in free form; the resulting apurinic site is then removed by short-patch excision repair (Fig. 1). The enzyme has properties typical of several different DNA glycosylases. Thus, it has no apparent cofactor requirement, it acts preferentially on double-stranded DNA, it is of relatively low molecular weight, and it seems highly specific for its particular substrate. While the E. coli 3-methyladenine-DNA glycosylase is able to excise also the closely similar ethylated derivative (3-ethyladenine), it shows no detectable activity against several other modified purine residues in DNA, such as 7-methylguanine, 7-ethylguanine, O^6-methylguanine, 3-methylguanine, 7-methyladenine, or N^6-methyladenine (Riazuddin and Lindahl, 1978).

The physiological role of the E. coli 3-methyladenine-DNA glycosylase has been clarified to some extent by the isolation of enzyme-deficient (tag) mutants (Karran et al., 1980). Such mutants have a strongly reduced ability to remove 3-methyladenine from their DNA after exposure to a high, but nonlethal, dose of methylating agent. Further, they are more sensitive to such agents than wild-type E. coli. These findings define the 3-methyladenine-DNA glycosylase as a DNA repair enzyme, active in the removal of a major alkylation lesion. One of the mutants investigated (PK432) has been shown to have a mutation within the structural gene for the enzyme since it produces an anomalously heat-sensitive 3-methyladenine-DNA glycosylase (Karran et al., 1980). Another enzyme-deficient, highly sensitive mutant (BK2012) with unusual mapping properties has been found on subsequent analysis to be a double mutant (tag ada). In addition to its 3-methyladenine-DNA glycosylase deficiency, the latter is also defective in the adaptive response against alkylating agents. Substrains of this mutant, containing one or the other mutation, have now been obtained by transduction experiments (G. Evensen and E. Seeberg, personal communication).

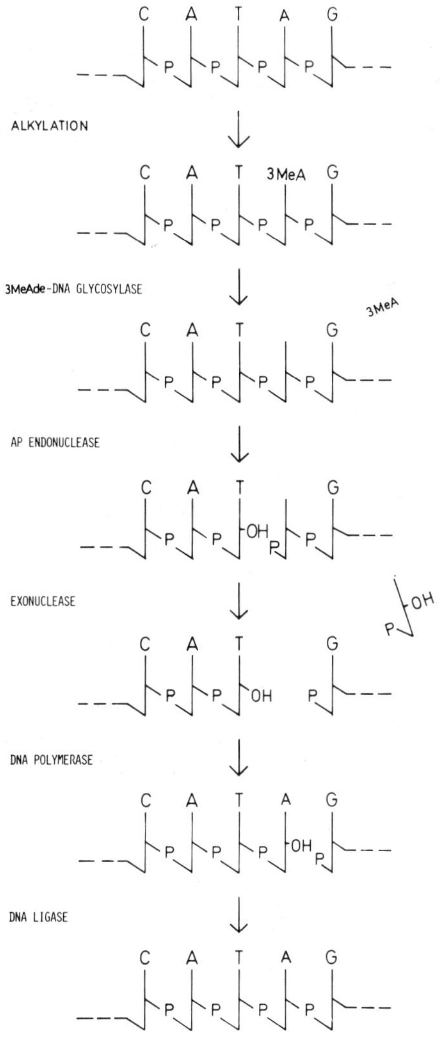

CELLULAR DEFENSE MECHANISMS

Fig. 1.
Removal of 3-methyladenine from alkylated double-stranded DNA by excision repair. The methylated purine is released as the free base by a 3-methyladenine-DNA glycosylase, and the DNA chain is then cleaved at the apurinic site by an AP endonuclease. In E. coli, the latter step is catalyzed by the endonucleolytic activity of exonuclease III (Rogers and Weiss, 1980; Verly et al., 1973) and probably also by the analogous endonuclease IV (Ljungquist, 1977). These endonucleases cleave at the 5' side of the apurinic site, but E. coli also contains at a lower level an endonuclease which acts at the 3' side of an apurinic site, endonuclease III (Warner et al., 1980). The latter may serve to remove subsequently the deoxyribose-phosphate residue from DNA. Alternatively, this residue may be excised by an exonuclease such as the 5' exonuclease activity of DNA polymerase I or exonuclease VII. Gap-filling of the double-stranded DNA by a polymerase, which may only add a single purine deoxynucleotide to the DNA, and ligation complete the repair event.

INDUCIBLE DNA GLYCOSYLASE ACTIVITIES FOR 3-METHYLADENINE AND 7-METHYLGUANINE RESIDUES

The evaluation of the role of the 3-methyladenine-DNA glycosylase in repair of alkylation damage is complicated by the recent finding that E. coli contains two different DNA glycosylases with the ability to remove 3-methyladenine: A major activity is the product of the tag$^+$ gene, while a different species accounts for 5 to 10% of the total level of activity towards this DNA lesion in a crude cell extract. The latter enzyme was discovered in extracts of E. coli tag mutants, which were observed to contain a residual small amount of activity for specifically releasing 3-methyladenine from alkylated DNA. On further analysis, this minor activity was shown not to be due to leakiness of the tag mutants but to the presence of a DNA glycosylase with distinctly different properties from the tag$^+$ gene product. This enzyme is present at similar levels in both wild-type cells and tag mutants (Karran et al., 1980) and differs from the tag$^+$ gene product in being considerably more heat stable, as well as being insensitive to product inhibition by free 3-methyladenine. It has not been extensively purified so far or characterized with regard to its detailed substrate specificity.

When E. coli cells are treated with low doses of an alkylating agent such as MNNG, they gain increased resistance to both the mutagenic and killing action of a subsequent challenge dose of the same agent (Jeggo et al., 1977; Samson and Cairns, 1977). At least part of this phenomenon, termed the adaptive response, is due to the induction of a protein that acts in the repair of O^6-methylguanine residues (Olsson and Lindahl, 1980; Robins and Cairns, 1979; Schendel and Robins, 1978). However, cell extracts of E. coli adapted by the procedure of Jeggo et al. (1977) also contain a twofold higher level of 3-methyladenine-DNA glycosylase activity than unadapted controls (P. Karran, T. Hjelmgren, and T. Lindahl, unpublished data). The enzymatic properties of this inducible activity indicate that it may be accounted for by a large increase (about 20-fold) in the level of the heat-stable 3-methyladenine-DNA glycosylase. Thus, it would appear that the tag$^+$ gene product is constitutively expressed and is present at similar levels in unadapted and adapted E. coli, while the minor 3-methyladenine-DNA glycosylase is an inducible enzyme.

Extracts of adapted E. coli contain yet another induced enzyme activity specific for alkylated DNA, that is, a 7-methylguanine-DNA glycosylase. Laval et al. (1981) have recently described a 7-methylguanine-DNA glycosylase in extracts of M. luteus and E. coli. We have confirmed that such an E. coli activity can be observed, although in our experiments only trace amounts seem to be present in nonadapted cells. This activity occurs at an about 20-fold higher level in extracts of adapted cells (T. Hjelmgren and T. Lindahl, unpublished data) so it is much more readily detected with such material. Nevertheless, even in extracts from adapted cells the 7-methylguanine-DNA glycosylase activity appears low, that is, about 15-fold lower than the total level of 3-methyladenine-DNA glycosylase activity.

It is presently unclear if the minor, heat-stable 3-methyladenine-DNA glycosylase activity and the 7-methylguanine-DNA glycosylase activity are due to two different enzymes, both of which are induced during the adaptive response, or if the two activities reside within a single inducible DNA glycosylase of relatively broad substrate specificity which removes 3-methyladenine more efficiently than 7-methylguanine. Gel chromatography experiments on Sephadex G-75 have shown that both activities appear to be associated with a protein of molecular weight 25,000 to 30,000. Since several clearly distinct DNA glycosylases are known to have similar molecular weights, these observations do not, however, resolve the question of whether one or two enzymes are present. Somewhat unexpectedly, an E. coli mutant strain, BS 21 (Adc), which possesses the ability to repair O^6-methylguanine in a constitutive fashion (Sedgwick and Robins, 1980), does not show the markedly increased levels of 3-methyladenine- and 7-methylguanine-DNA glycosylase activities that are detected after adaptation of wild-type E. coli B/r by exposure to a low dose of MNNG.

While 7-methylguanine is the quantitatively major lesion introduced by treatment of DNA with methylating agents in vivo or in vitro, this base modification has been assumed to be relatively harmless, since in double-stranded DNA the methylated N^7 site is not involved in hydrogen-bonding with the complementary chain, and cells exposed to alkylating agents can retain large amounts of this base for long time periods with little or no loss of viability. Even in adapted E. coli cells, active removal of 7-methylguanine is not readily observed after exposure to a high dose of MNNG (Schendel and Robins, 1978; Sedgwick and Robins, 1980), so the capacity to repair this form of damage would appear to be limited. It may be that the 7-methylguanine-DNA glycosylase activity is present at such a low level even after induction that only small amounts of 7-methylguanine are actively excised during the course of an analytical experiment. Another possibility might be that the torpid activity observed with regard to release of 7-methylguanine residues is a consequence of the enzyme removing more efficiently other guanine adducts with large substituent groups at the readily alkylated N^7 site. The relative rates of enzymatic release of 7-methylguanine and 7-ethylguanine could provide some information on this point, but this has not so far been investigated.

REMOVAL FROM DNA OF 7-METHYLGUANINE RESIDUES WITH OPENED IMIDAZOLE RINGS

7-Methylguanine in DNA is chemically a much less stable DNA component than unmodified guanine, and it is subject to two different kinds of hydrolytic reactions. In the absence of active repair, 7-methylguanine is released from DNA in free form by hydrolytic cleavage of the glycosyl bond with a half-life of about 110 h at 37°C and neutral pH (Lawley and Warren, 1976). In addition, 7-methylguanine is converted to a pyrimidine derivative by hydrolytic opening of the imidazole ring (Haines et al., 1962), and this alkali-catalyzed reaction happens readily in alkylated, double-stranded DNA incubated immediately below the pH at which alkaline denaturation of the secondary structure occurs (Chetsanga and Lindahl, 1979). Model experiments with 7-methyldeoxyguanosine solutions incubated at different pH and temperatures indicate that the ring-opening reaction takes place at a 200 to 400 times slower rate than glycosyl bond cleavage at pH 7.4 and 37°C (L. Breimer, unpublished data). This slow reaction may be of physiological significance since the substituted formamidopyrimidine generated by hydrolytic cleavage of the imidazole ring would presumably convert 7-methylguanine to a noncoding (or miscoding) base residue that remains in the DNA, while the more common hydrolytic event of glycosyl bond cleavage at a 7-methylguanine residue would result in the formation of an easily repairable apurinic site. When DNA that contained ring-opened 7-methylguanine residues was employed as a potential substrate in enzyme assays, effective release of these residues in free form was observed upon incubation with E. coli extracts. The active factor has been shown to be a distinct DNA glycosylase, which is unable to remove the species with an intact imidazole ring from DNA (Chetsanga and Lindahl, 1979). Thus, the activity is clearly due to an enzyme that is different from the inducible 7-methylguanine-DNA glycosylase described above. This formamidopyrimidine-DNA glycosylase may serve to remove a potentially dangerous, secondary alkylation product that is spontaneously generated in small amounts from the abundant, but less harmful, 7-methylguanine residues. Alternatively (or in addition), the enzyme may be active in the repair of damage caused by ionizing radiation since purines with opened imidazole rings are major base radiation products. A DNA glycosylase of this specificity has been found recently in rat and hamster liver extracts (Margison and Pegg, 1981) so it is present in bacteria and in mammalian cells.

THE UNUSUAL MECHANISM OF REPAIR OF O^6-METHYLGUANINE RESIDUES

Methylating agents which are strongly mutagenic cause the formation of considerable amounts of O-alkylated base residues in addition to the N-alkylated adducts. The most important DNA lesion in this regard appears to be O^6-methylguanine. This is a miscoding base because the methylation at the O^6 position locks guanine in an aberrant tautomeric form, which forms base pairs with thymine as well as with cytosine. Consequently, DNA replication over an O^6-methylguanine residue is associated with a very high risk of the introduction of a transition mutation. E. coli and mammalian cells can actively remove O^6-methylguanine residues, but the repair process seems more limited than the excision of 3-methyladenine residues (Medcalf and Lawley, 1981). In E. coli, and apparently in mammalian cells as well, the repair of O^6-methylguanine is an inducible process, triggered by exposure of cells to low concentrations of alkylating agents (Montesano et al., 1980; Samson and Cairns, 1977). The removal of O^6-methylguanine from DNA as a consequence of this adaptive response shows some unusual features. In particular, the repair activity initially is quite efficient but appears to be expended in its reaction with alkylated DNA in vivo (Cairns, 1980; Robins and Cairns, 1979).

The development of an in vitro assay for the removal of O^6-methylguanine has allowed a clarification of the biochemical mechanism of the reaction. When DNA containing a small proportion of O^6-methylguanine residues, radioactively labeled in the methyl groups, was incubated with a cell-free extract of adapted E. coli, the O^6-methylguanine could no longer be detected in DNA hydrolysates. Surprisingly, the reaction was not accompanied by any release of radioactive material in acid-soluble form (Karran et al., 1979). Instead, the methyl group was apparently transferred to another site while remaining associated with a macromolecule. Recent work on the nature of the methyl acceptor has shown that the methyltransferase responsible for the removal of the methyl group from the O^6 position covalently attaches this moiety to one of its own cysteine residues. Consequently, radioactive S-methylcysteine is found in a hydrolysate of the protein after its reaction with DNA (Olsson and Lindahl, 1980). The methyltransferase undergoes suicide inactivation in this process. That is, a defined amount of the protein has the ability to remove a stoichiometric amount of the O^6-methyl group very rapidly from DNA, but then the reaction ceases to occur and the now-methylated protein appears to be irreversibly inactivated (Lindahl, 1981). This inactivation is presumably due to permanent blocking of a sulfhydryl group since it is not accompanied by any detectable dissociation or cleavage of the protein. Thus, the methyltransferase shows a molecular weight of about 17,000 both in its native unmethylated form and in its inactive methylated form (Olsson and Lindahl, 1980, and unpublished data). The latter data, which are based on gel chromatography experiments of protein carrying a radioactively labeled methyl group in the presence of 6 M guanidium hydrochloride, also strongly indicate that the blocked sulfhydryl group is present in a cysteine residue within the actual protein chain rather than in a low-molecular weight cofactor, such as glutathione, which could conceivably have been associated with the methyltransferase. The formation of a stable, but inactive, covalent adduct of the protein by the attachment of a methyl group to the active site resembles the interaction between a "suicide enzyme inactivator" and the target enzyme (Abeles and Maycock, 1976).

A complementary experiment on the fate of DNA containing O^6-methylguanine during the adaptive response has been performed recently by Foote et al. (1980). These authors used the in vitro assay conditions defined by Karran et al. (1979) but employed a substrate containing O^6-methylguanine residues radioactively labeled in the purine rings instead of in the methyl groups. In this way it was possible to demonstrate that unmodified guanine was regenerated in DNA on removal of the O^6-methyl group. Taken together, these findings suggest that in this case there is no need for an excision-repair process involving the removal of nucleotide residues and subsequent gap-filling. Instead, a novel and apparently costly form of repair occurs in which an inducible "kamikaze protein" acts on its target, O^6-methylguanine in DNA, and undergoes suicide inactivation as a consequence of the methyl group transfer.

PRESENCE OF THE METHYLTRANSFERASE IN NONADAPTED CELLS

E. coli cells not previously exposed to MNNG or a similar alkylating agent show little or no detectable activity that repairs O^6-methylguanine in DNA (Schendel and Robins, 1978). Similarly, crude cell extracts from nonadapted cells do not contain quantities of the methyltransferase that are detectable in our standard in vitro assay (Karran et al., 1979). However, these observations do not exclude the possibility that E. coli cells could contain a small amount of the repair function, adequate for removal of small amounts of O^6-methylguanine from DNA but largely insufficient to deal with the damage introduced by a substantial challenge dose of an alkylating agent. We have developed a purification procedure for the methyltransferase (Olsson and Lindahl, 1980), and this has now been applied to cell extracts from wild-type E. coli B and E. coli K-12 grown in standard glycerol-yeast extract-salts media. These experiments have revealed the presence of small amounts of the methyltransferase in extracts of nonadapted cells. The levels are 100- to 200-fold lower than in fully adapted cells. Since adapted E. coli contains about 3000 molecules of the repair protein per cell (Cairns et al., 1981), the data imply that unadapted wild-type cells contain about 20 methyltransferase molecules per cell. This may represent a basal level of uninduced activity or a very low level of persistent adaptation to internal methylation events caused by S-adenosylmethionine and other intracellular methyl donors. It seems unlikely that the presence of the methyltransferase in nonadapted cells could be ascribed to the occurrence of alkylating agents in the bacterial growth medium because methylating agents such as MNNG which act directly on DNA are unstable in neutral aqueous solution and would be destroyed by autoclaving. Moreover, E. coli does not possess enzymes corresponding to the microsomal oxidases that cause activation of other (indirect acting) alkylating agents in mammalian cells, so it may be argued that even if traces of compounds such as dimethylnitrosamine were present in the growth media they would be unable to induce an adaptive response.

The presence of small amounts of the methyltransferase in non-adapted E. coli provides an explanation for the observation that MNNG treatment of phage λ particles yields few or no mutants while similar treatment of intracellular phage DNA in lytically infected E. coli is highly mutagenic (Yamamoto et al., 1978). In the latter case, the simultaneous and extensive methylation of the host chromosome would be expected to generate many more O^6-methylguanine residues in DNA than could be dealt with by available repair protein molecules. In contrast, the relatively few O^6-methylguanines within an incoming phage genome presumably could be repaired efficiently by the preexisting methyltransferase molecules of the host cell.

REPAIR OF O^6-ETHYLGUANINE

When DNA is treated with ethylating agents, lesions analogous to the methylated ones are introduced. Thus, a DNA substrate suitable for in vitro assays containing O^6-ethylguanine instead of O^6-methylguanine can be prepared simply by replacing N-methyl-N-nitrosourea with N-ethyl-N-nitrosourea. Employing such substrates, we have shown that the highly purified E. coli methyltransferase that acts on O^6-methylguanine also has the ability to remove ethyl groups from O^6-ethylguanine residues. The O^6-ethylguanine residues disappear from DNA with the concomitant transfer of ethyl groups to protein. Subsequent amino acid analysis of the transferase has shown that the radioactive ethyl groups are recovered in the form of S-ethylcysteine residues (A. Jacobsson, M. Olsson, and T. Lindahl, unpublished data). Similar conclusions about the ability of the induced repair function to react both with methylated and ethylated DNA have been drawn from in vivo data in which adapted E. coli cells have been challenged with an ethylating agent (B. Sedgwick, personal communication). In rat liver cells, a similar activity has been found. That is, O^6-methylguanine seems to be repaired in the same fashion in such cells (Montesano et al., 1980) as in E. coli, and in cell extracts an activity that transfers the ethyl group of O^6-methylguanine to a protein residue has been described recently (Renard et al., 1981). It is interesting to note in this context that the E. coli 3-methyladenine-DNA glycosylase also is able to release effectively 3-ethyladenine from ethylated DNA (Riazuddin and Lindahl, 1978). It therefore seems likely that these findings can be generalized and that methylation and ethylation damage can be repaired by the same enzymes. In addition, the bulkier ethyl adducts may in some cases be recognized and removed by uvr^+-dependent repair, while this is not the case for methyl adducts (Warren and Lawley, 1980).

CONCLUSIONS

E. coli cells appear well equipped to cope with low levels of alkylation damage. This might be related to the continuous DNA alkylation by S-adenosylmethionine and/or to a common occurrence of alkylating agents among environmental mutagens. A 3-methyladenine-DNA glycosylase is expressed constitutively in sufficient amounts to allow very rapid removal of 3-methyladenine from DNA. In addition, E. coli cells contain several other repair functions, including low levels of a second glycosylase activity for 3-methyladenine, a glycosylase activity for 7-methylguanine, and a distinct methyltransferase activity for O^6-methylguanine. The latter three functions are all inducible, although the induction of a repair pathway for 7-methylguanine appears much less striking and more limited than the response to O^6-methylguanine. Moreover, certain forms of alkylation damage might be subject to $recA^+$-dependent postreplication repair. It seems clear that E. coli has developed repair mechanisms that enable cells to counteract a low level of continuous alkylation, as well as allowing them to exist in an uncommon ecological niche in which considerable amounts of DNA alkylation would occur. Cells, on the other hand, seem more vulnerable to sudden exposure to a high dose of an alkylating agent, and this presumably accounts for the fact that N-nitroso compounds can act as efficient mutagens.

REFERENCES

Abeles, R. H., and Maycock, A. L., 1976, Suicide enzyme inactivators, Acc. Chem. Res., 9:313.

Brent, T. P., 1979, Partial purification and characterization of a human 3-methyladenine-DNA glycosylase, Biochemistry, 18:911.

Cairns, J., 1980, Efficiency of the adaptive response of E. coli to alkylating agents, Nature, 286:176.

Cairns, J., Robins, P., Sedgwick, B., and Talmud, P., 1981, Inducible repair of alkylated DNA, Prog. Nucleic Acid Res. Mol. Biol., 26:237.

Cathcart, R., and Goldthwait, D. A., 1981, Enzymatic excision of 3-methyladenine and 7-methylguanine by a rat liver nuclear fraction, Biochemistry, 20:273.

Chetsanga, C. J., and Lindahl, T., 1979, Release of 7-methylguanine residues whose imidazole rings have been opened from damaged DNA by a DNA glycosylase from E. coli, Nucleic Acids Res., 6:3673.

Foote, R. S., Mitra, S., and Pal, B. C., 1980, Demethylation of O^6-methylguanine in a synthetic DNA polymer by an inducible activity in E. coli, Biochem. Biophys. Res. Commun., 97:654.

Haines, J. A., Reese, C. B., and Lord Todd, 1962, The methylation of guanine and related compounds with diazomethane, J. Chem. Soc., IV:5281.

Ishiwata, K., and Oikawa, A., 1979, Actions of human DNA glycosylases on uracil-containing DNA, methylated DNA and their reconstituted chromatins, Biochim. Biophys. Acta, 563:375.

Jeggo, P., Defais, M., Samson, L., and Schendel, P., 1979, An adaptive response of E. coli to low levels of alkylating agent: Comparison with previously characterized DNA repair pathways, Mol. Gen. Genet., 157:1.

Karran, P., Lindahl, T., and Griffin, B. E., 1979, Adaptive response to alkylating agents involves alteration in situ of O^6-methylguanine residues in DNA, Nature, 280:76.

Karran, P., Lindahl, T., Ofsteng, I., Evensen, G. B., and Seeberg, E., 1980, E. coli mutants deficient in 3-methyladenine-DNA glycosylase, J. Mol. Biol., 140:101.

Laval, J., 1977, Two enzymes are required for strand incision in repair of alkylated DNA, Nature, 269:829.

Laval, J., Pierre, J., and Laval, F., 1981, Release of 7-methylguanine residues from alkylated DNA by extracts of Micrococcus luteus and Escherichia coli, Proc. Natl. Acad. Sci. U.S.A., 78: 852.

Lawley, P. D., and Orr, D. J., 1970, Specific excision of methylation products from DNA of E. coli treated with N-methyl-N'-nitro-N-nitrosoguanidine, Chem.-Biol. Interact., 2:154.

Lawley, P. D., and Thatcher, C. J., 1970, Methylation of DNA in cultured mammalian cells by N-methyl-N'-nitro-N-nitrosoguanidine, Biochem. J., 116:693.

Lawley, P. D., and Warren, W., 1976, Removal of minor methylation products 7-methyladenine and 3-methylguanine from DNA of E. coli treated with dimethyl sulphate, Chem.-Biol. Interact., 12:211.

Lindahl, T., 1976, New class of enzymes acting on damaged DNA, Nature, 259:64.

Lindahl, T., 1981, DNA methyltransferase acting on O^6-methylguanine residues in adapted E. coli, in: "Chromosome Damage and Repair," E. Seeberg, and K. Kleppe, eds., Plenum Publishing Corporation, New York.

Lindahl, T., and Nyberg, B., 1972, Rate of depurination of native DNA, Biochemistry, 11:3610.

Ljungquist, S., 1977, A new endonuclease from E. coli acting at apurinic sites in DNA, J. Biol. Chem., 252:2808.

Margison, G. P., and Pegg, A. E., 1981, Enzymatic release of 7-methylguanine from methylated DNA by rodent liver extracts, Proc. Natl. Acad. Sci. U.S.A., 78:861.

Medcalf, A. S. C., and Lawley, P. D., 1981, Time course of O^6-methylguanine removal from DNA of N-methyl-N-nitrosourea-treated human fibroblasts, Nature, 289:796.

Montesano, R., Bresil, H., Planche-Martel, G., Margison, G. P. and Pegg, A. E., 1980, Effect of chronic treatment of rats with dimethylnitrosamine on the removal of O^6-methylguanine from DNA, Cancer Res., 40:452.

Olsson, M., and Lindahl, T., 1980, Repair of alkylated DNA in E. coli: Methyl group transfer from O^6-methylguanine to a protein cysteine residue, J. Biol. Chem., 255:10569.

Paik, W. K., Lee, H. W., and Kim, S., 1975, Non-enzymatic methylation of proteins with S-adenosyl-L-methionine, FEBS Lett., 58:39.

Renard, A., Verly, W. G., Mehta, J. R., and Ludlum, D. B., 1981, Repair of O^6-ethylguanine in DNA by a chromatin factor from rat liver, Fed. Proc., 40:1763.

Riazuddin, S., and Lindahl, T., 1978, Properties of 3-methyladenine-DNA glycosylase from E. coli, Biochemistry, 17:2110.

Robins, P., and Cairns, J., 1979, Quantitation of the adaptive response to alkylating agents, Nature, 280:74.

Rogers, S. G., and Weiss, B., 1980, Exonuclease III of E. coli K-12, an AP endonuclease, Methods Enzymol., 65:201.

Samson, L., and Cairns, J., 1977, A new pathway for DNA repair in E. coli, Nature, 267:281.

Schendel, P. F., and Robins, P., 1978, Repair of O^6-methylguanine in adapted E. coli, Proc. Natl. Acad. Sci. U.S.A., 75:6017.

Sedgwick, B., and Robins, P., 1980, Isolation of mutants of E. coli with increased resistance to alkylating agents: Mutants deficient in thiols and mutants constitutive for the adaptive response, Mol. Gen. Genet., 180:85.

Singer, B., 1976, All oxygens in nucleic acids react with carcinogenic ethylating agents, Nature, 264:333.

Singer, B., and Brent, T. P., 1981, Human lymphoblasts contain DNA glycosylase activity excising N-3 and N-7 methyl and ethyl purines but not O^6-alkylguanines or 1-alkyladenines. Proc. Natl. Acad. Sci. U.S.A., 78:858.

Verly, W. G., Paquette, Y., and Thibodeau, L., 1973, Nuclease for apurinic sites may be involved in the maintenance of DNA in normal cells, Nature, 244:67.

Warner, H. R., Demple, B. F., Deutsch, W. A., Kane, C. M., and Linn, S., 1980, Apurinic/apyrimidinic endonucleases in repair of pyrimidine dimers and other lesions in DNA, Proc. Natl. Acad. Sci. U.S.A., 77:4602.

Warren, W., and Lawley, P. D., 1980, The removal of alkylation products from the DNA of E. coli cells treated with the carcinogens N-ethyl-N-nitrosourea and N-methyl-N-nitrosourea: Influence of growth conditions and DNA repair defects, Carcinogenesis, 1:67.

Yamamoto, K., Kondo, S., and Sugimura, T., 1978, Mechanism of potent mutagenic action of N-methyl-N'-nitro-N-nitrosoguanidine on intracellular phage lambda, J. Mol. Biol., 118:413.

CHAPTER 5

CELLULAR RESPONSES TO MUTAGENIC AGENTS:

A SUMMARY AND PERSPECTIVE

Sohei Kondo

Department of Fundamental Radiology, Faculty of
Medicine, Osaka University, Kita-ku, Osaka 530, Japan

Mutagenesis is a complicated process. Alteration of DNA is only one of the important factors involved. The final outcome of mutation depends on various biological factors; among them, DNA repair is now known to be as important as the initial DNA alteration. As partly discussed in this symposium, Escherichia coli has at least three seemingly independent repair systems for DNA damage.

Wild-type E. coli has a very efficient capacity to repair mismatched DNA bases as reviewed by Glickman (1981). The mismatch repair seems to be a constitutive function.

For bulky DNA damage such as pyrimidine dimers, E. coli has a well-known recA-dependent inducible repair. It is a surprise that the expression of the uvrA gene, which is indispensable for error-free excision repair, is subject to induction by the recA-lexA system (Walker, 1981) since this system had been supposed to control only error-prone repair in addition to induction of some other nonrepair functions (Witkin, 1976). Thus, we must conclude that the recA-lexA system controls the induction of not only error-prone, but also error-free, repair for bulky DNA damage. It should, however, be noted that the expression of the uvrA gene is partially constitutive. Thus, the two terms "inducible" and "constitutive" should not be used as exclusive to each other.

We now know that E. coli has an efficient repair capacity for at least some types of alkylated DNA bases, and that the repair enzymes seem to be specific for different sites of alkylation (Lindahl et al., 1981; Singer, 1981). In relation to the clear-cut work on methyltransferase by Lindahl et al. (1981), it may be appropriate to summarize the work done by Cairns and his colleagues (Jeggo, 1979; Jeggo et al., 1977; Samson and Cairns, 1977; Schendel et al., 1978; Sedgwick and Robins, 1980) on the adaptive response in E. coli.

Upon treatment with N-methyl-N'-nitro-N-nitrosoguanidine (MNNG), normal strains of E. coli show a typical dose-mutation frequency relationship composed of three parts, as illustrated in Fig. 1 (Jeggo, 1979; Schendel et al., 1978; Sedgwick and Robins, 1980). In the low-dose range, the mutation frequency increases linearly with dose (the first part). In the intermediate-dose range, the frequency increases with dose at a rate slower than linear, due to induced adaptive error-free DNA repair (the second part). In the high-dose range, the frequency increases sharply with dose (the third part) due to induced error-prone repair (Schendel et al., 1978; Yamamoto et al., 1978) or to induced saturation effects on the adaptive error-free repair. It should be stressed that both the error-prone (Schendel et al., 1978; Yamamoto et al., 1978) and the error-free repair (Jeggo et al., 1977; Schendel et al., 1978) are induced after treatment with alkylating agents in recA- or lexA-deficient strains as well as in normal strains. In Ada$^-$ (adaptive-response-deficient) strains, the dose-response relationship to MNNG mutagenesis consists of the first and third parts and lacks the second part of the normal strain response because adaptive error-free repair is lacking (Jeggo, 1979). Constitutive adaptive-response strains seem to give a response curve made of the third part only after a large threshold dose (Sedgwick and Robins, 1980).

Recently, Day and his colleagues (1980a,b) reported that many strains of human tumor cells are very sensitive to MNNG due to defective repair for MNNG-induced DNA damage; he named them Mer$^-$ strains. From the dose-response curves for sister chromatid exchanges (SCE) induced by MNNG (Fig. 1) (Day et al., 1980b), we find that the Mer$^-$ cell strains may correspond to Ada$^-$ strains of E. coli, and that Mer$^+$ strains of normal human cells correspond to constitutive-adaptive response strains of E. coli.

In my laboratory, we have recently found that N-hydroxyurethan seems to act as an inducer of a kind of error-free adaptive response but not of error-prone repair, because the response curve for NHU mutagenesis showed an increase with dose slower than a linear increase, and because uvrA, uvrB, and uvrC strains all showed about 10 times the sensitivity of their parental normal strains (Nagata and Kondo, 1979).

In summary, we may conclude that in E. coli there are at least two independent systems for inducible DNA repair: One, which is recA-dependent, is for bulky damage such as pyrimidine dimers, and the other is for alkylated bases and is recA-independent. In human cells the evidence is only suggestive, but there seem to exist also

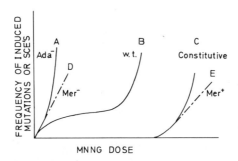

Fig. 1. Schematic dose-response curves of frequencies of MNNG-induced mutations or SCE. Mutations (———) in E. coli strains deficient (Ada⁻, curve A), normal (w.t., curve B) and constitutive (curve C) for adaptive response and SCE (— — —) in human cell strains defective (Mer⁻, curve D) and normal (Mer⁺, curve E). Constructed from Day et al. (1980b), Jeggo (1979), and Sedgwick and Robins (1980).

two independent DNA repair systems, which presumably work, at least partially, as constitutive functions, as suggested by the existence of xeroderma pigmentosum cells defective for excision repair of UV damage and Mer⁻ cell strains defective for repair of DNA damage caused by MNNG and other alkylating agents.

Finally, it should be stressed that cellular responses to mutagens are very complex. For example, mutagenesis and killing by alkylating agents depend not only on the alkylated-base-specific repair but also on repair for bulky damage as well as on mismatch repair. In E. coli, MNNG- or ethyl methanesulfonate-induced mutagenesis depend, at least under certain conditions, on recA, lexA, uvrA, polA (Ishii and Kondo, 1975; Kondo et al., 1970; Schendel et al., 1978) and dam-3 (mismatch-repair deficient) (Glickman et al., 1978) mutations. Killing adaptation does not occur after treatment with MNNG in polA or dam-3 strains (Jeggo et al., 1978). Further, in yeast (Lawrence et al., 1981) and Neurospora (de Serres, 1981) some types of DNA repair deficiency greatly affect mutagenesis induced by alkylating agents as well as UV mutagenesis, an indication that mechanisms of mutagenesis could, at least in some mutagenic pathways, substantially differ between E. coli and eukaryotes.

REFERENCES

Day, R. S., III, Ziolkowski, C. H. J., Scudiero, D. A., Meyer, S. A., and Mattern, M. R., 1980a, Human tumor cell strains defective in the repair of alkylation damage, Carcinogenesis, 1:21.

Day, R. S., III, Ziolkowski, C. H. J., Scudiero, D. A., Meyer, S. A., Lubiniecki, A. S., Girardi, A. J., Galloway, S. M., and Bynum, G. D., 1980b, Defective repair of alkylated DNA by human tumour and SV40-transformed human cell strains, Nature (London), 288: 724.

de Serres, F. J., 1981, Comparison of the induction of specific locus mutations in wild-type and repair-deficient strains of Neurospora crassa, this volume.

Glickman, B., van den Elsen, P., and Radman, M., 1978, Induced mutagenesis in dam⁻ mutants of Escherichia coli: A role for 6-methyladenine residues in mutation avoidance, Mol. Gen. Genet., 163:307.

Glickman, B. W., 1981, Methylation-instructed mismatch correction as a postreplication error avoidance mechanism in Escherichia coli, this volume.

Ishii, Y., and Kondo, S., 1975, Comparative analysis of deletion and base-change mutabilities of Escherichia coli B strains differing in DNA repair capacity (wild-type, uvrA⁻, polA⁻, recA⁻) by various mutagens, Mutat. Res., 27:27.

Jeggo, P., 1979, Isolation and characterization of Escherichia coli K12 mutants unable to induce the adaptive response to simple alkylating agents, J. Bacteriol., 139:783.

Jeggo, P., Defais, M., Samson, L., and Schendel, P., 1977, An adaptive response of E. coli to low levels of alkylating agent: comparison with previously characterized DNA repair pathways, Mol. Gen. Genet., 157:1.

Jeggo, P., Defais, M., Samson, L., and Schendel, P., 1978, The adaptive response of E. coli to low levels of alkylating agent: the role of polA in killing adaptation, Mol. Gen. Genet., 162:299.

Kondo, S., Ichikawa, H., Iwo, K., and Kato, T., 1970, Base-change mutagenesis and prophage induction in strains of Escherichia coli with different DNA repair capacities, Genetics, 66:187.

Lawrence, C. W., Christensen, R., and Schwarz, A., 1981, Mechanisms of UV mutagenesis in yeast, this volume.

Lindahl, T., Rydberg, B., Hjelmgren, T., Olsson, M., and Jacobsson, A., 1981, Cellular defense mechanisms against alkylation of DNA, this volume.

Nagata, H., and Kondo, S., 1979, N-Hydroxyurethan-induced mutagenesis in Escherichia coli (in Japanese). Abstracts of Japanese Environmental Mutagen Society Meeting, p. 65.

Samson, L., and Cairns, J., 1977, A new pathway for DNA repair in Escherichia coli, Nature (London), 267:281.

Schendel, P. F., Defais, M., Jeggo, P., Samson, L., and Cairns, J., 1978, Pathways of mutagenesis and repair in Escherichia coli exposed to low levels of simple alkylating agents, J. Bacteriol., 135:466.

Sedgwick, B., and Robins, P., 1980, Isolation of mutants of Escherichia coli with increased resistance to alkylating agents: Mutants deficient in thiols and mutants constitutive for the adaptive response, Mol. Gen. Genet., 180:85.

Singer, B., 1981, Mutagenesis from a chemical perspective: Nucleic acid reactions, repair, translation, and transcription, this volume.

Walker, G. C., Kenyon, C. J., Bagg, A., Elledge, S. J., Perry, K. L., and Shanabruch, W. G., Regulation and functions of Escherichia coli genes induced by DNA damage, this volume.

Witkin, E. M., 1976, Ultraviolet mutagenesis and inducible DNA repair in Escherichia coli, Bacteriol. Rev., 40:869.

Yamamoto, K., Kondo, S., and Sugimura, T., 1978, Mechanism of potent mutagenic action of N-methyl-N'-nitro-N-nitrosoguanidine on intracellular phage lambda, J. Mol. Biol., 118:413.

CHAPTER 6

MECHANISMS OF UV MUTAGENESIS IN YEAST

Christopher W. Lawrence, Roshan Christensen, and
Ann Schwartz

Department of Radiation Biology and Biophysics
University of Rochester, School of Medicine and
Dentistry, Rochester, New York 14642

SUMMARY

UV mutagenesis in yeast depends on the function of the RAD6 locus, a gene that is also responsible for a substantial fraction of wild-type resistance, suggesting that this eukaryote may possess a misrepair mechanism analogous to that proposed for Escherichia coli. The molecular mechanism responsible for RAD6 repair or recovery is not yet known, but it is different from either excision or recombination-dependent repair, processes carried out by the other two main repair pathways in yeast. RAD6-dependent mutagenesis has been found to have the following characteristics.

It is associated at best with only a small fraction of RAD6-dependent repair, the majority of the sensitivity of rad6 mutants being due to their lack of nonmutagenic repair. SRS2 metabolic suppressors restore a substantial fraction of UV resistance to rad6 mutants but do not restore their UV mutability. Strains containing mutations at loci (rev, umr) that are probably more directly involved in mutagenesis are only mildly sensitive, and there is a poor correlation between their sensitivity and mutational deficiency.

UV mutagenesis appears to require a large number of gene functions, perhaps ten or more. Where examined in detail, these genes have been found to be concerned in the production of only a specific range of mutational events, not all of them.

Mating experiments have shown that a substantial fraction, probably 40% or more, of UV-induced mutations are untargeted, that is, occur in lesion-free regions of DNA. UV irradiation, therefore, produces a general reduction in the normally high fidelity with which DNA is replicated on undamaged templates. It does not appear to be necessary for the causal lesion to be present in the same chromosome as the mutation it induces. The reduction in fidelity may be the consequence of the production of a diffusible factor in UV-irradiated cells, but definite evidence supporting this proposal has not yet been obtained.

INTRODUCTION

UV mutagenesis in the yeast Saccharomyces cerevisiae requires the function of the RAD6 gene (Lawrence et al., 1970, 1974), a function also involved in processes responsible for a substantial fraction of wild-type resistance to radiations and other DNA-damaging agents (reviewed in Lawrence, 1981). The molecular mechanisms that account for this resistance are not yet fully understood but do not appear to entail either excision (Prakash, 1977) or recombination (Hunnable and Cox, 1971; Kern and Zimmermann, 1978; Prakash, personal communication; Saeki et al., 1980). Since rad6 mutants are probably deficient in postreplication repair (DiCaprio and Cox, 1981), one mechanism may be transdimer synthesis though definitive evidence for this suggestion has not yet been obtained.

UV mutagenesis seems to be associated with only a minor fraction, at best, of these RAD6-dependent repair or recovery processes; rad6-1 mutants that carry an SRS2 metabolic suppressor are substantially more resistant to the lethal effects of UV light than the unsuppressed mutant but remain equally incapable of induced mutagenesis (Lawrence and Christensen, 1979a), while translational suppressors are also capable of dissociating mutagenesis and survival (Tuite and Cox, 1981). Similarly, strains containing mutations at the REV (Lemontt, 1971) or UMR (Lemontt, 1977) loci, which were identified on the basis of their deficiency with respect to UV mutagenesis and, therefore, are probably more immediately involved in this process, are either only moderately UV sensitive or virtually as resistant as the wild type. Such observations not only question whether mutagenesis is an intrinsic consequence of repair or recovery mechanisms, but, insofar as RAD6 repair may involve transdimer synthesis, they may also imply that this process can often be error free. This speculative inference obviously requires investigation and substantiation.

GENES INVOLVED IN UV MUTAGENESIS

Whatever the molecular mechanisms specifically responsible for UV mutagenesis, it is clear that they depend on the functions of a considerable number of genes. Lemontt (1971, 1977) has identified at least six genes, three REV and three or more UMR loci, and more recently we have isolated seven mutant strains of this kind that, on the basis of phenotype and complementation, appear to define four new genes, giving a total of at least ten. We embarked on this mutant hunt because it seemed unlikely that such mutations had been isolated exhaustively. The screening procedure employed by Lemontt detected deficiencies chiefly or exclusively with respect to the induction of base-pair substitutions by UV. As discussed below, rev mutations may exhibit a highly selective effect on the induction of different kinds of mutations, however, and it therefore appeared worthwhile to make use of a screening procedure based on the reversion of the frameshift allele his4-38 (Culbertson et al., 1977). This mutation reverts by frameshift events within the his4-38 locus itself and also by base-pair addition within certain tRNA loci.

Treatment of a his4-38 strain (kindly supplied by G. Fink) with 2% ethyl methanesulfonate, followed by the screening of 11,366 potentially mutant clones, yielded four reversion deficient strains containing mutations that complement rev1-1, rev2-1, rev3-1, and rad6-1, as well as two new rev1 alleles, one new rev2 allele, and six new rev3 alleles. These four strains fall into three complementation groups. In addition, screening of methyl methanesulfonate-sensitive strains, isolated for another purpose, yielded three reversion-deficient strains that fall into a single complementation group different from the other three. UV-induced reversion of his4-38 and survival curves for these strains are illustrated in Figs. 1 and 2. The new mutants appear to differ from the umr mutants in these respects, but an answer to the question of their possible allelism awaits the outcome of tests now in progress.

THE PHENOTYPES OF REVERSION-DEFICIENT MUTANTS

With the exception of rad6 mutants, none of the reversion-deficient strains that have been adequately studied are deficient with respect to all types of induced events; even rev3 mutants, which have the largest and most general deficiency, exhibit about half the normal frequency of the kind of base-pair substitutions required to revert cyc1-115 and cyc1-131 and lesser, though appreciable, frequencies of frameshift events (Lawrence and Christensen, 1979b). In keeping with this observation, three of the newly isolated mutations (those in RG-6 and RG-150, which are allelic, and in RG-114) have no effect on the UV-induced reversion of the ochre allele arg4-17 (Fig. 3), though they do, of course, reduce the frequency of his4-38 revertants (Fig. 1). Further work will be required to establish whether these mutations affect frameshift mutagenesis and not base-pair substitution in general.

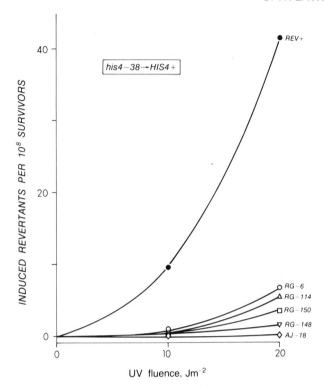

Fig. 1. UV-induced reversion frequencies of the frameshift allele his4-38 in wild-type (REV$^+$) and various rev$^-$ strains. The RG strains were isolated on the basis of their deficiency in his4-38 reversion induced by UV, and AJ-18 was isolated as sensitive to methyl methanesulfonate.

UNTARGETED MUTAGENESIS

In excision-deficient cells the activity of genes of the kind discussed above leads not only to the induction by UV of targeted mutations, those at the site of the causal lesion (Witkin and Wermundsen, 1979), but also, as shown by the results of experiments discussed below, to the production of a high frequency of untargeted mutations, events that occur in lesion-free stretches of DNA. A major consequence of UV irradiation to yeast cells, therefore, appears to be a general reduction in the normally high fidelity with which undamaged DNA templates are replicated. The occurrence of untargeted mutagenesis has been clearly established in experiments wit bacteriophage (e.g., Kondo and Ichikawa, 1973), but the relative pro portions of targeted and untargeted events have not been previously estimated. We have attempted to do this in mating experiments

MUTAGENESIS IN YEAST

conceptually similar (Fig. 4) to those devised by Fabre and Roman (1977) to study induced recombination. An excision-deficient (rad1-2) haploid strain carrying a nonrevertible deletion of the whole CYC1 locus (cyc1-363) is mated with an excision-deficient haploid of the opposite mating type that contains the UV-revertible ochre allele, cyc1-91). Estimates of the total number of diploid clones formed, and of the proportion of CYC1 revertants among them, can be obtained by plating the mating mixture on suitable selective media. Irradiation of the deletion, but not the cyc1-91 parent gives an estimate of the frequency of untargeted mutations, while the reciprocal treatment gives an estimate of the total frequency of both types of event. It is, of course, essential in this style of experiment to exclude all revertants from the score other than those resulting from changes within the CYC1 locus itself. Although the cyc1-91 allele is an ochre mutation, it is almost entirely nonsuppressible, and translationally suppressed revertants cannot be recovered from the selective medium (Prakash and Sherman, 1973). Metabolically suppressed revertants were excluded from the count by spectroscopic examination (Sherman

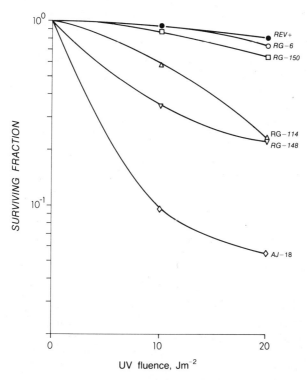

Fig. 2. UV-survival curves for the strains shown in Fig. 1.

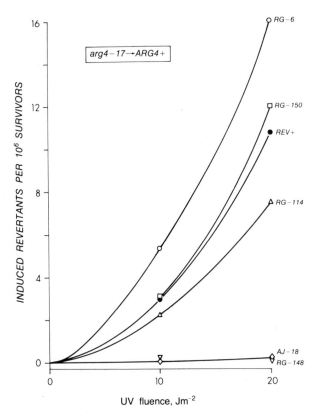

Fig. 3. UV-induced reversion frequencies of the ochre allele arg4-17 in wild-type (REV$^+$) and the various rev$^-$ strains shown in Fig. 1.

and Slonimski, 1964) of a sample of revertants from each experiment, particularly all slow-growing clones. This procedure ensures that virtually all revertants scored contain the high levels of iso-1-cytochrome c characteristic of intragenic reversion.

Data from mating experiments of this kind (Table 1) give estimates for the proportion of untargeted mutations that vary from 13 to 38%. These estimates, particularly the lower ones, are likely to be underestimates however. Untargeted mutations of the "near targeted" variety, for example, will not contribute to the untargeted category, because the end points of the cyc1-363 deletion are distant from the cyc1-91 site (greater than 20 base pairs on one side, greater than 300 base pairs on the other; Sherman and Stewart, 1978) Further, the lower estimates come from experiments with strains in which excision is blocked by a single mutation, which may be slightly

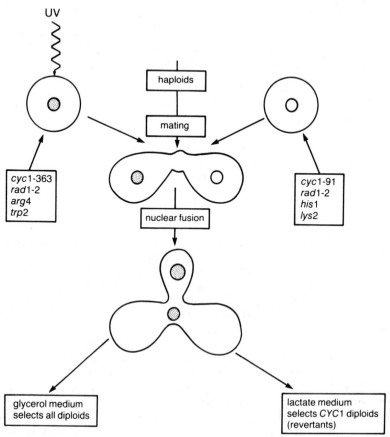

Fig. 4. <u>Indirect Mutagenesis</u>. Protocol for the mating experiments. Haploid strains of the opposite mating type that contain either the nonrevertible deletion <u>cyc1-363</u> (sometimes <u>cyc1-383</u>) or the ochre allele <u>cyc1-91</u> were grown to early stationary phase and samples of washed cells suspended in 0.9% KCl at a concentration of 1×10^7 cells/ml exposed to 0 or 4 J m^{-2} UV. Three matings, comprising irradiation to one or neither parent, were carried out by suspending equal numbers of parental cells in 1% yeast extract, 2% peptone, 10% dextrose medium for 5 h. About 1×10^8 cells of all types were plated on semisynthetic lactate medium to estimate the number of diploid <u>CYC1</u> revertants and suitable dilutions were plated on synthetic glycerol medium to estimate the number of diploid clones.

Table 1. Untargeted Mutagenesis in Yeast

Genotype of Strains	No. of Experiments	UV Fluence ($J\ m^{-2}$)	Induced Revertants/10^8 Survivors (% Survival) UV Irradiation to		Untargeted Revertants (%)	Spontaneous Mutation Frequencies $\times 10^8$		R/R_o^a	Mating (%)
			cyc1-91 Haploid	Deletion Haploid		Diploid	Haploid		
rad1-2 (cyc1-363)	2	4	1922 (11)	259 (14)	13	16	35	1.8	8
rad1-2 (cyc1-383)	2	4	2527 (13)	554 (13)	21	111	78	1.1	7
rad1-2 rad2-5	3	4	1680 (15)	632 (13)	38	52	67	1.8	12
rad1-2 rad6-1 SRS2-1	3	2^b	27 (32)	0 (53)	0	3	–	–	–
rad1-2 rev3-1	2	4	0 (16)	0 (14)	0	16	2	–	3
rad1-2 rad2-5 rad52-1	4	4	1839 (25)	681 (21)	37	159	74	1.3	12
rad1-2 kar1-1	2	8	668 (29)	0 (87)	0	–	53	–	–
rad1-2 kar1-1	3	30^c	1010 (6)	24 (74)	2	–	9	–	–
RAD^+	2	100	24600 (34)	10 (45)	0.04	14	5	1.2	8

[a] Number of revertant colonies (uncorrected for survival) when deletion is irradiated divided by number of revertants on control (unirradiated) plates.
[b] Exponential cells.
[c] Irradiation to 1×10^8 cells/ml.

leaky. No untargeted mutagenesis can be detected in excision proficient (RAD⁺) strains (Table 1), no doubt because lesions are excised from the deletion genome during the course of mating, which takes several hours. Data from strains in which excision is blocked by two mutations (rad1-2, rad2-5) give estimates that approach 40% (Table 1). Finally, relatively concentrated suspensions of the mating mixture were spread on the selective medium (about 1×10^8 cells of all kinds per plate) in order to restrict the amount of residual growth to about three to five cell generations (measured by washing cells from selective plates after 0, 1, and 2 days of incubation). This was done to minimize artifacts due to spontaneous mutation during residual growth on the plate, which can seriously inflate estimates of induced mutation frequency. In excision-deficient strains, such a procedure carries the penalty of preventing the expression of a fraction of the induced mutations, however (James et al., 1978; Kilbey and James, 1979), and this fraction appears to be larger following irradiation of the deletion than when the reciprocal treatment is given; spreading equal numbers of cells over a hundred plates containing selective medium rather than twenty of them gives a disproportionately greater increase in the number of revertant colonies when the deletion strain is irradiated than when the reciprocal treatment is given. Decreasing the plating density fivefold in this way allows a further two generations of residual growth.

It is likely, therefore, that the proportion of untargeted mutations exceeds 40%, but it is difficult to establish an upper limit to this value. Preliminary experiments in which low numbers of cells were plated on selective medium, whose results have been briefly reported (Lawrence and Christensen, 1980), appeared to indicate that more or less equal frequencies of cyc1-91 revertants could be induced by irradiating the deletion and ochre mutants, suggesting that virtually all mutations were untargeted; but it is difficult to exclude the possibility that this estimate is inflated by the occurrence of spontaneous mutations during residual growth. Nevertheless, even if only 40% of the induced mutations are untargeted, it is clear that UV irradiation of yeast cells produces a general and substantial reduction in the usually high fidelity with which undamaged DNA templates are replicated.

THE MECHANISM OF UNTARGETED MUTAGENESIS

UV-induced lesions in the deletion genome might give rise to untargeted mutations by one of two (or both) general mechanisms (Fig. 5): (1) by recombination between homologues and the introduction of lesions into the unirradiated chromosome, followed by some sort of error-prone "long patch" repair, or (2) by the induction (or release) of a diffusible agent that reduces replicational fidelity or otherwise promotes mutation. Data from experiments with excision-deficient strains that carry the rad52-1 allele, and are, therefore,

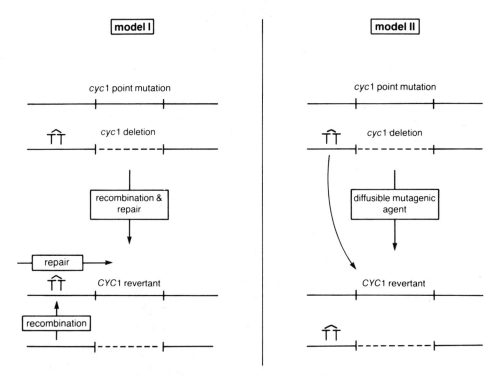

Fig. 5. <u>Indirect Mutagenesis</u>. Models for the mechanism of untargeted mutagenesis in the mating experiment.

also deficient with respect to induced and spontaneous recombination (Prakash et al., 1979; Resnick, 1975; Saeki et al., 1980), show that untargeted mutagenesis does not depend to any detectable extent on recombination (Table 1) and, therefore, suggest that a diffusible agent may be responsible. We have not yet been able to demonstrate the existence of such a factor, however, and experiments designed for this purpose (Fig. 6) have given negative results (Table 1). In these mating experiments, nuclear (but not cell) fusion is blocked by

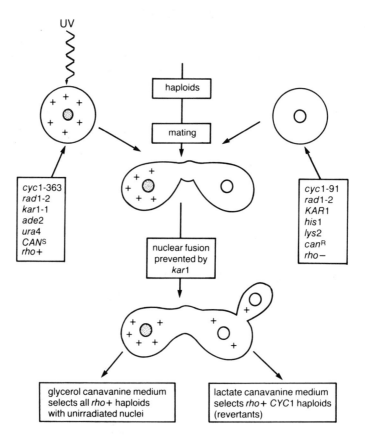

Fig. 6. _Indirect Mutagenesis._ Protocol for the mating experiments in which nuclear fusion is blocked by the kar1 mutation. Experimental procedure is as in Fig. 4, except that exponential cells are used and cells at a concentration of 1×10^8 cells/ml were exposed to 30 J m^{-2} in some experiments. Plating samples of the mating mixture on synthetic glycerol medium containing canavanine provides an estimate of the number of dikaryons that give rise to haploid clones carrying the cyc1-9 nucleus while plating on semisynthetic lactate medium containing canavanine provides an estimate of the proportion of these that are CYC1 revertants.

the karl mutation (Conde and Fink, 1976) so that mutagenesis in the unirradiated nucleus resulting from the sharing of a common cytoplasm with the irradiated nucleus could have been investigated. Whatever the outcome, it is likely that targeted and untargeted mutagenesis are the consequence of the same process since both depend alike on the function of the RAD6 and REV3 genes (Table 1).

ACKNOWLEDGEMENTS

This work was supported in part by U.S. Public Health Service Grant GM21858 and in part performed under Contract DE-AC02-76EV03490 with the U.S. Department of Energy at the University of Rochester, Department of Radiation Biology and Biophysics, and has been assigned Report No. UR-3490-1997.

REFERENCES

Conde, J., and Fink, G. R., 1976, A mutant of Saccharomyces cerevisiae defective for nuclear fusion, Proc. Natl. Acad. Sci. U.S.A., 73:3651.
Culbertson, M. R., Charnas, L., Johnson, M. T., and Fink, G. R., 1977, Frameshifts and frameshift suppressors in Saccharomyces cerevisiae, Genetics, 86:745.
DiCaprio, L., and Cox, B. S., 1981, The effect of UV irradiation on the molecular weight of pre-existing and newly synthesized DNA, Mutat. Res., 82:69.
Fabre, F., and Roman, H., 1977, Genetic evidence for inducibility of recombination competence in yeast, Proc. Natl. Acad. Sci. U.S.A., 74:1667.
Hunnable, E. G., and Cox, B. S., 1971, The genetic control of dark recombination in yeast, Mutat. Res., 13:297.
James, A. P., Kilbey, B. J., and Prefontaine, G. J., 1978, The timing of UV mutagenesis in yeast: Continuing mutation in an excision-defective (rad1-1) strain, Mol. Gen. Genet., 165:207.
Kern, R., and Zimmermann, F. K., 1978, The influence of defects in excision and error-prone repair on spontaneous and induced mitotic recombination and mutation in Saccharomyces cerevisiae, Mol. Gen. Genet., 161:81.
Kilbey, B. J., and James, A. P., 1979, The mutagenic potential of unexcised pyrimidine dimers in Saccharomyces cerevisiae rad1-1. Evidence from photoreactivation and pedigree analysis, Mutat. Res., 60:163.
Kondo, S., and Ichikawa, H., 1973, Evidence that pretreatment of Escherichia coli cells with N-methyl-N'-nitro-nitrosoguanidine enhances mutability of subsequently infecting phage λ, Mol. Gen. Genet., 126:319.
Lawrence, C. W., 1981, Mutagenesis in Saccharomyces cerevisiae, Adv. Genet., 21, in press.

Lawrence, C. W., and Christensen, R. B., 1979a, Metabolic suppressors of trimethoprim and ultraviolet light sensitivities of Saccharomyces cerevisiae rad6 mutants, J. Bacteriol., 139:866.

Lawrence, C. W., and Christensen, R. B., 1979b, Ultraviolet-induced reversion of cyc1 alleles in radiation-sensitive strains of yeast. III. rev3 mutant strains, Genetics, 92:397.

Lawrence, C. W., and Christensen, R. B., 1980, Undamaged DNA is replicated with low fidelity in UV-irradiated yeast, J. Supramol. Struct., Suppl. 4, 356, abstract.

Lawrence, C. W., Stewart, J. W., Sherman, F., and Christensen, R., 1974, Specificity and frequency of ultraviolet-induced reversion of an iso-1-cytochrome c ochre mutant in radiation-sensitive strains of yeast, J. Mol. Biol., 85:137.

Lawrence, C. W., Stewart, J. W., Sherman, F., and Thomas, F. L. X., 1970, Mutagenesis in ultraviolet-sensitive mutants of yeast, Genetics, 64:s36.

Lemontt, J. F., 1971, Mutants of yeast defective in mutation induced by ultraviolet light, Genetics, 68:21.

Lemontt, J. F., 1977, Pathways of ultraviolet mutability in Saccharomyces cerevisiae. III. Genetic analysis and properties of mutants resistant to ultraviolet-induced forward mutation, Mutat. Res., 43:179.

Prakash, L., 1977, Repair of pyrimidine dimers in radiation-sensitive mutants rad3, rad4, rad6, and rad9 of Saccharomyces cerevisiae, Mutat. Res., 45:13.

Prakash, L., and Sherman, F., 1973, Mutagenic specificity: Reversion of iso-1-cytochrome c mutants of yeast, J. Mol. Biol., 79:65.

Prakash, S., Prakash, L., Burke, W., and Montelone, B. A., 1979, Effects of the rad52 gene on recombination in Saccharomyces cerevisiae, Genetics, 94:31.

Resnick, M. A., 1975, The repair of double-strand breaks in chromosomal DNA of yeast, in: "Molecular Mechanisms for Repair of DNA," Part B, P. C. Hanawalt, and R. M. Setlow, eds., Plenum Press, New York.

Saeki, T., Machida, I., and Nakai, S., 1980, Genetic control of diploid recovery after λ-irradiation in the yeast Saccharomyces cerevisiae, Mutat. Res., 73:251.

Sherman, F., and Slonimski, P.P., 1964, Respiration-deficient mutants of yeast. II. Biochemistry, Biochim. Biophys. Acta, 90:1.

Sherman, F., and Stewart, J. W. 1978, The genetic control of yeast iso-1 and iso-2-cytochrome c after 15 years, in: "Biochemistry and Genetics of Yeasts, Pure and Applied Aspects," M. Bacila, B.L. Horecker, and A.O.M. Stoppani, eds., Academic Press, New York

Tuite, M. F., and Cox, B. S., 1981, The $RAD6^+$ gene of Saccharomyces cerevisiae codes for two mutationally separable deoxyribonucleic acid repair functions, Mol. Cell. Biol., 1:153.

Witkin, E. M., and Wermundsen, I. E., 1979, Targeted and untargeted mutagenesis by various inducers of SOS function in Escherichia coli, Cold Spring Harbor Symp. Quant. Biol., 43:881.

CHAPTER 7

SITE-SPECIFIC MUTAGENESIS: A NEW APPROACH FOR STUDYING

THE MOLECULAR MECHANISMS OF MUTATION BY CARCINOGENS

Robert W. Chambers

Department of Biochemistry, New York University
School of Medicine, New York, New York 10016

SUMMARY

Most chemical carcinogens are also mutagens. Usually, they are also electrophilic reagents that react at high electron density regions in DNA forming covalent adducts. It is generally believed that these adducts occasionally lead to mutations during replication and/or repair. There is no direct experimental evidence bearing on the molecular mechanisms that are responsible for producing mutations from these adducts. To complicate matters, there are at least 18 distinct sites where adducts can form. Not all sites react with a given carcinogen in vivo under a given set of conditions, but many different adducts are almost always formed. In order to gain a better understanding of how carcinogens produce mutations, and, hopefully, to obtain some insight into the relationship between mutagenesis and carcinogenesis, we need to answer the following questions: (1) Which of the covalent adducts that form when a carcinogen reacts with DNA actually produce mutations? (2) What kind of mutation does each different premutational lesion produce? (3) What role do the various DNA repair systems play in producing these mutations?

We have developed a site-specific mutagenesis system that is capable of answering these questions directly and unambiguously. The system involves gene G of bacteriophage ϕX174. Through a combination of chemical and enzymatic procedures we are able to introduce the covalent adduct to be studied at a single preselected site in this essential gene. The site-modified DNA produced is studied in vivo by transfection of spheroplasts. Since one of the strands in the modified RF DNA is wild type, all of the normal viral proteins are produced in the spheroplast and infectious mutant viruses are assembled even when the mutation is lethal. Mutants that are produced, regardless of their nature, are identified and propagated using a host cell carrying a functional copy of ϕX gene G on a plasmid. DNA isolated from different mutants is sequenced in the region that carried the original covalent adduct in order to identify the nature of the mutation unambiguously. By studying the same type of adduct (e.g., a methyl group) at different positions in the purine or pyrimidine rings, one at a time, it should be possible to determine which adducts are mutagenic and which are not. Furthermore, the same site-specific DNA adduct can be studied in spheroplasts derived from cells carrying mutations that produce defects in various DNA repair systems. By comparing mutant frequencies and mutant types produced in these different repair backgrounds, important information concerning the role of DNA repair and producing mutations from different kinds of lesions should be obtained.

INTRODUCTION

It is now established that a significant number of organic compounds, either by themselves or after metabolic activation, react with DNA to produce covalent adducts. These compounds fall into two groups. The first includes nucleophilic reagents such as HSO_3^- and $HONH_2$ that add to the 5,6-double bond of cytosine residues and lead, eventually, to C → U transitions (Hayatsu, 1976; Phillips and Brown, 1967). A second and much larger group includes a variety of electrophilic reagents that form stable adducts with DNA (Miller and Miller, 1977). Figure 1 shows three such reagents, each of which is carcinogenic though the adducts formed in each case are different. Most of the time these adducts are removed or bypassed by one of the DNA repair systems. Occasionally, however, an error occurs either during replication or repair, and a change in the DNA sequence occurs. This change (which may be a simple base substitution, a frameshift, a large deletion, an insertion, or a rearrangement) is passed along by transcription and translation and is recognized eventually as some phenotypic change.

The Ames test (Ames et al., 1975) and other similar mutagen tests (Hsie et al., 1979), are based on the extremes of this process. Virtually any chemical can now be tested in both prokaryotic and eukaryotic systems for mutagenicity by looking for well-characterized phenotypic changes. It is particularly interesting that out of 175 known carcinogens tested by Ames and his colleagues in the Salmonella system, 157 (90%) are also mutagens (McCann and Ames, 1976; McCann et al., 1975). This very high correlation between carcinogenicity and mutagenicity suggests that mutations represent the initial step in tumor production. There is a variety of evidence, however, that mutations per se are not sufficient to produce the complicated biological changes that seem to be involved in tumorigenesis (Boutwell, 1977; Weinstein and Pietropaulo, 1977).

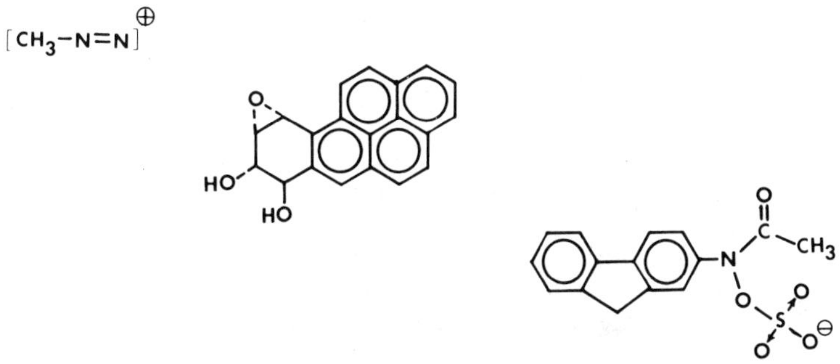

Fig. 1. Some electrophilic mutagens. Left to right: methyl diazonium ion, benzyprene diolepoxide-2, acetylaminofluorine sulfate.

It is relatively easy to study the chemistry and biochemistry involved in forming DNA adducts from electrophilic reagents, and a large literature has accumulated on this subject. Similarly, it is relatively easy to study the gross biological effects of these reactions by looking at phenotypic changes in a suitable test organism. With current technology, it is even possible to identify precisely the change in DNA sequence that has produced the phenotypic change that is observed. However, mutations are produced during replication and/or repair; very little is learned about the molecular mechanisms of mutation from covalent adducts by examining only the initial and final events. Although considerable effort has been devoted to studying DNA replication and repair, there is no direct evidence showing how various DNA adducts actually produce mutations. The problem is illustrated in Fig. 2.

DNA has 16 regions of high electron density (Kochetkov and Budovskii, 1971). While a given electrophilic reagent usually does not react with all of these sites under a given set of conditions, it always reacts with more than one. For example, ethyl nitrosourea, which is both a mutagen and a carcinogen, alklyates 11 of these possible sites (Singer, 1979). Representative data are shown in Fig. 2. At the moment there is no direct evidence indicating which of these lesions actually cause mutations. Even when the mutation is characterized at the molecular level and shown to be a result of a modification of one particular residue in DNA, as has been achieved in one case for a -2 frameshift produced by N-methyl-N nitrosoguanidine (Isono and Yourno, 1974), the precise adduct responsible for this mutation is still unknown.

One way to take the next important step in understanding the molecular mechanisms of mutation produced by carcinogens is to prepare site-specific adducts of a biologically active DNA and test

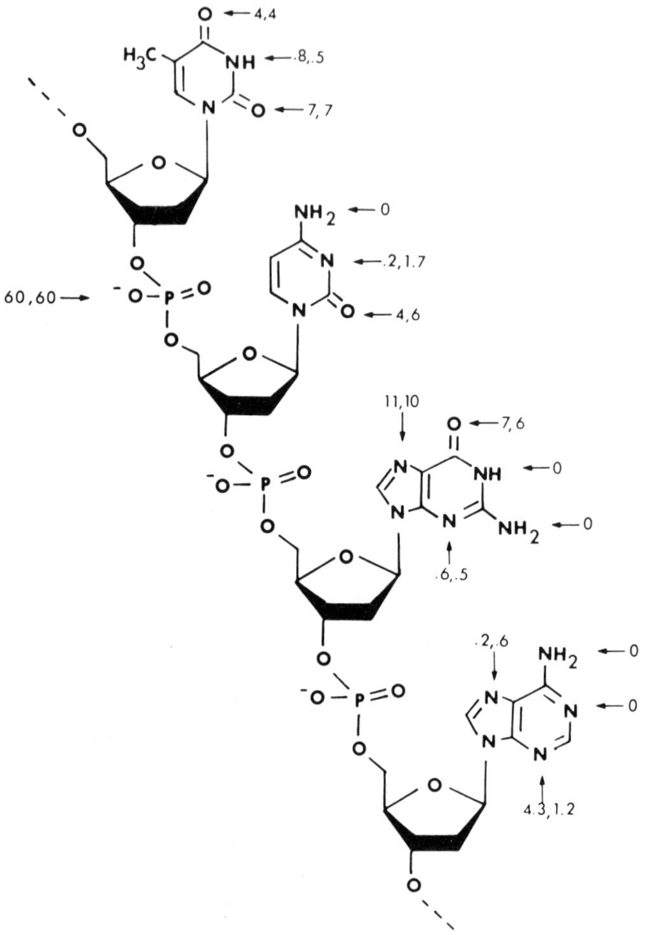

Fig. 2. Sites of high electron density in DNA. Numbers indicate the percentage of total alkylation with ethyl nitrosourea in vitro (Singer, 1979). First number is for double-stranded DNA; second number for single-stranded DNA.

them, one at a time, in vivo in order to answer the following questions. (1) Which of the carcinogen-induced covalent modifications of DNA actually produce mutations? (2) What kind of mutation(s) does each different premutational lesion produce? (3) What role do the various cellular DNA repair systems play in the mutation process?

RATIONALE OF OUR APPROACH

In order to answer the above questions in a direct manner, we have developed a system of site-specific mutagenesis. Our approach is summarized in Fig. 3. For our initial experiments we chose the bacteriophage ϕX174 because it was the only system available at the time we began our work that satisfied all of the technical requirements. Therefore, we will describe our approach in terms of this virus, but it should be kept in mind that the principles involved are general; and, if certain nontrivial technical problems can be solved, it should be possible to extend this approach to eukaryotic cells, hopefully to human cells in culture. Our initial work has focused on gene G of ϕX174 since the first sequence data from this virus described the early region of this gene (Robertson et al., 1973). Gene G is an essential gene coding for a viral spike protein (Sinsheimer, 1968). The gene product is essential for assembling infectious virus particles and for converting double-stranded RF DNA to single-stranded viral DNA that is packed into the virion. Therefore, we can anticipate that most changes in this gene will produce conditionally lethal or lethal effects.

Our approach begins with the chemical synthesis of a short, primer oligonucleotide corresponding to a preselected region of gene G. During this synthesis, the covalent adduct to be studied is introduced in an unambiguous manner. The site-modified primer is then annealed with circular DNA isolated from the virus. This partial duplex is converted to RF DNA enzymatically by the classical synthesis of Goulian and Kornberg (1967) but using the large fragment of Escherichia coli DNA polymerase, which lacks the 5' → 3' exonuclease activity (Klenow et al., 1971). The product is a double-stranded, closed circular DNA containing a single change at a preselected site. The effect of this single modification can be examined in vivo by transfection of spheroplasts. Replication of the normal template strand in the transfected cell will lead to a normal double-stranded RF DNA. Transcription of this DNA, followed by translation of the mRNAs, will produce a full complement of viral proteins. Replication of the complementary strand carrying the site-specific covalent adduct will lead to a mutant RF DNA if that particular covalent adduct is a premutational lesion. Eventually a mutant viral strand will be produced and converted to an infectious virus particle by combination with the appropriate viral proteins. Thus, two kinds of virus particles should be produced, wild type and mutant.

It should be noted that the original covalent modification can never be packaged in a virus particle since it resides in the minus strand of the RF DNA; only plus strands are packaged (Sinsheimer, 1968). Furthermore, as we will show by experiments described below, it makes no difference what kind of mutation is produced by the original covalent modification; the mutant DNA will be rescued by complementation and converted to an infectious particle.

The problem resolves to finding a simple method for distinguishing mutant virus from wild type and for propagating the mutant regardless of the nature of the mutation. We need a ϕX-sensitive host that is permissive for any kind of mutation that might occur. By comparing virus plaque formation on such a permissive host with that obtained using a host permissive only for wild-type or pseudo wild-type virus, we can easily distinguish mutants (see Fig. 3). The permissive host can then be used to propagate mutant virus and to isolate mutant DNA. Since we know the position of the original lesion, it is a simple matter to sequence the region of DNA that contained this covalent modification and to determine the exact nature of the mutation that has been produced.

Another important feature of this system needs emphasis. Since we are studying the biological effect of the covalent lesion by transfection of spheroplasts, we do not need a ϕX-sensitive host for this part of the experiment. Thus, all of the E. coli repair mutants that are known (Clark and Ganesan, 1975) can be used to study the effect of different repair systems on the mutation process. For example, after determining mutation frequency and characterizing the type of mutation produced by transfection of wild-type spheroplasts,

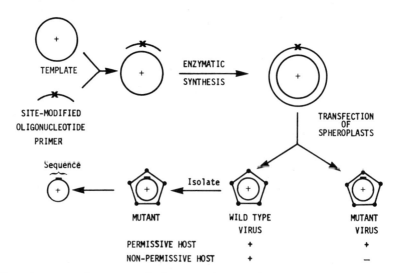

Fig. 3. A general approach for studying the mutagenic effect of individual DNA adducts at preselected sites in biologically active DNA.

SITE—SPECIFIC MUTAGENESIS

the experiment can be repeated with spheroplasts defective in excision repair (uvr mutants) to determine what effect this particular repair system has on the mutations observed. The experiments can then be repeated using a mutant deficient in recombination repair (recA) or one in which error-prone repair has been induced at 42°C (tif), etc. In this way, it should be possible to obtain valuable information concerning the role that various DNA repair systems play in the mutation process.

The entire φX gene G system diagrammed in Fig. 3 is operational in our laboratory. We will now describe the experimental system.

THE EXPERIMENTAL SYSTEM

Detection and Propagation of ΦX174 Gene G Mutants Carrying Various Types of Lethal or Conditionally Lethal Mutations

As we have stated, we will never know what kind of mutation a given covalent modification will produce ahead of time. In order to detect any kind of mutant that might be produced, we need a φX-sensitive host system that is permissive for all kinds of mutation. Complementation is the obvious way to achieve this. This problem was solved by constructing a plasmid carrying a functional copy of φX gene G. Briefly, this involved isolating gene G in a restriction fragment, adding poly(dT) "tails" and annealing this product with pMB9 that had been opened with EcoRI endonuclease and "tailed" with poly(dA). After transfection of calcium-shocked E. coli, transformants carrying the tetracycline resistance marker from the plasmid were isolated. These were then tested with a known φX gene G nonsense mutant (Gam9). Out of the first 11 tetracycline-resistant transformants isolated, nine were permissive for this gene G mutant. One of these, called pφXG, was isolated and thoroughly characterized (Humayun and Chambers, 1978).

Two important properties of cells carrying pφXG need to be emphasized. First, these pφXG-bearing hosts rescue only gene G mutants, as shown in Table 1. Wild-type virus grows on the host cells with or without the plasmid. φXGam9 grows only in those cells carrying pφXG. Viruses carrying mutations in gene E or gene A cannot be rescued by pφXG (Humayun and Chambers, 1978). Second, Humayun (unpublished data) has shown that mini-cells carrying pφXG produce gene G product without virus infection.

From these data, we conclude that pφXG rescues gene G mutations specifically by complementation. However, all of these experiments were conducted with a specific nonsense mutant (φXGam9), and we wanted to be sure that we could rescue lethal missense mutants. Since no such ΦX174 mutants had ever been isolated, we used pφXG to

Table 1. Plaque Formation by ΦX174 Mutants on E. coli Containing pϕXG Plasmids

ΦX174	Location of Mutation	Plaque Formation		
		H514	H514/ pϕXG105	H514/ pϕXG109
Wild type	–	+	+	+
am9	Gene G	–	+	+
am3	Gene E	–	–	–
am86	Gene A	–	–	–

isolate some examples. Accordingly, wild-type phage was mutagenized with nitrous acid and plated on a host carrying pϕXG. Individual plaques were picked and gridded sequentially onto two plates carrying prepoured lawns of the host with and without pϕXG, respectively. The results are shown in Fig. 4.

One expects HNO_2 to produce C → U = T, A → I = G, and G → X = A transitions. There are six codons in gene G (five CAG = Gln and one TGG = Trp) that can produce a nonsense codon (UAG or UGA) by one of these single change transitions. Six of the seven possible nonsense mutations (TGG can go to either TAG or TAA) produce the amber codon. All the remaining transitions produced in gene G by the above base changes give missense or silent (pseudo wild-type) mutations.

The screen shown in Fig. 4 was designed so that amber mutations that are derived from Gln codons are not recognized. Using this procedure, nine new mutants were isolated out of the first 455 plaques examined. One of these was found to be temperature sensitive. The other eight were neither temperature sensitive nor suppressible with the amber, ochre, or opal suppressors we had available. These eight mutants have been mapped by recombination (I. Kućan, unpublished data) and they scatter throughout gene G. Four have been sequenced (I. Kućan, unpublished data). In two cases, single-base changes producing missense mutations (G → A at position 2821, Gly → Ser; A → G at position 1678, Glu → Gly) have been identified. These represent lethal missense mutations. In two other cases no base changes could be identified in the coding sequence of gene G. Because of the way the sequencing was carried out, we regard it as unlikely that we have missed a base change in the coding sequence for these two mutants. Since these particular mutations map very early in the gene, we presume that the changes are located in the noncoding region between genes F and G where we have not yet sequenced. This is an intriguing finding since the gross properties

SITE—SPECIFIC MUTAGENESIS

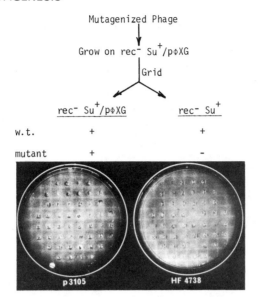

Fig. 4. A screen for ΦX174 mutants carrying a lethal missense mutation in gene G. E. coli HF 4738 is recA Su$^+$2; p3103 is HF 4738 carrying the plasmid pΦXG105. Fifty plaques are clearly visible on p3105; three, corresponding to mutants, are missing on HF 4738 (numbers 1, 21, and 44). These mutants were isolated from the p3105 plate.

of these two mutants are very similar to the missense mutants where the base change has been identified in the coding sequence (e.g., they grow only on host cells bearing pΦXG, and they have a burst size of about 40 compared to 200 for wild type on the same host (Z. Kućan, unpublished data).

These data show quite clearly that it is possible to rescue either nonsense or lethal missense mutations in ΦX gene G with pΦXG. It does not follow, however, that more complicated mutations, particularly frameshifts, can be rescued in this manner. The reason for this is as follows. It has been known for some time that nonsense mutations early in gene G are polar for the following gene, gene H, whose gene product is another spike protein (Benbow et al., 1972). Presumably this arises because of abnormal transcription termination as outlined in Fig. 5. Genes G and H are part of a polycistron whose promoter is located in gene D (for a summary see Fujimura and Hayashi, 1978). In order to assemble a virus particle, transcription from this promoter must proceed all the way through gene H since there is no promoter in either gene G or gene H. It has been known for some time that the mutation in ΦXGam9, a nonsense mutation in the early part of gene G, is polar. Complementation of gene H mutations by Gam9 is poor (Benbow et al., 1972). Examination of the viral proteins produced by Gam9 shows that the amount of gene H product is markedly reduced (Benbow et al., 1972). One can speculate that this occurs because the mutation in Gam9 occurs before the purported transcription termination site that has been identified by McMahon and Tinoco (1978). In such a case, the amount of completed message would presumably decrease and the amount of H protein produced could become limiting since a large amount of gene G product is being produced by the plasmid.

By examination of the gene G sequence (Sanger et al., 1978), we found that frameshift mutations in the early part of gene G generate in-frame nonsense codons, some of which occur before the purported transcription termination site. We were afraid that mutations of this kind might be so polar for gene H that we might be unable to rescue such a gene G mutant with cells carrying pɸXG. To test this, we set out to construct a site-specific deletion that would also produce a frameshift mutations. The rationale was as follows.

From a list of available restriction enzymes and of the known gene G sequence, we could identify three enzymes that would cut twice and only twice in gene G and that would not cleave single-stranded ɸX DNA. HincII was one of these. Armed with this information, we set out to construct a partial duplex that would contain only the gene G in the double-stranded region. Our strategy is outlined in Fig. 6. First, we isolated gene G in a restriction fragment. This was annealed to single-stranded viral DNA producing a partial duplex with two, and only two, restriction sites for HincII. This enzyme leaves blunt ends. These can be ligated together with T4 DNA ligase giving closed circular double-stranded DNA that has been specifically shortened, supposedly by 79 base pairs. This entire process was

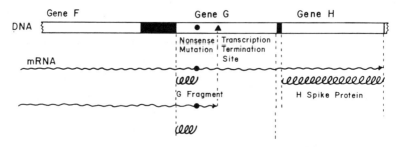

Fig. 5. A possible explanation for the polarity of nonsense mutations in the early part of ɸX174 gene G.

SITE—SPECIFIC MUTAGENESIS

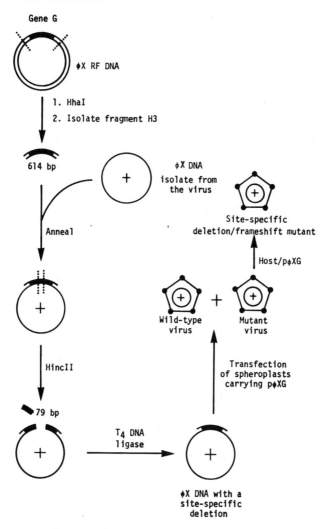

Fig. 6. Strategy for construction of a site-specific deletion/ frameshift mutation in ΦX174 gene G.

carried out without isolation of any intermediates, and the final mixture was used to transfect spheroplasts carrying pϕXG. Mutants were easily identified. One of these was isolated and thoroughly characterized (Humayun and Chambers, 1979). Sequence data showed clearly that 80 base pairs, instead of the expected 79, were excised. This produces a -2 frameshift and generates 13 in-frame nonsense codons, four of which occur upstream from the purported transcription termination site. This mutant has a very small burst size indicating, but not proving, that the mutation is polar. The fact that we were able to rescue such a mutant greatly increases our confidence that pɸXG will rescue any mutation in gene G.

More recently, Bhanot (unpublished data) has isolated a spontaneous mutant with pɸXG. It contains a deletion of 15 nucleotides starting at codon 143. It also contains an insertion of 50 nucleotide pairs at the deletion site. This insertion generates a -1 frameshift. The insertion sequence itself carries an in-frame termination codon, and the -1 frameshift generates an in-frame nonsense codon. It is interesting that the 50-base insertion is not derived from ɸX174. Presumably, it comes from recombination of the viral DNA with either the host chromosome or with the pMB9 portion of pɸXG. As expected, this extensive sequence change in this essential gene is lethal, yet this mutant grows happily in the presence of pɸXG.

Figure 7 summarizes the different kinds of mutations that have been rescued with pɸXG. From these data, we have every reason to believe that our system is completely permissive for any kind of mutation that might be produced from this specific covalent adduct in gene G.

Introduction of Site-Specific Changes into Gene G of ɸX174

In order to establish the enzymology of the ɸXG system, we have synthesized a site-specific mutant carrying a nonsense mutation in the third codon of gene G (Bhanot et al., 1979). The procedure that was used is outlined in Fig. 8.

First, we synthesized a short, minus-strand primer in which we changed the wild-type G at position 2401 to an A. It was clear that such a short primer with a single-base mismatch might prime at other positions in the genome. A computer search of the ɸX (Sanger et al., 1978) sequence indicated seven such sites. These are summarized in Fig. 9. We examined these possible sites experimentally by carrying out an elongation synthesis in the absence of the dCTP. As shown in Fig. 9, elongation of the primer will occur until the first G residue is encountered in the complete strand. The expected products are shown on the right-hand side of Fig. 9. The desired product is a 17-mer; the other products are all shorter. After annealing the terminally labeled primer to the template and carrying out such an elongation synthesis, the products were examined by electrophoresis. Results are shown in Fig. 10. The left lane shows the purity of the primer. The middle lane shows the degradative effect of the E. coli DNA polymerase-large fragment on the primer due to 3'-exonuclease activity that still resides in the protein. The right lane shows the complete reaction. Priming at four of the seven possible sites has been established; one site is uncertain. The desired product,

SITE—SPECIFIC MUTAGENESIS

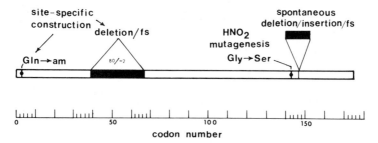

Fig. 7. Some ΦX174 Gene G Mutants. A summary of the kinds of ΦX174 mutants that have been rescued with a host carrying a functional copy of gene G on a plasmid, pɸXG.

a 17-mer, is formed in very small yield. Fortunately, enough material was present so that it could be eluted from the gel, sequenced, and used to reprime the synthesis of RF DNA as shown in the bottom portion of Fig. 8.

After transfection, the spheroplast lysate was searched for mutants. Out of 1355 plaques that were tentatively identified as possible mutants, 15 were confirmed. These fell into two groups. Eleven of the 15 mutant isolates grew on a suppressor host that inserts the wild-type amino acid glutamine at the nonsense codon. This group was designated OBI. A portion of gene G from one of the OBI isolates was sequenced and shown to have the expected C → T at position 2401 of the viral DNA.

The other four mutant isolates were not suppressed by any of the suppressor strains available to us. This group was designated OBII. Gene G of these four mutants has been sequenced (Bhanot, unpublished data). In three cases, a single base substitution was found. The fourth contains the deletion-insertion-frameshift mutation described above. Thus, each of the OBII mutants is different, and the change is not at the preselected site. From the limited data available, the frequency of their occurrence seems to be approximately 1 in 1300. We believe they are spontaneous mutants that were present in our template DNA stock. This is a potentially serious problem since 25% of the mutants we have isolated in this site-specific mutagenesis experiment did not have a change at the preselected site. However, we wish to emphasize that the transfection of spheroplasts was carried out with a reaction mixture that had been treated with S1 nuclease to reduce the amount of single-stranded template molecules in the mixture. Even though this procedure enriched the mixture some 200-fold for RF molecules, the remaining template molecules were still the predominate species. In subsequent experiments, we intend to purify the RF molecule before transfection.

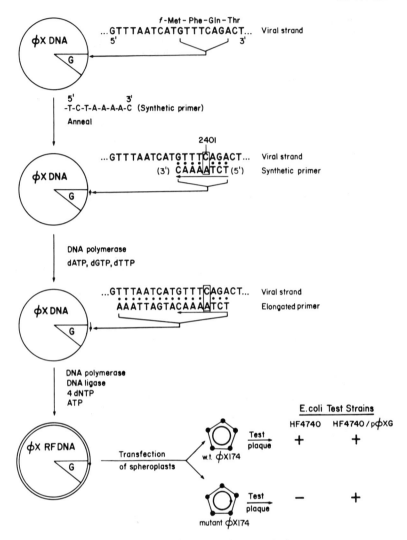

Fig. 8. Strategy for the synthesis of a ΦX174 mutant carrying an amber mutation is the third condon of gene G. HF 4740 is recA.

We hope this will drastically reduce this second class of mutations since the chance of using a mutant DNA as a template should be less than 10^{-3}.

SITE—SPECIFIC MUTAGENESIS   135

```
                  5'                             3'
Gene G    (+)···G T T T A A T C A T G T T T[C]A G A···       17-mer
(2401)    (-)   A A A T T A G T A C A A A[A]T C T            C→T; Gln→amber

Gene A    (+)···G T T T A T C G T T T T[T]G A···             14-mer
(4575)    (-)   A A A T A G C A A A A[T]C T                  T→A; Phe→Leu

Gene G    (+)···G C A C C T G T T T T A[C]A···               13-mer
(2491)    (-)   G T G G A C A A A A T[C]T                    C→G; Gln→Glu

Gene C    (+)···G A C T G[G]T T T A G A···                   11-mer
(276)     (-)   T G A C[A]A A A T C T                        G→T; Trp→Cys
                                                             not found

Gene D    (+)···G C C G T T T T[G]G A···                     10-mer
(470)     (-)   G G C A A A A[T]C T                          G→A; Leu→Leu

Gene F    (+)···G A A G T T T[A]A G A···                     10-mer
(2029)    (-)   T T C A A A[A]T C T                          A→T; Lys→amber
                                                             ?

Gene A    (+)···G[T]T T T A G A···                           not elongated
(4353)    (-)  [C]A A A A T C T                              in this experiment
                                                             T→G; Phe→Val
```

Fig. 9. Specificity of priming by d-(^{32}pT-C-T-A-A-A-A-C) on a viral
strand template. The sequences in ΦX174 DNA complementary
to the synthetic primer with no more than one base mismatch
were located with a computer. The gene involved and the
position of the mismatch are shown at the left. The synthetic primer is underlined and its polarity is indicated
by the arrowhead. The mismatched bases are boxed. The
reading frame is indicated by the bracket over the template
strand. The nature of the mutation expected from the mismatch is shown on the right. Priming was measured by
enzymatic elongation of the ^{32}P-labeled primer with E. coli
DNA polymerase I (large fragment) in the absence of dCTP.
The sequence of each elongated product is shown; the length
of the product is indicated at the right.

Calibration of the Biological Effect Produced by Base-substitution
Mutations in the Third Codon of ΦX174 Gene G

These experiments demonstrate that the ΦXG system is operational. However, we can obtain some further information about the
biological effects that mutations in the third codon of gene G will
have by studying the suppression of the site-specific mutant
ΦXGam2401, which we have synthesized (Fig. 8). Ideally, we would
like to use a simple plaque assay to distinguish base substitutions
from other kinds of mutation. We would also like to distinguish
different kinds of base substitutions.

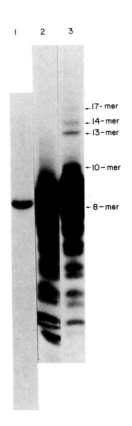

Fig. 10
Electrophoretic separation of the elongation products formed from an octanucleotide primer and a viral strand template in the absence of dCTP. Lane 1, the synthetic primer, d-^{32}pT-C-T-A-A-A-A-C). Lane 2, primer + E. coli DNA polymerase I (large fragment). Lane 3, primer + ΦX174 DNA + dATP, dGTP, dTTP + DNA polymerase I (large fragment). The bands were located by radioautography.

Using the approach pioneered by Miller and his co-workers (1979), we have examined the permissiveness of a variety of suppressor strains of E. coli (without pΦXG) for ΦXGam2401. The results are shown in Table 2. There are nine possible base substitutions at an amber codon. These are summarized in Fig. 11. As shown in Table 2, we verified that an Su2$^+$ suppressor, which inserts the wild-type amino acid, glutamine, was permissive for ΦXGam2401. Similarly, we showed that an ochre suppressor that inserted glutamine was permissive (data not shown). Insertion of tyrosine, on the other hand, was lethal (Table 2). The control, ΦXGamH116 (a ΦX gene H amber mutant), showed that there was nothing wrong with the suppressor host itself. Insertion of serine produced a temperature-sensitive phage (Table 2). The data with the suppressors that insert leucine and lysine are less certain because no positive controls have been obtained as yet (Table 2). Thus, we cannot be sure whether insertion of these amino acids is lethal because non-functional gene G protein is made or whether the suppression efficiency is so poor that an insufficient amount of this essential protein is made.

Table 2. Plaque-forming Units Produced by Suppression of a Site-specific Nonsense Mutation in the Third Codon of φX Gene G

Plaque	Plaque-forming Units Produced by Infection or Transfection									
	Su^+(Gln)		Su^+(Tyr)		Su^+(Ser)		Su^+(Leu)		Su^+(Lys)	
	32°C	38°C	32°C	38°C	32°C	38°C	32°C	38°C	32°C	38°C
Wild type	0.93	1.00	0.51	0.50	0.45	0.61	1.00	—	0.80	—
φGam2401	0.93	1.00	0	0	0.42	0	0	—	0	—
φXGamH116	1.00	1.00	0.39	0.42	0.39	0.40	0	—	0	—

```
                          T A T^Tyr

                          T A C^Tyr

                          T A A^Oc

TTG^Leu   TCG^Ser   T A G^Am   TGG^Trp

                          C A G^Gln

                          A A G^Lys

                          G A G^Glu
```

Fig. 11. Codons that are produced by a single base change in an amber codon.

FUTURE WORK

This is the φXG system for studying the molecular mechanisms of mutation by electrophilic carcinogens as it now stands. Though the third codon of gene G of ΦX174 does not have all the properties we would like for an ideal system, it is adequate for our initial studies. To illustrate the kind of experiments we are now attempting, let us consider two examples.

The Mutagenic Effects of O^6-Methylguanine Residues

There is considerable interest in the lesion O^6-methylguanine (O^6-MeG) as an initiating event for tumor formation. This is based on the observation that there is a strong correlation between the amount of O^6-MeG that is produced in vivo by the reaction of various methylating agents with DNA and the carcinogenic index of the methylating agent (Loveless, 1969). Thus, S_N2 reagents such as dimethylsulfate, which are noncarcinogenic, produce very little O^6-MeG, but S_N1 reagents such as N-methylnitrosourea, which form significant quantities of O^6-MeG, are carcinogenic (Lawley, 1976).

O^6-MeG should produce transition mutations by mispairing. This is illustrated in Fig. 12. With the φXG system, we should be able to measure the frequency of this transition mutation in vivo by introducing O^6-MeG into the minus strand opposite the first position of the third codon in a wild-type template. A transition will convert the wild-type Gln codon to amber giving the site-specific mutant φXGam2401 that we have been discussing. It should be possible to measure the mutation frequency quantitatively in vivo by a simple plaque assay using an $Su2^+$ suppressor. Furthermore, it should be possible to study the effect of various repair systems by repeating the experiment with various DNA repair mutants.

There is no evidence that simple base substitutions are pre-carcinogenic lesions, and there are some arguments that they are not (Cairns, 1981). More than half of the carcinogens that have been shown to be mutagens have frameshift activity. Even simple methylating agents such as N-methyl-N-nitro-nitrosoguanidine, which forms significant amounts of O^6-MeG, have frameshift activity (Isono and Yourno, 1974). Therefore, we would like to know whether O^6-MeG produces mutations other than simple base substitutions. It may be possible to examine this by isolating a few representative phages from the lethal mutation group and sequencing the region around the third codon of gene G. If frameshifts, deletions, or rearrangements occur with approximately the same frequency as each other and the same or greater frequency as transversions (the only base substitutions that can be in the lethal group), then we should detect these interesting mutations. This information is particularly important in view of the recent suggestion that the initiating event in carcinogenesis is a transposition rather than a simple base substitution (Cairns, 1981).

The Mutagenic Effects of Apurinic/Apyrimidinic (AP) Sites

As a final example, consider the following. It has been known for some time that the 7- position of guanine residues is a major site for alkylation of DNA with simple methylating agents (Lawley, 1966; Lawley and Wallick, 1957; Reiner and Zamenhof, 1957). The 7-MeG adduct breaks down spontaneously to give a so-called AP site

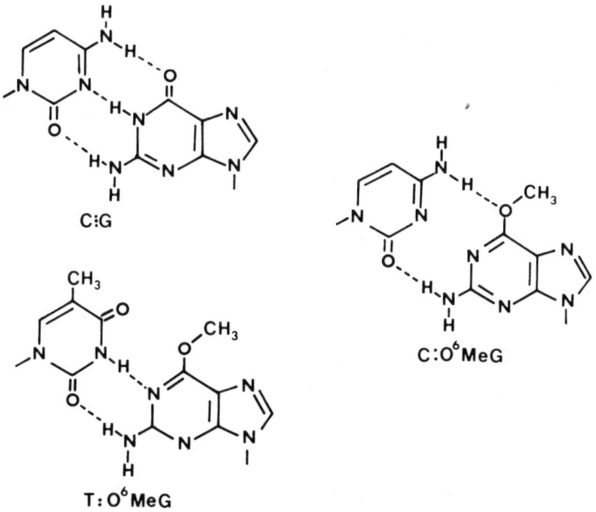

Fig. 12. Pairing and mispairing of O^6-methylguanine residues in DNA.

(Lindahl, 1979). A similar reaction occurs from 3-methyladenine (3-MeA) residues (Lindahl, 1979). In addition, 3-MeA is converted to an AP site enzymatically by a specific glycosylase (Lindahl, 1979). As we have already indicated, certain nucleophilic reagents produce deamination of cytosine residues in DNA leading to deoxyuridine residues in DNA. Because of the prevalence of this kind of damage, it is not surprising that the cell contains a specific glycosylase to remove the uracil residue, again producing an AP site (Lindahl, 1979). Finally, in certain bacteria the first step in removing a thymine dimer involves cleavage of a glycosyl linkage to produce an AP site (Haseltine et al., 1980). These reactions are summarized in Fig. 13.

Many years ago, Bautz and Freese (1960) suggested that AP sites give rise to transversions, but there has never been any direct evidence to substantiate this hypothesis. Therefore, we have designed an experiment to see whether or not a transversion can arise from such a lesion, and whether or not mutations other than base substitutions also occur. This is shown in Fig. 14. First, a site-specific ochre mutant is prepared in exactly the same way that the site-specific amber mutant was obtained. This is used to isolate template DNA. Deoxyuridine is introduced into the second position of the third codon during the synthesis of the primer molecule. The uracil moiety can presumably be removed either at the level of the primer or at the level of the completed RF DNA using the uracil glycosylase (Lindahl, 1979). Upon transfection of wild-type E. coli spheroplasts, several things may occur. First, the AP site may be repaired by excision or by insertion of the "correct" base, T, leading to ochre virus. Alternatively, a mutation may occur by insertion of the "wrong" base. The transition A → G gives the codon TGA, which is conditionally lethal. The transversion A → C gives a codon that inserts serine instead of the wild-type amino acid glutamine. We know from the calibration of the codon that this transversion will produce a temperature-sensitive phage. The transversion A → T leads to a replacement of glutamine with leucine; the suppression data suggest that this is lethal. Therefore, if the spheroplast lysate is grown on wild-type E. coli at 32°C, the only phage that can grow is the temperature-sensitive mutant produced from an A → C transversion. Even if this mutation occurs at very low frequency, it should be possible to find it since we have a positive screen. Furthermore, it should be possible to determine whether or not other kinds of mutation are produced by examining a few representatives from the lethal group.

CONCLUSIONS

These examples illustrate the potential of in vivo site-specific mutagenesis for studying the molecular mechanisms of mutation. In

SITE—SPECIFIC MUTAGENESIS

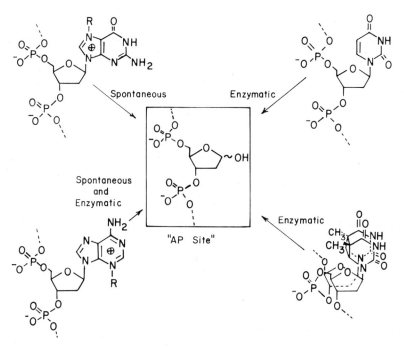

Fig. 13. Some reactions that give rise to AP sites in DNA.

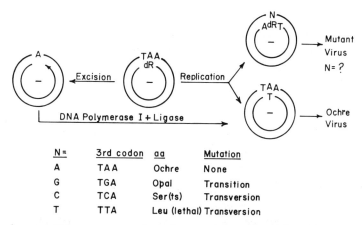

N=	3rd codon	aa	Mutation
A	TAA	Ochre	None
G	TGA	Opal	Transition
C	TCA	Ser(ts)	Transversion
T	TTA	Leu (lethal)	Transversion

Fig. 14. A sensitive screen for detecting a low frequency transversion from a site-specific AP site in the third codon of ΦX174 gene G. dR = a deoxyribose residue; aa = codon specificity. See text for further discussion.

principle, any lesion that can be introduced into a preselected site in gene G of ϕX174 can be studied in this system. We believe that these experiments will provide a better understanding of the role played by different covalent modifications in mutagenesis. Furthermore, we see no reason why this approach cannot be extended to eukaryotic systems. It is essential that this be done since there is no guarantee that the results will be the same in prokaryotic and eukaryotic organisms, particularly if different DNA repair systems play a critical role in the mutation process (Radman, 1975; Witkin, 1975). It is, of course, still a quantum jump from mutagenesis to carcinogenesis. But even if this approach tells us nothing about cancer, it almost surely will yield results that are of fundamental importance in understanding the molecular mechanisms of mutation.

ACKNOWLEDGMENTS

I am especially indebted to Opinder S. Bhanot and M. Zafri Humayun for the major contributions they have made to the development of the system described in this paper. I also thank Morton C. Schneider for his expert assistance and Ira Kućan and Zeljko Kućan for allowing me to mention their results with the lethal missense mutants prior to publication. This work was supported by Grant 2 R01 CA16319 awarded by the National Cancer Institute, and Grant BC-252B awarded by the American Cancer Society.

REFERENCES

Ames, B. N., McCann, J., and Yamasaki, E., 1975, Methods for detecting carcinogens and mutagens with the Salmonella/mammalian-microsome mutagenicity test, Mutat. Res., 31:347.

Bautz, E., and Freese, E., 1960, On the mutagenic effect of alkylating agents, Proc. Natl. Acad. Sci. U.S.A., 46:1585.

Benbow, R. N., Mayol, R. F., Picchi, J. C., and Sinsheimer, R. L., 1972, Direction of translation and size of bacteriophage ϕX174 cistrons, J. Virol., 10:99.

Bhanot, O. S., Khan, S. A., and Chambers, R. W., 1979, A new system for studying molecular mechanisms of mutation by carcinogens, J. Biol. Chem., 254:12684.

Boutwell, R. K., 1977, The role of the induction of ornithine decarboxylase in tumor promotion, in: "Origins of Human Cancer, Book B, Mechanisms of Carcinogenesis," H. H. Hiatt, J. D. Watson, and J. A. Winsten, eds., Cold Spring Harbor Laboratory, New York

Cairns, J., 1981, The origin of human cancers, Nature, 289:353.

Clark, A. J., and Ganesan, A., 1975, Lists of genes affecting DNA metabolism in Escherichia coli, in: "Molecular Mechanisms of Repair of DNA," Part B, P. C. Hanawalt, and R. B. Setlow, eds., Plenum Press, New York and London.

Fujimura, F. K., and Hayashi, M., 1978, Transcription of isometric single-stranded DNA phage, in: "The Single-Stranded DNA Phages," D. T. Denhardt, D. Dressler, and D. S. Ray, eds., Cold Spring Harbor Laboratory, New York.

Goulian, M., and Kornberg, A., 1967, Enzymatic synthesis of DNA. XXIII. Synthesis of circular replicative form of phage ΦX174 DNA, Proc. Natl. Acad. Sci. U.S.A., 58:1723.

Haseltine, W. A., Gordon, L. K., Lindan, C. P., Grafstrom, R. H., Shaper, N. L., and Grossman, L., 1980, Cleavage of pyrimidine dimers in specific DNA sequences by a pyrimidine dimer DNA-glycosylase of M. luteus, Nature, 285:634.

Hayatsu, H., 1976, Bisulfite modification of nucleic acids and their constituents, Prog. Nucleic Acid Res. Mol. Biol., 16:75.

Hsie, A. W., O'Neill, J. P., and McElheny, V. K., 1979, "Banbury Report 2, Mammalian Cell Mutagenesis: The Maturation of Test Systems," Cold Spring Harbor Laboratory, New York.

Humayun, M. Z., and Chambers, R. W., 1978, Construction and characterization of an Escherichia coli plasmid bearing a functional gene G of bacteriophage ΦX174, Proc. Natl. Acad. Sci. U.S.A., 75:774.

Humayun, M. Z., and Chambers, R. W., 1979, Construction of a site-specific, deletion-frameshift mutation in an essential gene of bacteriophage ΦX174, Nature, 278:524.

Isono, K., and Yourno, J., 1974, Chemical carcinogens as frameshift mutagens: Salmonella DNA sequence sensitive to mutagenesis by polycyclic carcinogens, Proc. Natl. Acad. Sci. U.S.A., 71:1612.

Klenow, H., Overgaard-Hansen, K., and Patkar, S. A., 1971, Proteolytic cleavage of native DNA polymerase into two different catalytic fragments. Influence of assay conditions on the change of exonuclease activity and polymerase activity accompanying cleavage, Eur. J. Biochem., 22:371.

Kotchetkov, N. K., and Budovskii, E. I., 1971, Electronic structure and reactivity of the monomer components of nucleic acids, in: "Organic Chemistry of Nucleic Acids," Chapter 3, Plenum Press, London and New York.

Lawley, P. D., 1966, Effects of some chemical mutagens and carcinogens on nucleic acids, Prog. Nucleic Acid Res. Mol. Biol., 5:89.

Lawley, P. D., 1976, Carcinogenesis by alkylating agents, in: "Chemical Carcinogens," C. E. Searle, ed., ACS Monograph 173, American Chemical Society, Washington, D.C.

Lawley, P. D., and Wallick, C. A., 1957, The action of alkylating agents on deoxyribonucleic acid and guanylic acid, Chem. Ind. (London) 633.

Lindahl, T., 1979, DNA glycosylases, endonucleases for apurinic/apyrimidinic sites, and base excision-repair, Prog. Nucleic Acid Res. Mol. Biol., 22:135.

Loveless, A., 1969, Possible relevance of O-6 alkylation of deoxyguanosines to the mutagenicity and carcinogenicity of nitrosamines and nitrosoamides, Nature, 223:206.

McCann, J., and Ames, B. N., 1976, Detection of carcinogens as mutagens in the Salmonella/microsome test: Assay of 300 chemicals: Discussion, Proc. Natl. Acad. Sci. U.S.A., 73:950.

McCann, J., Choi, E., Yamasaki, E., and Ames, B. N., 1975, Detection of carcinogens as mutagens in the Salmonella/microsome test: Assay of 300 chemicals, Proc. Natl. Acad. Sci. U.S.A., 72:5135.

McMahon, J. E., and Tinoco, I., Jr., 1978, Sequences and efficiencies of proposed mRNA terminators, Nature, 271:275.

Miller, J. A., and Miller, E. C., 1977, Ultimate chemical carcinogens as reactive mutagenic electrophiles, in: "Origins of Human Cancer, Book B, Mechanisms of Carcinogenesis," H. H. Hiatt, J. D. Watson, and J. A. Winsten, eds., Cold Spring Harbor Laboratory, New York.

Miller, J. H., Coulondre, C., Hofer, M., Schmeissner, U., Sommer, H., Schmitz, A., and Lu, P., 1979, Genetic studies of the lac repressor. IX-Generation of altered proteins by the suppression of nonsense mutations, J. Mol. Biol., 131:191.

Phillips, J. H., and Brown, D. M., 1967, The mutagenic action of hydroxylamine, Prog. Nucleic Acid Res. Mol. Biol., 7:349.

Radman, M., 1975, SOS repair hypothesis: Phenomenology of an inducible DNA repair which is accompanied by mutagenesis, in: "Molecular Mechanisms for Repair of DNA," Part A, P. C. Hanawalt, and R. B. Setlow, eds., Plenum Press, New York and London.

Reiner, B., and Zamenhof, S., 1957, Studies on the chemically reactive groups of deoxyribonucleic acids, J. Biol. Chem., 228: 475.

Robertson, H. D., Barrell, B. G., Weith, H. L., and Donelson, J. E., 1973, Isolation and sequence analysis of a ribosome-protected fragment from bacteriophage ΦX174 DNA, Nature New Biol., 241:38.

Sanger, F., Coulson, A. R., Friedman, T., Air, G. M., Barrell, B. G., Brown, N. L., Fiddes, J. C., Hutchison, C. A. III, Slocombe, P. M., and Smith, M., 1978, The nucleotide sequence of bacteriophage ΦX174, J. Mol. Biol., 125:225.

Singer, B., 1979, N-nitroso alkylating agents: Formation and persistence of alkyl derivative in mammalian nucleic acids as contributing factors in carcinogenesis, J. Natl. Cancer Inst., 62:1329.

Sinsheimer, R. L., 1968, Bacteriophage ΦX174 and related viruses, Prog. Nucleic Acid Res. Mol. Biol., 8:115.

Weinstein, I. B., and Pietropaolo, C., 1977, The action of tumor-promoting agents in cell culture, in: "Origins of Human Cancer, Book B, Mechanisms of Carcinogenesis," H. H. Hiatt, J. D. Watson, and J. A. Winsten, eds., Cold Spring Harbor Laboratory, New York.

Witkin, E. M., 1975, Relationships among repair, mutagenesis, and survival: Overview, in: "Molecular Mechanisms for Repair of DNA," Part A, P. C. Hanawalt, and R. B. Setlow, eds., Plenum Press, New York and London.

CHAPTER 8

SINGLE-STRANDED GAPS AS LOCALIZED TARGETS

FOR IN VITRO MUTAGENESIS

        David Shortle and David Botstein

        Department of Biology, Massachusetts Institute of
        Technology, Cambridge, Massachusetts 02139

SUMMARY

    Short single-stranded gaps in circular DNA molecules can be generated enzymatically, often at predetermined sites. These can serve as targets for in vitro mutagenesis procedures that result in alterations in nucleotide sequence within or very near the gap. Deamination of unpaired cytosine residues with sodium bisulfite has been used to induce mutations in the BglI restriction site of SV40 DNA and within defined regions of the β-lactamase gene on pBR322. A new method of induction of mutations at gaps, called "gap misrepair," has been developed; it was used to cause changes at the HindIII and ClaI restriction sites on pBR322 DNA. Gap misrepair reactions using DNA polymerase I of Micrococcus luteus in the presence of T4 DNA ligase and three of the four deoxynucleoside triphosphates yielded all three possible substitutions for adenine and cytosine residues in the DNA.

INTRODUCTION

    A stretch of single-stranded DNA (a gap) in an otherwise duplex molecule has a number of structural features that render the nucleotide sequence in the gap specifically susceptible to mutagenesis in vitro by two methods. First, cytosine residues in the single-stranded segment become accessible to deamination by sodium bisulfite, a reaction resulting in $C \cdot G \rightarrow T \cdot A$ transition mutations (Shortle and Nathans, 1978). Second, a gap can serve as a primer-template for mutagenic "misrepair" reactions with DNA polymerase, either by incorporation of nucleotide analogues (Muller et al., 1978) or, as shown below, by misincorporation of the standard four nucleotides. These two reactions can be used for site-specific mutagenesis by enzymatically generating a gap, the target of mutagenesis, at a predetermined site on a circular DNA molecule. In the following discussion, the enzymatic reactions used to generate gaps at specific sites will be briefly reviewed, and then some of the mutants that have been recovered specifically at single-stranded gaps will be described.

GENERATION OF SINGLE-STRANDED GAPS

The basic route by which a single-stranded gap is generated enzymatically in a duplex DNA molecule is outlined in Fig. 1. First, an endonuclease is used to introduce a break or nick in one of the two DNA strands; then, an exonuclease that can initiate hydrolysis at a nick is allowed to remove a few nucleotides, proceeding either in the $5' \rightarrow 3'$ or $3' \rightarrow 5'$ direction away from the nick. Once the site of the nick is specified in the first reaction, the direction and extent of hydrolysis by the exonuclease determine the final structure of the gap. To generate single-stranded gaps at uniquely defined positions, therefore, requires methods for nicking DNA at specific sites plus exonucleolytic reactions that can be controlled.

Nicking at Restriction Sites

When incubated with a negatively supercoiled circular DNA containing one or more restriction sites, type II restriction endonucleases cleave at (or near) their recognition sequences by generating two nicks, one in each strand. With some type II enzymes (e.g., BglI, HindIII, and ClaI), this cleavage reaction can be inhibited after the first nick has been induced by including in the reaction mixture the intercalating compound ethidium bromide (Parker et al., 1977). Using purified supercoiled DNA and ethidium bromide at a

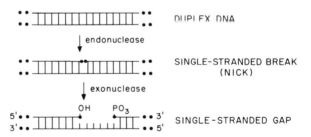

Fig. 1. The two enzyme-catalyzed reactions used in the generation of single-stranded gaps.

concentration determined to be optimal by titration, it is often possible to nick between 50 and 90% of the input DNA, yielding open circular molecules with a single nick located at the restriction enzyme's normal cleavage site (Shortle and Nathans, 1978). After conversion of the nick to a gap that exposes the restriction recognition sequence, such DNA molecules provide convenient substrates for assaying mutagenesis by reactions specific for single-stranded gaps, since mutations affecting the recognition sequence confer the easily scorable phenotype of loss of a particular restriction site.

Nicking Within Specific DNA Segments

The number of sites at which single-stranded gaps can be generated via restriction enzyme-catalyzed nicking is, of course, limited by the availability of usable restriction sites. When the objective is to construct mutations within a defined nucleotide sequence interval carried on a circular DNA, a two-step, segment-specific nicking procedure can be used (Shortle et al., 1980). In principle, with this procedure a nick can be introduced into any segment of a circular DNA, provided that a single-stranded fragment corresponding to that segment can be isolated. With ATP serving as cofactor, the recA protein is used to catalyze the annealing of a unique, single-stranded fragment to the complementary sequence on a covalently closed circular DNA (McEntee et al., 1979; Shibata et al., 1979). The annealed fragment displaces one strand of the circular DNA, creating a single-stranded D-loop. In a second reaction, this D-loop structure becomes a substrate for the single-strand specific endonuclease S1. Once the first nick is induced by S1 nuclease attack on the displaced strand, the D-loop rapidly breaks down by spontaneous displacement of the fragment, yielding an open circular DNA molecule with a nick located within the DNA segment corresponding to the single-stranded fragment.

When negatively supercoiled plasmid pBR322 DNA (4.3 kilobases in length) was subjected to nicking with this procedure, nicks were efficiently induced in defined segments as short as 130 base pairs in length (Shortle et al., 1980). Before use of this nicked DNA as a substrate for gapping and mutagenesis, the specificity of the nicking reaction was assayed by incorporation of α-^{32}P-labeled nucleotides at the site of the nick in a limited nick translation, followed by determination of the site of the incorporated label by restriction analysis and gel electrophoresis. From this biochemical analysis and from fine-structure genetic mapping of the mutants subsequently constructed by using this DNA as substrate for mutagenesis, it was concluded that greater than 90% of nicks were localized to the 130 base-pair segment specified by the single-stranded fragment.

Exonuclease Reactions

DNA polymerase I from Micrococcus luteus has both the 3' → 5' and 5' → 3' exonuclease activities typical of this class of bacterial polymerase (Miller and Wells, 1972). The intrinsic activity of both exonuclease functions is relatively low in comparison to Escherichia coli DNA polymerase I. Consequently, this enzyme (available free of endonuclease activity from commercial sources) can be used in a time-controlled reaction to remove an average of approximately five to six nucleotides predominantly in the 5' → 3' direction (Shortle and Nathans, 1978), although gaps as long as 20 nucleotides may be generated. Alternatively, the exonuclease activity of T4 DNA polymerase can be used for hydrolysis in the 3' → 5' direction (Rawlins and Muzyczka, 1980). This enzyme has the advantage that the extent of hydrolysis is controllable by the addition of one or more deoxyribonucleotide triphosphates in the reaction mixture (Englund et al., 1974).

BISULFITE MUTAGENESIS

Sodium bisulfite catalyzes the deamination of cytosine under mild conditions of temperature and pH to form uracil. Although cytosine residues in single-stranded polynucleotides react at nearly the same rate as the free mononucleotide, cytosine residues within the duplex structure of double-stranded DNA are essentially inert to bisulfite deamination (for discussion, see Hayatsu, 1976). Therefore, sodium bisulfite is in effect a single-strand specific mutagen and can be used to specifically deaminate cytosine residues exposed in a gap. On filling in with DNA polymerase, bisulfite-treated gapped DNA molecules which have suffered C·G → U·A changes would be expected to result in C·G → T·A transition mutations on replication and/or repair.

The BglI Restriction Site of SV40

That sodium bisulfite can be used for site-specific mutagenesis of single-stranded gaps was demonstrated at the single BglI restriction site on SV40 DNA (Shortle and Nathans, 1978). As shown in Fig. 2, BglI-induced nicks can occur on either strand, and thus gapping in the 5' → 3' direction with the M. luteus DNA polymerase generates two types of gapped molecules. After incubation of BglI site-gapped DNA with sodium bisulfite under conditions that yield approximately 30% deamination of accessible cytosine residues, the gap was filled in with DNA polymerase in vitro. Molecules that lost the BglI site were recovered by digesting the modified DNA with BglI and isolating the circular DNA resistant to cleavage. This BglI-resistant DNA was transfected onto permissive tissue culture cells to obtain individual SV40 plaques. Analysis of 23 independent SV40

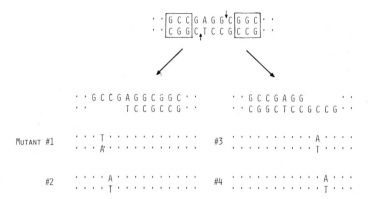

Fig. 2. The nucleotide sequences of the BglI restriction site (the recognition sequence is enclosed in boxes) in SV40 DNA and four viral mutants induced with bisulfite. The probable structures of the two types of single-stranded gaps generated by nicking with BglI plus ethidium bromide and gapping with the exonuclease activity of M. luteus DNA polymerase I are shown in the second line.

isolates revealed that 19 contained viral DNA that had lost the BglI site. These BglI-resistant SV40 mutants could be grouped into four phenotypic classes on the basis of plaque size at three different temperatures, and the nucleotide sequence surrounding the missing BglI site determined for one member of each class is shown in Fig. 2 (Shortle and Mathans, 1979). Three mutants had the base substitution expected of bisulfite-induced deamination events - a C·G → T·A transition, and a fourth SV40 mutant had a C·G → A·T transversion mutation. A second C·G → A·T transversion appeared in one or two other SV40 mutants induced with bisulfite at a short gap generated by very limited exonuclease III hydrolysis from the BglI nick (Shortle and Nathans, 1979). The molecular mechanism responsible for these "nonstandard" bisulfite-induced base substitutions is not known.

The β-lactamase Gene: Codons 1 to 46

To obtain point mutants in the "signal sequence" of the β-lactamase (bla) gene, which could be used for biochemical analysis of transport of this enzyme into the periplasmic space, specific nicks were induced in a segment of the gene (carried on the plasmid pBR322) that spans codon 1 to codon 46 (Shortle et al., 1980). After converting the nick into a gap with the M. luteus polymerase, reacting with sodium bisulfite, and filling in the gap in vitro, the modified plasmid DNA was used to transform an E. coli strain by selection for

the tetracycline resistance marker. Of the 22 bla⁻ mutants recovered
from 800 transformants, 14 mapped genetically within the signal sequence (codons 1 to 23) and the remaining eight mutants mapped within
the amino terminal end of the mature protein (codons 24 to 129).
Nucleotide sequence analysis of some of these bla⁻ mutants and others
induced in this same DNA segment in subsequent experiments has identified the four single C·G → T·A transition mutations shown in
Fig. 3; the mutants in codon 4 and in codon 20 have been independently isolated three times each (Shortle, Grisafi, Koshland, and
Botstein, unpublished data). A majority of the remaining mutants
have, in addition to one or a few tightly clustered C·G → T·A transition mutations, frameshift mutations consisting of either a single T
residue added to a run of four or more T residues (+1 frameshift) or
a deletion (-1 frameshift) of a single C residue within a run of
three C residues.

Experiments to determine the mechanism responsible for these
frameshift mutations are in progress; at this point one can only
speculate. One reasonable hypothesis on the origin of the +1 frameshifts is that the extra nucleotide was inserted in vitro during the
gapping reaction. In this reaction, dATP was included at high concentrations to stimulate the 5' → 3' exonuclease activity of the
M. luteus DNA polymerase (Miller and Wells, 1972). If the initial

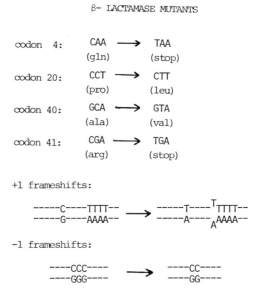

Fig. 3. The nucleotide sequences of β-lactamase mutants recovered
after segment-directed mutagenesis of codons 1 to 46.

nick occurred in a run of A residues, the polymerase would be expected to nick translate to the end of such a run and stop. Slippage of the primer strand, however, might then allow the polymerase to add on the extra A residue. Tending to support this hypothesis is the failure to find this type of frameshift among mutants that were induced with bisulfite reaction on pBR322 DNA molecules gapped in this same segment of the bla gene but without dATP present and from which the in vitro gap filling was omitted. The other type of frameshift mutation (CCC → CC), however, has been recovered regardless of the presence of dATP or the subsequent gap filling. This lesion might arise at a low frequency in vivo after transformation at sites of uracil residues in single-stranded DNA. Noteworthy in this regard is that the E. coli strain used for all transformations has an active DNA-uracil N-glycosylase.

NUCLEOTIDE MISINCORPORATION: THE GAP MISREPAIR REACTION

When polymerizing a complementary strand on a primer-template in vitro, purified DNA polymerases are observed to incorporate noncomplementary nucleotides at frequencies orders of magnitude higher than the spontaneous mutation rate measured in vivo. Furthermore, the frequency of misincorporation can be substantially increased by addition of manganese (II) ion to the reaction mixture or by creating large imbalances in the ratio of the four deoxyribonucleoside triphosphates (for discussion and references, see Loeb et al., 1979). Since a short, single-stranded gap in a duplex DNA molecule is a good substrate for many DNA polymerases, nucleotide misincorporation during repair (misrepair) of such a gap in vitro would generate a base substitution in the newly synthesized strand. If DNA synthesis is terminated by ligation of the nick once the polymerase has completely filled in the gap, this mutagenic reaction will be confined to those nucleotides within the single-stranded gap; i.e., the short stretch of single-stranded DNA becomes the target for mutagenesis. Ligation also serves to trap misrepaired molecules, since closed circular DNA is no longer a substrate for DNA polymerase's repair functions. In the absence of mismatch repair, the misincorporated nucleotide should segregate from the wild-type sequence at the first round of replication after transformation into an appropriate host cell, resulting in a mutant carrying a base substitution.

In preliminary experiments to examine the feasibility of in vitro mutagenesis by gap misrepair, short gaps have been constructed in pBR322 DNA at the single cleavage sites for the restriction enzymes HindIII and ClaI (Shortle and Botstein, in preparation). When purified circular DNA molecules gapped at the HindIII site were incubated with DNA polymerase I from M. luteus in the presence of all four deoxyribonucleoside triphosphates plus T4 ligase and ATP, the gap was repaired in greater than 90% of molecules as measured by

conversion of the DNA from an open to a covalently closed circular
form. Restriction enzyme analysis of the plasmid progeny appearing
after transformation of this repaired DNA revealed that all 40 independent plasmid isolates screened retained the HindIII restriction
site. When the first nucleotide required to repair the gap (dATP)
was omitted from the reaction mixture and manganese (II) ion was
added in addition to magnesium ion (see Fig. 4), again greater than
90% of the circular DNA molecules were eventually converted to a
covalently closed form. However, five out of 36 plasmid isolates
(14%) recovered on transformation with this DNA were resistant to
cleavage by the HindIII enzyme. The nucleotide sequence changes
identified in these five HindIII site mutants (one of which has two
additional mutations adjoining the restriction site) are consistent
with a noncomplementary nucleotide being incorporated in place of
the missing nucleotide dATP during gap filling [A → G(5); A → C(1);
A → T(1)].

A higher rate of mutagenesis was observed at the ClaI site when
the gap misrepair reaction was carried out in the absence of dCTP.
Twelve of 36 plasmid isolates (33%) in this experiment had lost the
ClaI restriction site, and the nucleotide sequences of six of these
mutants have been determined. Again several isolates had additional
base-substitution mutations flanking the restriction site, but in
each case the pattern of base substitution [C → T(6); C → G(4);
C → A(3)] is consistent with misincorporation at sites where the
missing nucleotide (dCTP) would have been expected.

From these preliminary results, it would appear that the in
vitro misrepair of specifically placed single-stranded gaps can be
used for efficient site-specific mutagenesis. Since in the experiments described above all three base substitutions for A and for C
have been induced, it is possible that, with the proper ratios of

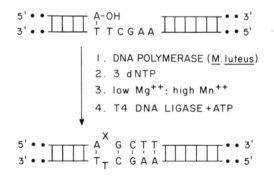

Fig. 4. Schematic diagram of the gap misrepair reaction. The DNA
substrate is a short gap at the HindIII site in pBR322 DNA.

the three added deoxynucleoside triphosphates, the noncomplementary nucleotide that is inserted in lieu of the missing nucleotide can be controlled. The use of other DNA polymerases, such as the error-prone avian myeloblastosis virus reverse transcriptase, in this gap misrepair reaction may provide another device for controlling the pattern of misincorporation.

CONCLUSIONS

Single-stranded gaps constitute local disruptions of the uniform, highly ordered structure of DNA. Gaps can be enzymatically constructed in regions specified in advance. Two mutagenic reactions that depend on structural features unique to single-stranded gaps can thus be used for site-specific mutagenesis. In the future it seems reasonable to expect that additional, single-strand specific mutagenic reactions will be developed. As methods for precisely placing single-stranded gaps are further refined, it may become possible to routinely construct virtually any desired base substitution or frameshift mutation.

ACKNOWLEDGMENTS

This work was supported by grants from the American Cancer Society (MV-90) and from the National Institutes of Health (GM18973 and GM21253). D.S. was supported by a Postdoctoral Research Fellowship from the Helen Hay Whitney Foundation.

REFERENCES

Englund, P. T., Price, S. S., and Weigel, P. N., 1974, The use of the T4 DNA polymerase in identification of 3' terminal nucleotide sequences of duplex DNA, Methods Enzymol., 29:273.

Hayatsu, H., 1976, Bisulfite modification of nucleic acids and their constituents, Prog. Nucleic Acid Res. Mol. Biol., 16:75.

Loeb, L. A., Weymouth, L. A., Kunkel, T. A., Gopinathan, K. P., Beckman, R. A., and Dube, D. K., 1979, On the fidelity of DNA replication, Cold Spring Harbor Symp. Quant. Biol., 43:921.

McEntee, K., Weinstock, G. M., and Lehman, I. R., 1979, Initiation of general recombination catalyzed in vitro by the recA protein of Escherichia coli, Proc. Natl. Acad. Sci. U.S.A., 76:2615.

Miller, L. K., and Wells, R. D., 1972, Properties of the exonucleolytic activities of the Micrococcus luteus deoxyribonucleic acid polymerase, J. Biol. Chem., 247:2667.

Muller, W., Weber, H., Meyer, F., and Weissmann, C., 1978, Site-directed mutagenesis in DNA: Generation of point mutations in cloned β globin complementary DNA at the positions corresponding to amino acids 121 to 123, J. Mol. Biol., 124:343.

Parker, R. C., Watson, R. M., and Vinograd, J., 1977, Mapping of closed circular DNAs by cleavage with restriction endonucleases and calibration by agarose gel electrophoresis, Proc. Natl. Acad. Sci. U.S.A., 74:851.

Rawlins, D. R., and Muzyczka, N., 1980, Construction of a specific amber codon in the simian virus 40 T-antigen gene by site-directed mutagenesis, J. Virol., 36:611.

Shibata, T., DasGupta, C., Cunningham, R. P., and Radding, C. M., 1979, Purified Escherichia coli recA protein catalyzes homologous pairing of superhelical DNA and single-stranded fragments, Proc. Natl. Acad. Sci. U.S.A., 76:1638.

Shortle, D., Koshland, D., Weinstock, G. M., and Botstein, D., 1980, Segment-directed mutagenesis: Construction in vitro of point mutations limited to a small predetermined region of a circular DNA molecule, Proc. Natl. Acad. Sci. U.S.A., 77:5375.

Shortle, D., and Nathans, D., 1978, Local mutagenesis: A method for generating viral mutants with base substitutions in preselected regions of the viral genome, Proc. Natl. Acad. Sci. U.S.A., 75:2170.

Shortle, D., and Nathans, D., 1979, Regulatory mutants of simian virus 40: Constructed mutants with base substitutions at the origin of DNA replication, J. Mol. Biol., 131:801.

CHAPTER 9

MUTAGENESIS AT SPECIFIC SITES: A SUMMARY AND PERSPECTIVE

Michael Smith

Department of Biochemistry, Faculty of Medicine
2146 Health Sciences Mall, University of British
Columbia, Vancouver, B.C. V6T 1W5, Canada

Recent developments in molecular genetics, principally molecular cloning of segments of genomic DNA and the rapid ladder methods of DNA sequence determination, have provided new opportunities for construction of specific mutants and for definition of the specific locations and nature of point mutations. The four papers in this Section of the symposium describe important investigations in these two areas.

The work of Miller on the lac repressor gene of Escherichia coli has provided an exceedingly detailed and interesting catalogue of mutational changes (Miller and Schmeissner, 1979). This detailed study of a gene provides a most productive route to defining the function of amino acids in a protein (Miller, 1979; Miller et al., 1979). In addition it provides a very important tool for defining the nature of changes in DNA induced by mutagens (Miller and Schmeissner, 1979). In his symposium paper, Miller discussed the sequence specificity and the changes induced by spontaneous mutations, by UV light and by chemical mutagens. It should also be noted that the lacI gene provides a very useful tool for the study of the enzymatic mechanisms of error correction (Glickman, this volume).

The analogous gene to the E. coli lacI gene for eukaryotes (in terms of detailed genetic analysis) is the CYC1 locus of Saccharomyces cerevisiae. This gene, which codes for the apoprotein of iso-1-cytochrome c, has been subjected to detailed genetic analysis by Sherman's group (Sherman and Stewart, 1971; Sherman et al., 1975). Recently, a synthetic oligodeoxyribonucleotide probe has been constructed (Gillam et al., 1977) which has made possible the isolation of the iso-1-cytochrome c gene (Montgomery et al., 1978) and the determination of the sequence of the gene and its adjacent regions (Smith et al., 1979). Thus, the CYC1 locus provides an ideal substrate for studying the molecular changes introduced by mutagens in a eukaryotic DNA. In this session, Lawrence described his studies on the mechanisms of UV mutagenesis in yeast by analysis of changes induced at the CYC1 locus.

As was noted in the first paragraph of these remarks, clones of characterized DNA fragments provide substrates for the in vitro construction of defined mutants. This method of producing mutants has developed into a powerful and widely employed tool of modern genetics (Peden et al., 1980; Smith and Gillam, 1981a,b). Particularly important to the detailed understanding of the function of DNA are methods for construction of defined point mutants. One of the most useful strategies in this area is that which uses bisulfite to deaminate deoxycytidine residues in a short single-stranded segment of DNA (Shortle and Nathans, 1978; Shortle et al., 1980). In his paper, Shortle described his studies in this area and also a new approach to in vitro mutation, namely enzymatic misrepair.

In the last paper of the session, Chambers described a procedure, involving genetic complementation using a clone of the target gene, for detecting mutants in gene G of bacteriophage ØX174. The most specific of the in vitro methods for mutant construction uses a synthetic oligodeoxyribonucleotide, with a defined point change relative to its wild-type complement in a DNA, as a specific mutagen (Hutchison et al., 1978; Smith and Gillam, 1981b). This method has been used to induce both types of transition mutations (Hutchison et al., 1978) and both types of transversion mutations (Gillam et al., 1979). Under appropriate conditions, these mutations can be obtained with very high efficiencies (Gillam and Smith, 1979a). Single base-pair deletions can also be produced (Gillam et al., 1980). In addition, the mutagenic oligonucleotide can be used to purify the mutant DNA by genotypic selection (Gillam and Smith, 1979b; Gillam et al., 1980). Chambers described the production of a mutant using this strategy. In addition, he described how the method could be used to introduce specific modified bases into a defined position in a DNA as a prelude to defining action of the modified base.

REFERENCES

Gillam, S., Astell, C. R., and Smith, M., 1980, Site-specific mutagenesis using oligodeoxyribonucleotides: isolation of phenotypically silent ØX174 mutant, with a single nucleotide deletion, at very high efficiency, Gene, 12:129.

Gillam, S., Jahnke, P., Astell, C., Phillips, S., Hutchison, C. A. III, and Smith, M., 1979, Defined transversion mutations at a specific position in DNA using synthetic oligodeoxyribonucleotides as mutagens, Nucleic Acids Res., 9:2973.

Gillam, S., Rottman, F., Jahnke, P., and Smith, M., 1977, Enzymatic synthesis of oligonucleotides of defined sequence; synthesis of a segment of the yeast iso-1-cytochrome c gene, Proc. Natl. Acad. Sci. U.S.A., 74:96.

Gillam, S., and Smith, M., 1979a, Site-specific mutagenesis using synthetic oligodeoxyribonucleotide primers: I. Optimum conditions and minimum oligodeoxyribonucleotide length, Gene, 8:81.

Gillam, S., and Smith, M., 1979b, Site-specific mutagenesis using synthetic oligodeoxyribonucleotide primers: II. In vitro selection of mutant DNA, Gene, 8:99.

Glickman, B. W., 1981, Methylation-instructed mismatch correction as a postreplication error-avoidance mechanism in Escherichia coli, this volume.

Hutchison, C. A. III, Phillips, S., Edgell, M., Gillam, S., Jahnke, P., and Smith, M., 1978, Mutagenesis at a specific position in a DNA sequence, J. Biol. Chem., 253:6551

Miller, J. H., 1979, Genetic studies of the lac repressor. XI. On aspects of lac repressor structure suggested by genetic experiments, J. Mol. Biol., 131:249.

Miller, J. H., Coulondre, C., Hofer, M., Schmeissner, U., Sommer, H., Schmitz, A., and Lu, P., 1979, Genetic studies of the lac repressor. IX. Generation of altered proteins by the suppression of nonsense mutations, J. Mol. Biol., 131:191.

Miller, J. H., and Schmeissner, V., 1979, Genetic studies of the lac repressor. X. Analysis of missense mutations in the lacI gene, J. Mol. Biol., 131:223.

Montgomery, D. L., Hall, B. D., Gillam, S., and Smith, M., 1978, Identification and isolation of the yeast cytochrome c gene, Cell, 14:673.

Peden, K. W. C., Pipas, J. M., Pearson-White, S., and Nathans, D., 1980, Isolation of mutants of an animal virus in bacteria, Science, 209:1392.

Sherman, F., Jackson, M., Liebman, S. W., Schweingruber, M., and Stewart, J. E., 1975, Deletion map of CYC1 mutants and its correspondence to mutationally altered iso-1-cytochrome c of yeast, Genetics, 81:51.

Sherman, F., and Stewart, J. W., 1971, Genetics and biosynthesis of cytochrome c, Annu. Rev. Genet., 5:157.

Shortle, D., Koshland, D., Weinstock, G. M., and Botstein, D., 1980, Segment-directed mutagenesis: Construction in vitro of point mutations limited to a small predetermined region of a circular DNA molecule, Proc. Natl. Acad. Sci. U.S.A., 77:5375.

Shortle, D., and Nathans, D., 1978, Local mutagenesis: A method for generating viral mutants with base substitutions in preselected regions of the viral genome, Proc. Natl. Acad. Sci. U.S.A., 75:2170.

Smith, M., and Gillam, S., 1981a, In vitro construction of specific mutants, in: "Developmental Biology Using Purified Genes," D. D. Brown, and C. F. Fox, eds., Academic Press, New York, in press.

Smith, M., and Gillam, S., 1981b, Constructed mutants using synthetic oligodeoxyribonucleotides as site-specific mutagens, in: "Genetic Engineering, Principles and Methods," Vol. 3, J. K. Setlow, and A. Hollaender, eds., Plenum Press, New York.

Smith, M., Leung, D. W., Gillam, S., Astell, C. R., Montgomery, D. L., and Hall, B. D., 1979, Sequence of the gene for iso-1-cytochrome c in Saccharomyces cerevisiae, Cell, 16:753.

CHAPTER 10

POLYMERASE INFIDELITY AND FRAMESHIFT MUTATION

Lynn S. Ripley and Nadja B. Shoemaker

Laboratory of Molecular Genetics, National Institute
of Environmental Health Sciences, Research Triangle
Park, North Carolina 27709

SUMMARY

Mutant T4 DNA polymerases which alter mutation rates in vivo have been used to approach questions of replication fidelity. Most studies have characterized "mutator" or "antimutator" polymerases by their influence upon base-pair substitution mutation, particularly transitions. We are extending the characterization of mutant polymerases to include the role of T4 DNA polymerase in frame fidelity.

We conclude that T4 DNA polymerase indeed plays a major role in frame fidelity from the observation that among 26 ts DNA polymerase alleles more than half increased or decreased the revertant frequency of an rII frameshift between 3- and 400-fold. Furthermore, the revertant frequency at this frameshift site was not altered by some polymerases known to increase base-pair substitution frequencies. Thus, polymerase mutants may provide a means to separate fidelity elements which are uniquely important to either frameshift or to base-pair substitution mutations.

Our approach provides a detailed description (including frequency, location, and addition or deletion character) of frameshifts produced in the presence of different polymerases. The frameshifts arise in the T4 rIIB gene and are recognized as "revertants" or "suppressors" of other T4 rIIB frameshifts having a defined genetic sign (+ or -). From several initial tester frameshifts, spectra of suppressor frameshifts have been determined for a number of polymerases. The frameshift spectra of mutant polymerases are distinct from the spectrum of the wild-type polymerase. Furthermore, the mutant polymerases have individually distinct propensities for producing frameshifts at particular genetic loci. This specificity has allowed us to correctly predict the influence of several mutant polymerases on frameshifts other than those originally tested and suggests an important correlation with particular fidelity defects and frameshifts at particular sites. We believe it particularly noteworthy that some polymerase mutants exhibit a distinct bias in their influence upon the frequency of frameshifts having different genetic signs. We plan to sequence these frameshifts to determine their addition or deletion nature and thereby gain a more precise molecular view of the frame fidelity defects of these mutant polymerases.

FRAMESHIFT MECHANISMS

Frameshift mutations were first shown to arise preferentially in repeated DNA sequences by Streisinger et al. (1966) by an elegant combination of genetics and protein sequencing of the T4 lysozyme gene. A model was offered suggesting that such DNA repeats allowed the misannealing of DNA strands, resulting in the formation of looped structures from which further DNA metabolism generated either deletion or addition mutations.

Studies in a variety of systems have now confirmed that frameshift mutations (and deletions) do indeed often occur in DNA sequences predicted to be prone to frameshifting by the Streisinger model. In bacteriophage T4, repeating A's and T's are frameshift hot spots (Okada et al., 1972; Pribnow et al., 1981). In the Salmonella hisD gene, one frameshift hot spot was found to be a DNA sequence of 4 tandem CG's (Isono and Yourno, 1974). In the Escherichia coli lacI gene, a frameshift hot spot accounting for approximately two-thirds of the spontaneous mutations detected in that gene also conforms to the model in perhaps a more complex form: the frameshift site in wild-type DNA is composed of a tandem repeat of a four-base sequence three times in succession. Direct sequencing of frameshift mutations at the site demonstrated that the two classes of frameshift mutations generated at that site resulted from the deletion or the addition of a four-base unit within the site (Farabaugh et al., 1978). Tandemly repeated DNA sequences may also be important in the generation of deletion mutations, since many deletions isolated in the yeast iso-1-cytochrome c gene (Stewart and Sherman, 1974) and in the E. coli lacI gene (Farabaugh et al., 1978) are flanked by such repeats. These tandemly repeated sequences may also be important in producing mutations in mammalian DNAs. Deletions identified in the sequences of human β globin genes and in noncoding sequences of mammalian β-like globin sequences arose at DNA sites consistent with the interpretation that they were previously flanked by tandem repeats (Efstratiadis et al., 1980).

While the general model of frameshift mutation proposed by Streisinger appears to be widely applicable as a definition of a class of DNA sequences at which additions or deletions of base pairs might be expected at relatively high frequencies, we do not understand the more precise enzymological processes involved in frameshift production. The proposed initial event of DNA misalignment suggests that nicked DNA molecules (or the physical terminus of a linear molecule) could be involved. Many DNA metabolic processes produce nicked DNA molecules as intermediates and thus might contribute structural precursors to spontaneous frameshift mutation. A theoretical case can be made for the role of the DNA structural intermediates of replication, recombination, and repair as frameshift precursors. However, the particular contribution of any of those DNA metabolic processes to frameshift mutation remains equivocal (see Roth, 1974).

An approach to defining elements of DNA metabolism important to frameshift mutation has been to try to identify a possible role for DNA metabolic proteins by measuring the effects of mutations in the genes specifying these proteins on the frequency of frameshift mutation. Previous studies in bacteriophage T4 concentrated on the influence of mutations in genes 30 (DNA ligase) and 32 (helix-destabilizing protein) on the reversion of frameshift mutations (Koch and Drake, 1973; Koch et al., 1976). The influence of mutant alleles on frameshift mutation frequencies was small and thus not very useful in defining any particular role for these enzymes in the frameshift process. A broader survey of the effects of mutant alleles of DNA replication proteins (genes 30, 32, 36, 38, 42, 43, 44, 45, 46, 47, 49, and 56) on frameshift frequencies revealed large effects only for the L56 allele of the T4 DNA polymerase (Bernstein, 1971).

Since we wished to define the role of an enzyme in the frameshift process with the intention that frameshift models might later be tested in vitro, an enzyme producing large effects on frameshift frequencies seemed necessary. Thus our attention focused upon the DNA polymerase of T4. The polymerase has a profound influence upon the fidelity of DNA with respect to base-pair substitution mutation, but very little is known about its influence on frameshift mutation. We have asked the question, "Does the DNA polymerase play a major role in the maintenance of proper frame?" Anticipating an affirmative answer, we hoped to gain some insights into the mechanism(s) of the frameshift process, and into the role of the DNA polymerase in frame fidelity, through a detailed analysis of frameshift mutations produced in the presence of a variety of mutant DNA polymerase enzymes.

MEASURING FRAMESHIFT MUTATION IN THE T4 rIIB GENE

Because it appears that DNA sequences play an intimate role in the probability of frameshift mutation, we have chosen to examine frameshift mutations in a context which allows the manipulation of the DNA target size and site. The rIIB gene provides such a target. The direct isolation of frameshift mutations can be accomplished in a target which may be selected to be as large as approximately 200 base pairs or as small as approximately 20 simply by isolating phenotypic revertants of appropriately located frameshift mutations within that gene.

A rather large portion of the N-terminus of the rIIB protein has the exceptional property of being insensitive to amino acid sequence changes, as deduced from the lack of missense mutations in this region, and from the existence of a deletion mutation in which this region is absent while the remaining polypeptide is able to provide rIIB$^+$ function (Champe and Benzer, 1962). When amino acid sequence changes per se do not produce a mutant phenotype, one frameshift can be suppressed to the wild phenotype by any second frameshift mutation which restores the reading frame to 0, if no nonsense codons are generated between the two frameshift mutations.

Approximately 20 years ago the rIIB gene was used by Crick et al. (1961) to deduce a coding ratio of three for the genetic code. In the course of that study more than 70 different frameshift mutations, serially defined as frameshift suppressors (i.e., frameshift suppressors of frameshift suppressors and so on), were isolated and mapped (Barnett et al., 1967). The suppressors isolated in this system are themselves frameshift mutations having fully mutant phenotypes when separated from the initial frameshift mutation.

Genetic signs were assigned to frameshifts in the gene in the following manner. A single frameshift mutation, FCO, was given the genetic sign "+." All mutations suppressing FCO are defined as genetically "-." Likewise, frameshifts suppressing the "-" frameshifts are defined as "+" frameshifts. The combination of a "+" and a "-" frameshift produces a "0" change in frame, although intervening codons may change. The assignment of genetic signs to frameshift mutations does not define their addition or deletion character. Frameshifts which are genetically "+" might, for example, be additions of +1, +4, or +7 bases but could also be deletions of -2, -5, or -8 bases. Conversely genetically "-" frameshift mutations could be deletions of -1, -4, or -7 bases or additions of +2, +5, or +8 bases.

All pairs of "+" and "-" frameshifts are expected to produce a wild phenotype except when the shift in frame between a pair of mutants generates an in-frame termination codon. The location of such termination codons, "barriers" to suppression between frameshifts otherwise expected to suppress, was deduced genetically by Crick et al. (1961). The wild-type DNA sequence of this portion of the rIIB gene has been determined by Pribnow et al. (1981). The barriers can of course be recognized directly in the DNA sequence. The position of frameshift mutations with respect to barriers can be determined genetically and thus the approximate location of many frameshifts in the sequence can be deduced. Furthermore, the sequence confirms that the genetic sign of FCO is actually "+", and thus that rIIB frameshift mutations with this sign are altered by $+1 \pm 3N$ bases (where N is an integer).

Figure 1 shows the portion of the rIIB gene in which we are measuring frameshift mutation. "AUG" marks the beginning of the translated portion of the gene. Barriers 1^B and M^B do not permit suppression between a "-" frameshift to the left of the barrier and a "+" frameshift to its right. The remaining barriers have the opposite polarity not permitting the suppression between a "+" frameshift to the left of the barrier and a "-" frameshift to its right. The DNA target within which suppressing frameshifts may be expected is determined from the location of the barriers in the sequence, the sign of the initial frameshift mutation and its location. The DNA targets for the suppression of the frameshift mutations in rIIB used in this paper are also shown in Fig. 1.

THE ISOLATION AND CHARACTERIZATION OF rIIB FRAMESHIFT MUTATIONS

Frameshift mutations of sign "+" were isolated as phenotypic revertants (suppressors) of a "-" frameshift mutation. Similarly, "-" frameshift mutations were isolated as suppressors of "+" frameshift mutations. Each phenotypic revertant was crossed against the wild type. When the revertant contained a suppressing frameshift, two different frameshift mutations were present among the progeny of the cross, along with occasional spontaneously arising forward R mutations. The frameshift mutations have the mutant (R) phenotype when grown on an appropriate host (E. coli B) in contrast to the wild phenotype (R^+) exhibited by the parents of the cross. The frameshift mutations are identified by selecting R progeny from each backcross to the wild type and determining (by recombination spot tests) whether these progeny have the frameshift mutation initially present in the stock or a different R mutation which may be a suppressor of the initial frameshift. The suppressing frameshift is recognized as an R mutation which arises frequently among R progeny of the backcross at a location distinct from the site of the initial frameshift mutation. From each backcross R's were isolated and ten were characterized. If several R's mapped to the position of the initial frameshift mutation and several R's mapped to a location different from the initial frameshift mutation but to the same location as each other, the latter class of R mutants was considered to

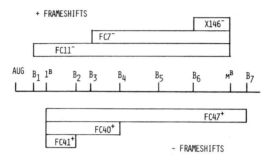

Fig. 1. A genetic map of the promoter-proximal end of the rIIB gene. The locations of frameshift mutations with respect to barriers and the predicted target sizes for suppressing frameshift mutations are shown. The targets for suppressing "+" frameshifts from each initial "-" frameshift are shown above the barrier map, while the targets for suppressing "-" frameshifts from initial "+" frameshifts are shown below the map. The mutants are described by Barnett et al. (1967). The size of the DNA targets in base pairs (including the codons containing the barriers which define each end of the target) have been determined from the wild-type DNA sequence of Pribnow et al. (1981). No correction has been made for the number of bases which may be added or deleted within the mutant sequence of each target. The approximate target sizes in base pairs are as follows: X146 = 23, FC7 = 86, FC11 = 122, FC47 = 124, FC40 = 49, and FC41 = 22.

represent the suppressor frameshift responsible for the phenotypic reversion of the original frameshift mutation. In those cases where two distinct classes of R mutations were not isolated, 10 to 20 additional R's were characterized. When no R's were found to be the same as the original frameshift mutation, it was presumed that the frameshift mutation occurred at the site of the original frameshift or within a few base pairs of it. Because of the high level of recombination exhibited by T4, mutations lying within several base pairs of each other can usually be resolved by such recombination tests.

Once the frameshift mutations, identified as suppressors of the initial frameshift mutations, have been separated, they are characterized by a number of genetic tests. The frameshift is separated from the ts gene 43 allele and then mapped into the target region more precisely. This region is densely marked with deletion, frameshift, and nonsense mutations which we have used to determine the genetic locations of the mutations. Furthermore, as we have collected mutants, many independently isolated examples of mutants have been found which map very close together, all failing to recombine with a marker frameshift in our recombination test. When we tested the ability of these frameshift mutations to recombine with each other, some crosses resolved previously apparently homogeneous groups of mutations into subclasses.

THE INFLUENCE OF MUTANT DNA POLYMERASES ON FRAMESHIFT MUTATION

The gene 43 mutant alleles used in this study were originally isolated after mutagenesis in a T4D genetic background (Edgar and Lielausis, 1964). Since the gene 43 alleles had never been crossed out of their mutagenized backgrounds and since all the rII mutations used in this study lie in a T4B genetic background, all the alleles were crossed to T4B four times, selecting phage which had retained temperature sensitivity. The final backcrossed versions of these alleles failed to recombine with the nonbackcrossed version for the temperature-sensitive phenotype. Those alleles known to demonstrate strong effects on the frequency of base-pair substitution mutations were tested to determine that this property was retained in the backcrossed version.

Our initial test for the possible role of polymerases in frame fidelity consisted of measuring the influence of each of 26 mutant alleles mapping into 19 recombinationally distinguishable gene 43 sites (see Allen et al., 1970), on the reversion of an rII frameshift mutation, r131. This rIIA mutation maps in a hot spot for frameshift mutation. More than one-third of spontaneously arising rII mutants occur at this site in the presence of the wild-type polymerase (Benzer, 1961). The position of this mutant can be approximated in the rII sequence of Pribnow et al. (1981) and probably lies in a run of 6A's in the wild type. Based on revertant frequency measurements of the r131 allele and of other frameshift mutants mapping at the same site having much higher revertant frequencies, an argument has been made that the r131 mutant is likely to be a deletion of an A, producing a 5A sequence (Pribnow et al., 1981). The open bars in Fig. 2 demonstrate the increases or decreases in revertant frequencies seen in the presence of each gene 43 mutant allele when compared to the wild-type allele. The spontaneous revertant frequency of the r131 allele in the presence of the wild-type enzyme was 25×10^{-8}. Just over half of the tested gene 43 alleles produced greater than threefold increases or decreases in revertant frequency at the site. This, we believe, confirms a major role for DNA polymerase in frame fidelity. Most importantly, a number of mutant polymerases produced very large effects upon frameshift mutations: L56 and L98 produced large increases in the revertant frequency and L88 produced a more modest increase. These three alleles also promote base-pair substitution mutatenesis (Drake et al., 1969; Ripley, 1975; Speyer et al., 1966). The L42 and L141 alleles decreased the reversion of r131

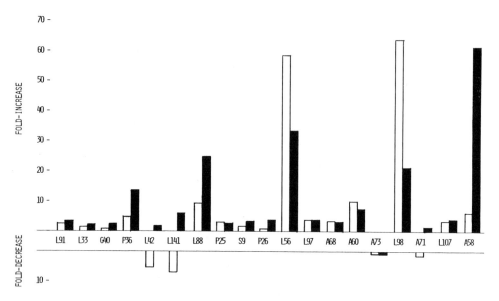

Fig. 2. The influence of T4 DNA polymerase alleles on frameshift mutation. The influence of ts gene 43 alleles on the revertant frequency of r131 (open bars) or on the average of FC47 and FC11 (closed bars) is compared to the wild-type enzyme. The mutant sites are presented in a linear array consistent with the map order determined by Allen et al. (1970). Relative increases in revertant frequencies are shown above the gene 43 allele; relative decreases are shown below. Revertant frequencies were determined from measurements of independent stocks, inoculated with approximately 10^3 phage particles and grown for 4 h at 30°C in E. coli BB cells. Final titers were approximately 2×10^{10}/ml. At least 4 stocks were measured for each combination of rII mutant and gene 43 allele. The median value was used. The revertant frequency of the r131 allele in the presence of the wild-type enzyme was 25×10^{-8}; the revertant frequency for FC47 was 55×10^{-8}, and for FC11 it was 33×10^{-8}. Increases or decreases of at least threefold were considered significant in these measurements.

compared to the wild type. These polymerase alleles also decrease revertant frequencies for spontaneously arising A·T to G·C transition mutations but do not decrease G·C-site transitions or transversion mutations arising spontaneously (Drake et al., 1969; Ripley, 1975).

THE IMPORTANCE OF DNA TARGETS IN POLYMERASE-ALTERED FRAMESHIFT FIDELITY

Because of the strong dependence of frameshift mutation on DNA sequence, it seemed important to determine whether any of the polymerases might also depend upon the exact DNA sequence for their influence upon mutation frequency. Thus, the frequency of revertants produced in the presence of each gene 43 allele was measured for both the FC47 (sign "+") and the FC11 (sign "-") frameshift mutations in the rIIB gene. Figure 1 demonstrates the mutational target for the suppression of these frameshift mutations. The solid bars in Fig. 2 demonstrate the average increase or decrease in suppressor frequency produced by the mutant polymerase alleles compared to the wild type for these two frameshifts. It is clear that in a number of cases the influence of mutant polymerase is profoundly different on these rIIB targets than on the r131 mutation. The A58 allele produced a much higher level of frameshift mutation in this region compared to that of the wild-type enzyme. Larger increases were seen in this region for L88 and P36 as well. The L42, L141, and A71 alleles, which produced lower revertant frequencies than the wild-type enzymes at the r131 site, did not lower revertant frequencies at the rIIB sites. In the case of the L141 allele, an increase of sevenfold compared to the wild-type enzyme was found.

The influence of DNA sequence upon frameshift frequencies produced in the presence of gene 43 mutant alleles was further investigated by subdividing the mutational targets into smaller subtargets. The targets for frameshift suppressors of the alleles used are shown in Fig. 1. Table 1 describes both the frequency of mutation in the presence of the wild-type enzyme and the relative increase in revertant frequency produced in the presence of five different polymerase alleles when compared to the wild type (no relative decreases being observed in these tests). A comparison between the different frameshift targets in the presence of the wild-type enzyme demonstrates the strong influence of DNA sequence on frameshift frequency. In the presence of the wild-type polymerase, the highest frequency of "-" frameshifts is seen for the smallest of the three targets. Except for the change in DNA sequence produced by the FC41 mutation itself, this target is totally contained within the larger FC40 and FC47 targets. Yet, the smaller target exhibits a three- to fivefold larger revertant frequency. Thus, a large fraction of frameshifts which suppress the FC41 frameshift appear to be due to the presence of the FC41 frameshift in the sequence.

Within all of the rIIB targets we have tested, the L56 and L98 alleles produce larger increases in genetically "-" than in genetically "+" frameshifts compared to the wild-type enzyme. Whether this reflects biases among these enzymes for creating particular frameshifts at certain DNA sequences or whether it might also

Table 1. The Frequency of "−" and "+" Frameshifts as a Function of T4 DNA Polymerase Allele and the DNA Target[a]

Polymerase	"−" Frameshifts				"+" Frameshifts		
Wild Type	(FC47)	(FC40)	(FC41)		(FC11)	(FC7)	(X146)
	55×10^{-8} (16)	36×10^{-8} (4)	156×10^{-8} (8)		33×10^{-8} (20)	28×10^{-8} (4)	5×10^{-8} (8)
L56	47(4)	99(4)	115(12)		18(11)	3(4)	1(4)
L98	34(4)	27(4)	93(8)		7(14)	2(4)	2(4)
L88	33(4)	8(4)	19(8)		15(8)	26(4)	25(4)
A58	31(4)	59(4)	59(8)		85(4)	30(4)	45(4)
L141	7(4)	1(4)	6(8)		4(17)	9(4)	13(4)

[a] The frequency of frameshift mutations in each target in the presence of the wild-type polymerase is shown at the top of the Table. The ratio of the median frameshift frequency for each mutant polymerase to the wild type is then given for each target. The number of stocks used to determine each median is given in parentheses. Target sizes and locations are given in Fig. 1. The method of measuring revertant frequencies is described in Fig. 2.

reflect a bias for the addition or deletion of particular numbers of bases will require the determination of the DNA sequence of the frameshift mutations. The L88 and A58 alleles produced large frequencies of frameshifts in all of these targets when compared to the wild-type enzyme. Although the specificity of infidelity of the A58 allele has not been well studied, it did not influence the revertant frequencies of two rII mutants capable of reverting by A·T and G·C site transition mutations, respectively (Drake et al., 1969). We plan to determine the specificity of mutational pathways promoted in the presence of this polymerase. The enzyme may represent a class of T4 DNA polymerase which is altered primarily in frame fidelity without a concomitant change in base-pair substitution fidelity. The L141 enzyme produced moderate increases within most targets. However, since the enzyme actually decreased frequencies of frameshifting in the r131 target (and at other targets, data not shown), this enzyme may nonetheless have strongly different properties of frameshift fidelity at different DNA sites. That this is indeed the case is confirmed below.

THE FINE-SCALE GENETIC IDENTIFICATION OF FRAMESHIFT TARGETS

We have determined the genetic locations of approximately 50 frameshift mutations arising in each of two DNA targets (FC11 and FC47) in the presence of the wild-type polymerase. In addition, we have determined the genetic locations of additional sets of approximately 50 mutants produced in each of three mutant polymerase backgrounds: L56, L88, and L141. Four DNA regions stand out as being frequent sites of frameshift mutation for both genetically "+" and genetically "-" frameshifts, since representative frameshifts were found in each of these four regions for nearly every combination of mutants tested.

Figure 3 demonstrates the frequency of frameshift mutation in each of these four DNA regions. On this fine-scale map of frameshift targets, we again find dramatically different effects of these polymerase alleles on frameshift mutations. Region 1 is defined by the inability of a suppressor frameshift isolated from FC47 to recombine with FC11 or by our inability to isolate a suppressor frameshift from FC11. Similarly, Region 4 is defined by the inability of a suppressor frameshift isolated from FC11 to recombine with FC47 or to be separated from FC47 by recombination. Regions 2 and 3 are defined by the inability of suppressor frameshifts to recombine with FC88 or with FC36, respectively.

The mutations mapping into any one of these four particular DNA regions are still not identical to each other. Those isolated from FC11 have the opposite genetic sign as those isolated from FC47. Even among mutations having a common sign, some subclasses can be

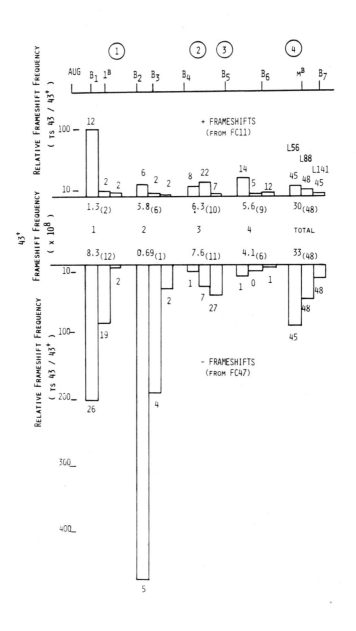

recognized by their genetic properties (e.g., spontaneous revertant frequency or recombination frequency with other nearby markers). However, the distance between the mutations in each region is quite small. The recombination tests that we are using easily resolve recombination values of 10^{-3} and lower. In T4, a recombination value of 10^{-3} is approximately that expected between two point mutations separated by 10 base pairs (Comer, 1977; Stahl et al., 1964), although site-to-site variation and marker effects can play either a positive or negative role, particularly at distances of less than ten base pairs. The four genetic regions described above account for a considerable fraction of the mutants we have isolated: 55 to 65% for the wild-type polymerase, 75 to 90% for the L56 allele, 65% for the L88 allele, and 50 to 70% for the L141 allele.

Each of the polymerase mutants compared to Fig. 3 demonstrates a unique mutational pattern. The L56 allele produced a very large (200-fold) increase in "-" frameshift mutations in Region 1, while increasing frameshift mutations in Region 3 by only ninefold compared to the wild-type enzyme. The wild-type enzyme produced frameshift mutations at similar frequencies in these two regions (8.3 x 10^{-8} and 7.6 x 10^{-8}, respectively). Thus, it might be surmised that Region 1 contains a DNA target particularly prone to L56-mediated infidelity.

Genetically "+" frameshifts produced in Regions 1 and 4 by the L88 allele are increased only slightly in frequency compared to the wild-type enzyme and at a much lower frequency than the production of frameshift mutations by L56 at those sites. However, the pro-

←―――――――――――――――

Fig. 3. Frameshift mutation frequencies in four rIIB regions. The influence of mutant polymerase alleles on frameshift mutation is shown for four regions. The map at the top of the figure indicates the approximate locations of these four regions. The central portion of the figure shows the frequency of frameshift mutations arising in the presence of the wild-type polymerase for each region; this frequency was determined by multiplying the total revertant frequency by the proportion of isolates in a given region among the total frameshifts characterized. The bars represent the relative increases in frameshift frequency measured in each region in the presence of the L56, L88, and L141 alleles, respectively, when compared to the wild type. The "+" frameshifts arising in the FC11 mutant are shown above the central area; the "-" frameshifts arising in FC47 are shown below. The frameshift frequencies for the entire target areas are labeled "total." The numbers in parentheses, or above bars, represent the number of characterized mutations.

production of "-" frameshifts in Region 1 is relatively very large compared to the wild-type enzyme. Thus, there appears to be a site in Region 1 at which L88 is particularly prone to create "-" frameshift mutations, but the site may not exist for the creation of "+" frameshift mutations due to the presence of the FC11 allele or perhaps there is a site in the region at which L88 creates only genetically "-" frameshifts.

One of the most striking examples of site specificity is seen for the L141 enzyme and was predicted on the basis of the experiments in Table 1. This enzyme increases the frequency of "-" frameshift mutations in Regions 1 and 4 by only threefold compared to the wild type (an insignificant increase in each individual case, since only one or two frameshifts were isolated). In contrast, in Region 3, 27 frameshifts were isolated from L141. The frequency of mutation for L141 in this region was 43-fold greater than was the wild-type polymerase. Furthermore, when the 27 frameshifts were mapped, two distinct classes of frameshifts were found, which accounted for 25 of the frameshifts. When the "-" frameshifts produced by L88 in this region were mapped, we found that among seven isolated, six mapped into just one of the two subclasses (the 7th mutant mapped at a 3rd location). The fact that there are at least two distinct mutational outcomes to the rather unique stimulation of frameshift mutations by the L141 allele may give us a clue as to some common step between the two mutational classes. The fact that L88 appears to stimulate only one of these events may help us deduce the common and unique aspects of infidelity exhibited by the L88 enzyme compared to the L141 and wild-type polymerases.

WILD-TYPE DNA SEQUENCES AND FRAMESHIFT TARGETS

While we have described, using genetic techniques, the approximate locations of frameshift mutations and their frequencies, a more precise description of their qualities depends upon determining their DNA sequences. A description of the specific physical addition or deletion nature of the frameshift and the precise DNA sequence should suggest clues from which further experiments may determine the mechanisms by which defects in DNA polymerase are responsible for increased frameshift mutation.

Thus far, we have examined the wild-type DNA sequences in the vicinity of the four genetically defined regions of DNA responsible for more than half of the mutants isolated. As discussed earlier, the FC11 frameshift, which fails to recombine with Region 1 frameshift mutations, is likely to be located in a run, or very near to a run, of 5A's in the wild-type DNA sequence. Region 2 frameshifts fail to recombine with FC88 which lies between barriers B_4 and B_5 (Barnett et al., 1967). Pribnow et al. (1981) have suggested a run of 3 C's as a possible location for that frameshift. Frameshift mutations mapping into Region 3 lie in or near a complex DNA sequence containing a tandem repeat of six bases separated by two bases. Such a sequence is an obvious candidate for the production of frameshift mutations via misannealed DNA intermediates. However, among the frameshifts which we have isolated, there appear to be at least four classes of "+" frameshifts and four or more classes of "-" frameshifts. If all of these different mutations lie within the 14 base pairs bounded by the tandem repeat, then a large number of different mutagenic outcomes are possible in this region. The stimulation of only two of these classes by L141 may allow us to identify a common step between two of the frameshift outcomes. Region 4 frameshifts fail to recombine with FC47. From genetic arguments, this mutation is almost surely located in the run of 4T's adjacent to the M^B barrier (Pribnow et al., 1981). Many of the mutations in this region may map into this sequence. One base pair to the right of the T's is also a run of 5 A's but on the right side of the M^B barrier. The A run would be a target for "-" frameshifts but not for "+" frameshifts (unless they remove the M^B barrier).

We have recently begun to investigate the possibility that potential DNA secondary ("hairpin") structures may define a class of DNA targets having high probabilities of producing addition or deletion mutations. Our hypothesis developed from the observation that the frameshift sites in Regions 2 and 3 discussed above were brought into juxtaposition by the formation of such a structure, and that the structure, at least if formed in RNA, would have a net free energy of approximately -14 Kcal (Tinoco et al., 1975) (see Fig. 4). Several potential structures might actually form, each having somewhat different, but still substantial, negative free energies. Such a diversity of potential structures might be the structural precursors to the diversity of frameshift mutations detected in Region 3.

DNA secondary structure might potentiate the addition or deletion of bases through a number of DNA metabolic mechanisms. The potential for deletion and/or addition by means of nicking followed by either excision and/or gap filling becomes immediately obvious. Furthermore, unpaired DNA loops might be particularly prone to single-strand invasion as a precursor to recombination; thus mutational mechanisms involving illegitimate recombination might be enhanced. In this particular DNA secondary structure, a tandem 6-base repeat separated by two bases at barrier 5 has its ends in unpaired regions. We are particularly interested in investigating the role that such secondary structure might play in the frameshift fidelity of polymerase mutants, since there is some indication from in vitro experiments that at least one polymerase mutant is likely to be defective in its replication of polynucleotides containing secondary structure (Gillin and Nossal, 1976). We intend to explore the importance of secondary structures in frameshift mutation. If a role for secondary structure is confirmed, it would constitute a

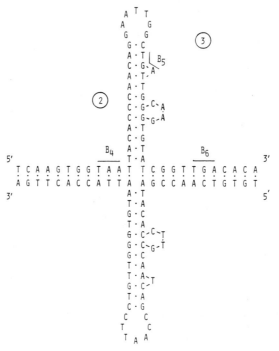

Fig. 4. One isomer of potential secondary structure in a frameshift-prone DNA sequence in the rIIB gene of bacteriophage T4. Using the wild-type DNA sequence of Pribnow et al., (1981) the structure shown in the figure can be drawn between the B_4 and B_6 barriers in the rIIB gene. Frameshifts in Region 2 lie to the right of B_4 and fail to recombine with the frameshift mutation FC88 which may be in the run of 3C's (Pribnow et al., 1981). Region 3 is identified by the inability of frameshifts to recombine with FC36 which is likely to be located in a tandemly repeated sequence ATTGGC falling on both sides of B_5.

novel approach to the identification of the structural precursors to addition and deletion mutations and furthermore would provide evidence for the actual in vivo existence of such secondary DNA structure.

ACKNOWLEDGMENTS

The authors appreciate the assistance of Mr. Gary Bayliss in performing these experiments. We are especially grateful to

Drs. David Pribnow and Britta Singer for sharing their DNA sequence and a number of rII mutants used in these experiments.

REFERENCES

Allen, E. F., Albreght, I., and Drake, J. W., 1970, Properties of bacteriophage T4 mutants defective in DNA polymerase, Genetics, 65:187.

Barnett, L., Brenner, S., Crick, F. H. C., Shulman, R. G., and Watts-Tobin, R. J., 1967, Phase-shift and other mutants in the first part of the rIIB cistron of bacteriophage T4, Philos. Trans. R. Soc. London, Ser. B, 252:487.

Benzer, S., 1961, On the topography of the genetic fine structure, Proc. Natl. Acad. Sci. U.S.A., 47:403.

Bernstein, H., 1971, Reversion of frameshift mutations stimulated by lesions in early function genes of bacteriophage T4, J. Virol., 7:460.

Champe, S. P., and Benzer, S., 1962, An active cistron fragment, J. Mol. Biol., 4:288.

Comer, M. M., 1977, Correlation between genetic and nucleotide distances in a bacteriophage T4 transfer RNA gene, J. Mol. Biol., 113:267.

Crick, F. H. C., Barnett, L., Brenner, S., and Watts-Tobin, R. J., 1961, General nature of the genetic code for proteins, Nature, 192:1227.

Drake, J. W., Allen, E. F., Forsberg, S. A., Preparata, R.-M., and Greening, E. O., 1969, Spontaneous mutation, Nature, 221:1128.

Edgar, R. S., and Lielausis, I., 1964, Temperature-sensitive mutants of bacteriophage T4D: Their isolation and genetic characterization, Genetics, 49:649.

Efstratiadis, A., Posakony, J. W., Maniatis, T., Lawn, R. M., O'Connell, C., Spritz, R. A., DeRiel, J. K., Foget, B. G., Weissman, S. M., Slighton, J. L., Blechl, A. E., Smithies, O., Baralle, F. E., Shoulders, C. C., and Proudfoot, N. J., 1980, The structure and evolution of the human β-globin gene family, Cell, 21:653.

Farabaugh, P. J., Schmeissner, U., Hofer, M., and Miller, J. H., 1978, Genetic studies of the lac repressor. VII. On the molecular nature of spontaneous hotspots in the lacI gene of Escherichia coli, J. Mol. Biol., 126:847.

Gillin, F. D., and Nossal, N. G., 1976, Control of mutation frequency by bacteriophage T4 DNA polymerase. I. The CB120 antimutator DNA polymerase is defective in strand displacement, J. Biol. Chem., 251:5219.

Isono, K., and Yourno, J., 1974, Chemical carcinogens as frameshift mutagens: Salmonella DNA sequence sensitive to mutagenesis by polycyclic carcinogens, Proc. Natl. Acad. Sci. U.S.A., 71:1612.

Koch, R. E., and Drake, J. W., 1973, Ligase-defective bacteriophage T4. I. Effects on mutation rates, J. Virol., 11:35.

Koch, R. E., McGaw, M. K., and Drake, J. W., 1976, Mutator mutations in bacteriophage T4 gene 32 (DNA unwinding protein), J. Virol., 19:490.

Okada, Y., Streisinger, G., Owen, J. (Emrich), Newton, J., Tsugita, A., and Inouye, M., 1972, Molecular basis of a mutational hot spot in the lysozyme gene of bacteriophage T4, Nature, 236:338.

Pribnow, D., Sigurdson, D. C., Gold, L., Singer, B. S., Brosius, J., Dull, T. J., and Noller, H. F., 1981, The rII cistrons of bacteriophage T4: DNA sequence around the intercistronic divide and positions of genetic landmarks, J. Mol. Biol., 149:337.

Ripley, L. S., 1975, Transversion mutagenesis in bacteriophage T4, Mol. Gen. Genet., 141:23.

Roth, J. R., 1974, Frameshift mutations, Annu. Rev. Genetics, 8:319.

Speyer, J. F., Karam, J. D., and Lenny, A. B., 1966, On the role of DNA polymerase in base selection, Cold Spring Harbor Symp. Quant. Biol., 31:693.

Stahl, F. W., Edgar, R. S., and Steinberg, J., 1964, The linkage map of bacteriophage T4, Genetics, 50:539.

Stewart, J. W., and Sherman, F., 1974, Yeast frameshift mutations identified by sequence changes in iso-1-cytochrome c, in: "Molecular and Environmental Aspects of Mutagenesis," L. Prakash, F. Sherman, M. W. Miller, C. W. Lawrence, and H. W. Taber, eds., C. C. Thomas, Springfield, Illinois.

Streisinger, G., Okada, Y., Emrich, J., 1966. Frameshift mutations and the genetic code, Cold Spring Harbor Symp. Quant. Biol., 31:77.

Tinoco, I., Borer, P. N., Dengler, B., Levine, M. D., Uhlenbeck, O. C., Crothers, D. M., and Gralla, J., 1973, Improved estimation of secondary structure in ribonucleic acids, Nature (London), New Biol., 246:40.

CHAPTER 11

IN VITRO REPLICATION OF MUTAGEN-DAMAGED DNA:

SITES OF TERMINATION

>Peter D. Moore, Samuel D. Rabkin, and Bernard S. Strauss
>
>Department of Microbiology, The University of Chicago
>Chicago, Illinois   60637

SUMMARY

We have examined the effect of DNA lesions, which in vivo are potentially mutagenic, on in vitro DNA synthesis carried out by a number of purified DNA polymerases using a ØX174 template. Both acetyl aminofluorene (AAF) adducts and UV-induced pyrimidine dimers are blocks to elongation by DNA polymerases. On UV-irradiated DNA templates synthesis terminates one nucleotide before the sites of pyrimidine dimers with all of the enzymes tested: Pol I and Pol III holoenzyme from <u>Escherichia coli</u>, T4 DNA polymerase, avian myeloblastosis virus reverse transcriptase and a mammalian DNA polymerase α. With AAF, which reacts at the C-8 position of guanine, differences are observed between the above enzymes, with the latter two inserting a nucleotide opposite the site of the lesion. Substitution of Mn^{2+} for Mg^{2+} as the cation in the Pol I reactions causes changes in the termination pattern on both UV-irradiated and AAF-reacted templates. The significance of these results to the process of inducible error-prone repair and the possible bypass of lesions in the DNA is discussed.

INTRODUCTION

Error-Prone Repair

While some mutagens cause relatively subtle changes in the bases of DNA, leading to mutation directly via mispairing during replication, others, e.g., UV-induced pyrimidine dimers, result in lesions that effectively destroy coding capacity and block DNA synthesis. The production of mutations by this latter type of mutagen requires the involvement of an error-prone repair process, one of several "SOS" functions that are coordinately induced after the introduction of DNA damage, require protein synthesis, and are regulated by the lexA and recA genes (Radman, 1974; Witkin, 1976). At least two other SOS functions, an inducible inhibitor of exonuclease V (Volkert et al., 1976) and increased production of the uvrA and uvrB gene products (Fogliano and Schendel, 1981; Kenyon and Walker, 1980) may be involved in DNA repair. It is currently thought that error-prone repair involves the induction of a protein(s) that either independently or in conjunction with DNA polymerase III allows DNA synthesis to bypass pyrimidine dimers and other noncoding lesions that are normally blocks to the cell DNA polymerases. A strong candidate for such a protein is the umuC gene product, which is induced after UV irradiation, regulated by lexA, and the only SOS functions in which it participates are UV-induced mutagenesis and enhancement of UV survival (Kato and Shinoura, 1977).

How Much Repair?

Bacteria have several error-free mechanisms for repairing pyrimidine dimers (Hanawalt et al., 1979). Photoreactivation enzyme plus visible light simply reverses the dimer chemically, while enzymatic excision of dimers from DNA is accurate so long as an intact complementary DNA strand exists. Replication of unexcised dimers leads to the formation of daughter-strand gaps, opposite the dimers, in the newly synthesized DNA. These can be repaired, again accurately, by recombination with the homologous sister DNA molecule. The involvement of the error-prone repair system becomes necessary when the damage results in complete loss of informational capacity at a particular point in the chromosome. Such situations may arise (Fig. 1) when the gap formed by the excision of one pyrimidine dimer extends past another dimer in the opposite strand forming an overlapping excision gap (OEG) or when the postreplication gaps opposite two closely spaced dimers overlap (ODSGs or overlapping daughter-strand gaps). Similarly, the replication of single-stranded DNA, such as ØX174 phage, is blocked by a single dimer (Doubleday et al., 1981) that cannot be circumvented with accurate conservation of information.

While repair of most daughter-strand gaps is constitutive, dependent on the $recA^+$ gene product, and involves recombination (Rupp et al., 1971; Sedgwick, 1975), the repair of some gaps is dependent on postirradiation protein synthesis and the $lexA^+$, as well as the $recA^+$, genotype (Sedgwick, 1975; Youngs and Smith, 1976). The number of gaps that require protein synthesis for repair increases in a dose-squared manner and correlates with the frequency that might be expected for ODSGs (Sedgwick, 1976). Similarly, there appear to be two components of excision repair: the major,

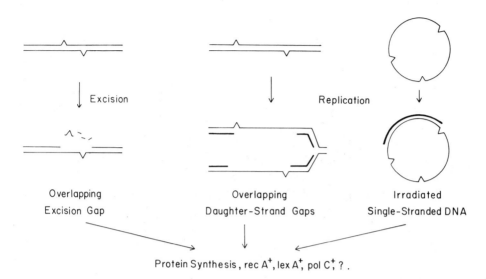

Fig. 1. Possible substrates for error-prone repair activity. The physical integrity of the DNA cannot be restored without losing informational content at the site of the lesion.

constitutive one having a short excision patch size of 5 to 20 nucleotides, and a second component with a patch size of 200 or more that is responsible for repair of only a small fraction of dimers. This "long-patch excision" repair requires protein synthesis, is $lexA^+$ and $recA^+$ dependent, and utilizes DNA polymerase III (Hanawalt et al., 1979). The similarity of the requirements for long-patch excision and the rejoining of ODSGs to those for the production of UV-induced mutagenesis suggest strongly that OEGs and ODSGs are indeed the sites of action of the error-prone repair system. Further, although the error-prone repair gene umuC has not yet been shown to be necessary for repair of ODSGs, the introduction of a umuC mutation into an excision-deficient bacterium results in a degree of sensitivity of UV compatible with the prediction that a single ODSG would be lethal in this strain (Table 1). A similar level of sensitivity is observed for a uvrA strain carrying lexA rnm, a derivative of lexA deficient in error-prone repair but normal for a different repair-associated SOS function (Volkert et al., 1976).

If ODSGs are indeed the lethal lesion repaired by the error-prone repair system (at least in excision-deficient bacteria), the cell has a substantial capacity for their biochemical restoration: at least 50 to 100 strand breaks (25 to 50 ODSGs) per chromosome, as measured by sucrose gradients, at doses between 9 to 12 J m^{-2}

Table 1. D_{37} of E. coli Repair Deficient Strains

Strain	D_{37} (J m^{-2})[a]	Pyrimidine Dimers[b]	Overlapping Excision Gaps[c]	Overlapping Daughter-Strand Gaps[c]
Wild type	50	2800	5.1	—
uvrB	3	168	—	6.6
uvrB umuC	0.8	45	—	0.5
uvrA recA	0.02	1.1	—	—

[a] Survival data from Howard-Flanders (1968) and Kato and Shinoura (1977).
[b] Number of dimers calculated as in Unrau et al. (1973).
[c] Overlapping gaps calculated from d^2g/m, where $\underline{d}$ is the number of dimers, $\underline{m}$ is the number of nucleotides in the E. coli genome (7.7 x 10^6), and $\underline{g}$ is the size of the gap, assumed to be 5 nucleotides for excision gaps and 1800 for daughter-strand gaps (Sedgwick, 1976).

(Sedgwick, 1976; Youngs and Smith, 1976). However, the biological capacity (i.e., in terms of killing protection) of the error-prone repair system appears much more limited. From the increase in survival due to induced repair, the average number of lethal hits removal per cell (ΔN) can be calculated (Table 2). It can be seen that in an excision-deficient bacterium the inducible error-prone repair system can remove at most two to three lethal lesions from either its own DNA or from λ phage DNA. While at low doses almost all the lesions are repaired, the number of lesions removed does not increase at higher doses. For irradiated single-stranded $\emptyset$X174 DNA, the number of lesions repaired is only one or two. For excision-proficient strains, the increase in survival due to inducible repair again appears to be limited and a maximum capacity of five to six lesions removed can be estimated from published data (Ganesan and Smith, 1972; Sedliakova et al., 1978). It would therefore appear that while the cell has an extensive capacity for the inducible repair of daughter-strand gaps, the concomitant enhancement of survival is much more limited.

How Error Prone is Error-Prone Repair?

Both transitions and transversions are found among the base-substitution mutations induced by UV light (Coulondre and Miller, 1977). The GC → AT transition is the most common change found although the AT → GC transition (which cannot be detected in mutations to amber or ochre) has not been adequately studied. The relatively high frequency of transversions (35 to 40%) induced by UV implies a greater loss of coding specificity for pyrimidine dimers than for damage introduced by mutagens such as ethyl methanesulfonate or 2-aminopurine, which do not act via the error-prone repair pathway. UV induces base substitutions ten times more frequently at positions of potential pyrimidine dimer formation than at other sites (Coulondre et al., 1978); however, a substantial amount of untargeted mutagenesis also occurs (Coulondre et al., 1978; Doubleday et al., 1981). Tandem double-base substitutions occur with UV and, where it is possible to compare within the same codon, such doubles account for about 10% of all base-substitution mutation (Coulondre and Miller, 1977). However, this figure may be an underestimate since, in scoring mutation to amber at a particular codon, one in three possible single substitutions may be detected while only one in nine tandem doubles can be recognized. Some untargeted tandem doubles (Doubleday et al., 1981) and some nontandem doubles also occur (Coleman et al., 1979).

With a clearly high rate of base substitution resulting from error-prone repair of pyrimidine dimers, it is of interest to know how frequently these result in phenotypically detectable mutation and, in particular, mutation leading to death of the organism. An examination of the genetic code indicates that elimination of base-

Table 2. The Number of Lethal Lesions Removed by Error-Prone Repair

Genome	Dose (J m^{-2})	N^a (No Repair)	ΔN^b	Reference
E. coli uvrB Postirradiation incubation in absence vs presence of chloramphenicol	3.8 7.6 11.4	2.5 5.8 8.6	2.2 2.7 3.1	Sedliakova et al. (1978)
λ Phage Assayed on UV-induced vs uninduced uvr$^-$ E. coli	50	7.8	2.3	Defais et al. (1976)
ØX174 DNA Assayed on UV-induced vs uninduced E. coli spheroplasts	30 60 90	2.3 4.4 5.8	0.9 1.6 1.5	Doubleday et al. (1981)

^{a}Average number of lethal hits per genome (N) calculated from surviving fraction (F) in absence of error prone repair (i.e., with addition of chloramphenicol or without an inducing irradiation of the host): N = -ln F̄.

^{b}Number of lesions removed, ΔN = N - N' = ln F' - ln F, where N' and F' are, respectively, the number of lethal hits remaining and surviving fraction under error-prone repair conditions.

pairing constraints at the site of a lesion results in a high probability (especially for tandem doubles) of amino acid alteration due to base substitution (Table 3). A minimum estimate of the deleterious effects of a codon alteration can be taken as 0.05 (3/63, the frequency of nonsense codons). The overall extent of deleterious effects due to amino acid alteration will vary with the composition and functional flexibility of particular proteins; however, for the lacI gene nonsense mutations were found to represent about 10% of all base substitutions that inactivated the protein (Eisenstadt et al., 1981). Since the number of codons that can mutate to nonsense in this gene is about that expected at random, a frequency of 0.5 may be a reasonable estimate for deleterious base substitutions.

From the estimates in Table 3 it is not hard to imagine not only that the probability of generating phenotypically detectable mutations is high, but, depending on the organism, mutations leading to loss of viability might be expected from repair of as few as four or five dimers. This could account for the discrepancy between the small effect of error-prone repair on survival and the high capacity for inducible repair of daughter-strand gaps. Thus the error-prone system, while being an efficient mechanism for restoring the physical integrity of the DNA, may exact such a high price in loss of informational accuracy that it may handle only a few lesions with reasonable chances of survival for the organism.

We have set out to investigate error-prone repair by studying the behavior of various purified DNA polymerases on encountering noncoding lesions in the DNA, attempting to distinguish the properties of the enzymes that determine termination at the site of a lesion, and seeking conditions that might allow an enzyme to bypass lesions. This approach, we hope, may lead to a method for directly assaying error-prone repair activity in extracts from induced bacteria.

GENERAL PROCEDURES

Enzymes

Pol I, large fragment (Klenow) was purchased from Boehringer-Mannheim. T4 DNA polymerase was a gift from Nicholas Cozzarelli, Pol III holoenzyme was a gift from U. Hübscher, and avian myeloblastosis virus (AMV) reverse transcriptase was provided by J. W. Beard. DNA polymerase α from Daudi human lymphoma cells (Bose et al., 1978) was prepared by K. Bose.

Table 3. Estimation of Phenotypically Detectable Mutation

	Probability of Amino Acid Change[a]		
	None	Single	Double
Single-base random selection	0.433	0.567	—
Tandem double-base random selection	0.166	0.727	0.107
	Estimate of Probability of Protein Loss of Function[b]	0.834	
Mutation to nonsense codon	0.050		
All missense mutations	0.50		

[a] Probability of mutational change to triplet specifying a different amino acid (or nonsense): averaged for all amino acid coding triplets. Calculation assumes error-prone repair inserts a nucleotide(s) at random at the site of one (or two adjacent) mutilated bases.

[b] Assuming (i) mutation to nonsense (3 out of 63 possible mutant codons) causes loss of function, and (ii) nonsense mutations can account for as few as 10% of lacI⁻ mutations caused by base substitution (Eisenstadt et al., 1981).

DNA Polymerase Reactions

Construction of the primed ØX174 DNA templates has been described (Moore and Strauss, 1979; Moore et al., 1980). Reactions for sequencing gels were carried out as described (Moore and Strauss, 1979). The following amounts of enzyme, chosen to give maximum incorporation of label, were used in 10 µl of reaction mix: Pol I (0.2 units), Pol III (52 units), T4 DNA polymerase (0.36 units), AMV reverse transcriptase (19 units) or DNA polymerase α (2.4 units). Sequence standards were synthesized by Pol I using chain-terminating nucleotides (Sanger et al., 1977).

Sequence Gels

One microgram of sonicated DNA was added to the samples, which were diluted 1:1 with 8 M urea/0.2 N NaOH/0.005% bromophenol blue/ 0.005% xylene cyanol FF, and analyzed on a 20% polyacrylamide gel (Moore and Strauss, 1979).

TERMINATION OF SYNTHESIS BY POL I AT DNA LESIONS

Our strategy to investigate the behavior of DNA polymerases at lesions such as pyrimidine dimers or adducts of acetyl aminofluorene (AAF) has been to use an in vitro system with primed ØX174 DNA as a template. A single, purified restriction fragment is annealed to the single-stranded phage DNA so that synthesis starts at a unique site on the template, enabling the newly synthesized products to be analyzed by sequencing techniques (Moore and Strauss, 1979). On an unreacted template DNA, polymerase I from Escherichia coli (Pol I) completely replicates the single-stranded region to give a fully double-stranded circular product (Moore et al., 1980). Reaction of the DNA with N-acetoxy-AAF leads to the formation of adducts of AAF at the site of guanine in the DNA (Kriek, 1974), and under our conditions of reaction more than 95% of the product is N-acetyl-N-(guan-8-yl)-2-aminofluorene (P. Moore and A. Osborn, unpublished data). These adducts are blocks to synthesis by Pol I and termination is found to take place primarily one nucleotide before the site of a reacted guanine (Moore and Strauss, 1979). Similarly, on UV-irradiated templates Pol I terminates one nucleotide before the first of two adjacent pyrimidines, presumably dimerized in the template. All three types of dimer, cystosine:cytosine, cytosine:thymine, and thymine:thymine, block synthesis, and the intensities of the termination bands at the different types of dimer appear to reflect their relative rates of dimerization in the DNA (Unrau et al., 1973). Since reaction with AAF appears to be equally efficient at all the guanines in the DNA, it has been possible to use AAF-reacted templates to quantitate termination. We have concluded (Moore et al.,

1980) that termination is essentially 100% both for Pol I and the DNA polymerase III holoenzyme from E. coli.

THE PATTERN OF TERMINATION OF DIFFERENT DNA POLYMERASES

We have investigated the termination of several DNA polymerase enzymes, from both bacterial and mammalian sources. On a UV-irradiated template (Fig. 2) a series of termination bands, not found with unirradiated DNA, are generated by all five enzymes: Pol I, Pol III holoenzyme, T4 DNA polymerase, AMV reverse transcriptase, and DNA polymerase α purified from human lymphoma cells. While there are differences in intensity of bands between the enzymes, which simply reflect the polymerizing activities of each enzyme, termination bands are observed at identical sites for all five enzymes. The sites of termination occur, as previously reported for Pol I (Moore and Strauss, 1979), one nucleotide before the sites of pyrimidine dimers.

On AAF-reacted DNA (Fig. 3) termination bands are observed with all five enzymes with the bands corresponding to the sites of guanine (presumably guanyl-AAF) in the template sequence. For Pol III holoenzyme and T4 polymerase, synthesis terminates one nucleotide before the guanyl-AAF adduct. For Pol I also, the principal site of termination is one nucleotide before the reacted guanine, but there is also a slight secondary band opposite the guanine, e.g., positions 34 and 43 (Moore et al., 1981). With AMV reverse transcriptase, termination bands are found at every guanine (the absence of bands at positions 19 and 20 was peculiar to this one out of many AMV experiments), but with this enzyme termination has occurred with an additional nucleotide inserted opposite the guanyl-AAF. The position of the termination bands with the α polymerase varies. At some sites termination occurs one nucleotide before the AAF adducts and at others an additional nucleotide is inserted. This variation may be related to the specific sequence at different sites of the adducts (Moore et al., 1981).

In order to determine the specificity, if any, of the nucleotide inserted by AMV reverse transcriptase at the site of a guanyl-AAF

Fig. 2. Sequence gel analysis of DNA synthesized on either untreated or UV-irradiated templates by Pol I, AMV reverse transcriptase, Pol III holoenzyme, DNA polymerase α, and T4 DNA polymerase. The templates were primed with HaeII restriction fragment 5 and UV irradiation was c. 1500 J m^{-2} Lanes A, C, G, and T are sequence standard synthesized by Pol I with chain-terminating nucleotides. The sequence given is of the synthesized strand and is numbered from the site of the HaeII recognition cut.

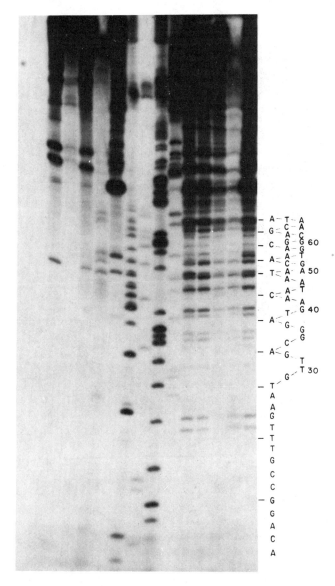

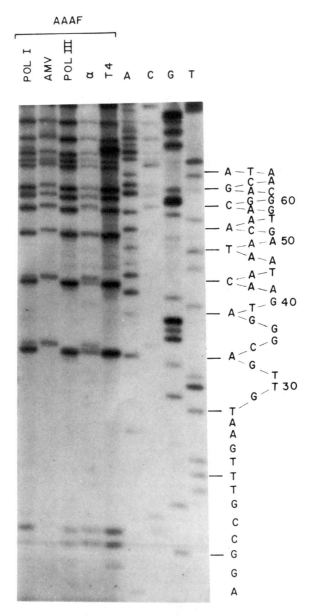

Fig. 3. Sequence gel analysis of DNA synthesized on a template containing 113 AAF residues per ØX174 molecule. Other details as for Fig. 2.

the polymerization reaction was carried out in two stages. In the
first stage of the reaction, synthesis was carried out with T4 DNA
polymerase under standard reaction conditions. The product DNA was
then isolated and purified free of contaminating nucleotides by gel
filtration on a 1-ml Sephadex G50 column. In the second stage of
the reaction, reverse transcriptase was reacted with aliquots of the
DNA either without or with added deoxynucleoside triphosphates
(Fig. 4). In the absence of a second polymerization step (Lane 1),
termination bands are found, as expected for the T4 reaction, one
nucleotide before the sites of guanyl-AAF, while a second reaction
with AMV reverse transcriptase plus all four deoxynucleoside triphosphates leads insertion of an extra nucleotide (Lane 7). When
added individually, only deoxycytidine triphosphate (Lane 4) allows
reverse transcriptase to elongate. Examination of Lanes 2-6 of this
gel also indicates unexpectedly that this preparation of reverse
transcriptase, in the absence of deoxynucleoside triphosphates, is
capable of removing one to three nucleotides from the 3' terminus
of the DNA.

CHANGES IN THE PATTERN OF TERMINATION BY POL I

Substituting Mn^{2+} for Mg^{2+} can result in reduced fidelity of
DNA synthesis by Pol I and other polymerases on intact DNA templates
(Sirover et al., 1979; Weymouth and Loeb, 1978). When Pol I-
directed synthesis on an AAF-reacted template takes place in the
presence of Mn^{2+} (0.5 or 1.0 mM) instead of Mg^{2+} (8 mM), termination
occurs but predominantly with a nucleotide incorporated opposite the
guanyl-AAF (Moore et al., 1981). Although the termination pattern
superficially resembles that obtained with AMV reverse transcriptase,
there are significant differences, notably a reduction in intensity
or possibly complete loss of termination bands at a few specific
sites (Moore et al., 1981). Another difference is that Mn^{2+} also
changes the pattern of termination by Pol I on UV-irradiated DNA
(Fig. 5). In contrast to the AMV reverse transcriptase pattern,
Pol I in the presence of Mn^{2+} terminates with a nucleotide inserted
opposite the first of two pyrimidines in a dimer. Interestingly,
the use of Mn^{2+} produces no changes in the normal AMV reverse transcriptase pattern.

CONCLUSIONS

It has been suggested (Villani et al., 1978) that termination
of DNA replication at pyrimidine dimers may be due to the 3' → 5'
proofreading exonuclease activity associated with some DNA polmyer-
ases. To test this hypothesis we have examined termination of synthesis by several DNA polymerases in an in vitro system. Both UV-
induced pyrimidine dimers and guanyl-AAF adducts block DNA synthesis.

All the enzymes tested, including DNA polymerase III holoenzyme from E. coli and a human DNA polymerase α, terminate synthesis one nucleotide before the first of two pyrimidines in a UV-induced dimer. No 3' → 5' exonuclease activity has been detected in the α polymerase preparation, even on a substrate that would allow removal of a single nucleotide to be observed (Moore et al., 1981). On AAF-reacted DNA, again termination is observed for all five enzymes but here interesting differences are found. Specifically, in contrast to Pol III holoenzyme and T4 DNA polymerase which terminate abruptly one nucleotide before the AAF adduct, AMV reverse transcriptase inserts cytosine opposite the AAF-reacted guanine but does not elongate further. These results imply not only that 3' → 5' exonuclease is not essential for termination at lesions but that termination can be independent of correct base selection by the polymerase. Thus under certain circumstances the inability to continue synthesis may be a function of the specific spatial conformation of the stalled polymerase and damaged base with regard to the template and primer DNA strands. The possible effect of neighboring bases on the pattern of termination by DNA polymerase α on AAF-treated DNA could also result from conformation differences due to stacking interactions (Moore et al., 1981).

Substitution of Mn^{2+} for Mg^{2+}, which changes the termination pattern of Pol I allowing insertion of an extra nucleotide, is accompanied by an increase in the exonuclease/polymerase activity ratio for this enzyme (Moore et al., 1981). This result, together with the observation that the AMV reverse transcriptase preparation may have a small amount of 3' → 5' exonuclease activities, suggests that the presence or absence of a proofreading nuclease does not play a critical role in determining even the position of termination, i.e., opposite or one nucleotide before a lesion. While the results on UV-irradiated DNA clearly distinguish Pol I synthesizing in the presence of Mn^{2+} from the AMV reverse transcriptase reaction, it is not apparent whether either situation bears any relation to bypass of lesions. Quantitative studies to determine whether termination

Fig. 4. The specificity of terminal nucleotide insertion by AMV reverse transcriptase on an AAF-treated template. Synthesis was first carried out under normal reaction conditions by T4 DNA polymerase. The template was then isolated and purified free of nucleotides by gel filtration through a Sephadex G50 column. In the second stage of the reaction, the DNA was incubated in polymerase reaction buffer at 37°C for 15 min. Lane 1 is the control with no enzyme. Lanes 2-7 all contained 19 units of AMV reverse transcriptase with the following nucleotides added at 50 µM: Lane 2-none; Lane 3-dATP; Lane 4-dCTP; Lane 5-dGTP; Lane 6-TTP; Lane 7-all four deoxynucleoside triphosphates.

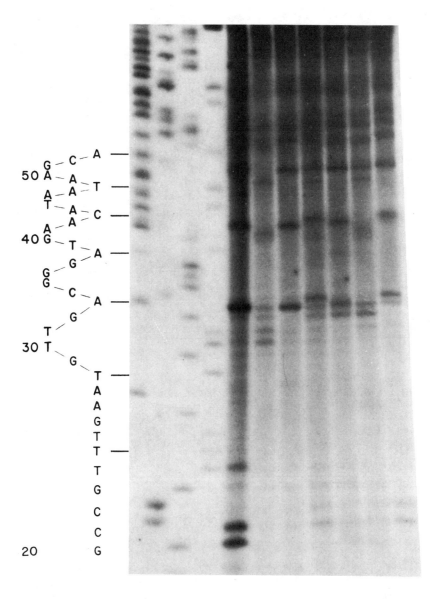

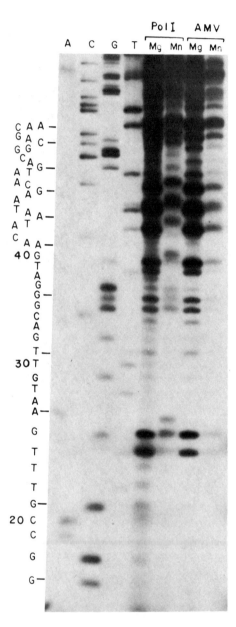

Fig. 5. The effect of divalent cation on sites of termination by Pol I and AMV reverse transcriptase on UV-irradiated DNA. Conditions were as described except for replacement of 8 mM $MgCl_2$ by 0.5 mM $MnCl_2$ when indicated.

in these situations is 100% have not yet been completed   Ultimately we expect that this system will prove successful in detecting "bypass synthesis" (if it occurs) in fractions or extracts of cells induced for error-prone repair. We are currently adapting our techniques to these ends.

ACKNOWLEDGMENTS

This work was supported by National Institutes of Health grants GM 07816, CA 14599, and CP 85669, and by Department of Energy grant EV 765-02-2040. We are grateful to Ann Osborn for skilled assistance and to Kallol Bose for the preparation of DNA polymerase α.

REFERENCES

Bose, K., Karran, P., and Strauss, B. S., 1978, Repair of depurinated DNA in vitro by enzymes purified from human lymphoblasts, Proc. Natl. Acad. Sci. U.S.A., 75:794.

Coleman, R. D., Dunst, R. W., and Hill, C. W., 1979, UV-induced double base change in a glycine tRNA gene, Mol. Gen. Genet., 177:213.

Coulondre, C., and Miller, J. H., 1977, Genetic studies of the lac repressor. IV. Mutagenic specificity in the lacI gene of Escherichia coli, J. Mol. Biol., 117:577.

Coulondre, C., Miller, J. H., Farabaugh, P. J., and Gilbert, W., 1978, Molecular basis of base substitution hot spots in Escherichia coli, Nature, 274:775.

Defais, M., Caillet-Fauquet, P., Fox, M. S., and Radman, M., 1976, Induction kinetics of mutagenic DNA repair activity in E. coli following ultraviolet irradiation, Mol. Gen. Genet., 148:125.

Doubleday, O. P., Brandenberger, A., Godson, G. N., Wagner, J. R., and Radman, M., 1981, DNA repair and mutagenesis of single-stranded bacteriophages, in: "Gann Monograph on Cancer Research," C. Heidelberger, N. Inui, T. Kuroki, and M. Yamada, eds., University Park Press, Baltimore, in press.

Eisenstadt, E., Warren, A. J., Atkins, D., Porter, J., and Miller, J., 1981, Mutational specificity of carcinogens in Escherichia coli, J. Supramol. Struct. Suppl. 5 (Cell. Biochem.), 204.

Fogliano, M., and Schendel, P. F., 1981, Evidence for the inducibility of the uvrB operon, Nature, 289:196.

Ganesan, A. K., and Smith, K. C., 1972, Requirement for protein synthesis in rec dependent repair of DNA in Escherichia coli after UV or X irradiation, J. Bacteriol., 111:575.

Hanawalt, P. C., Cooper, P. K., Ganesan, A. K., and Smith, C. A., 1979, DNA repair in bacterial and mammalian cells, Annu. Rev. Biochem., 48:783.

Howard-Flanders, P., 1968, DNA repair, Annu. Rev. Biochem., 31:175.

Kato, T., and Shinoura, Y., 1977, Isolation and characterization of mutants of Escherichia coli deficient in induction of mutations by UV light, Mol. Gen. Genet., 156:121.

Kenyon, C. J., and Walker, G. C., 1980, DNA damaging agents stimulate gene expression at specific loci in Escherichia coli, Proc. Natl. Acad. Sci. U.S.A., 77:2819.

Kriek, E., 1974, Carcinogenesis by aromatic amines, Biochim. Biophys. Acta, 355:177.

Moore, P. D., Bose, K. K., Rabkin, S. D., and Strauss, B. S., 1981, Sites of termination of in vitro DNA synthesis on UV and AAF treated ØX174 templates with prokaryotic and eukaryotic DNA polymerases, Proc. Natl. Acad. Sci. U.S.A., 78:110.

Moore, P. D., Rabkin, S. D., and Strauss, B. S., 1980, Termination of in vitro DNA synthesis at AAF adducts in the DNA, Nucleic Acids Res., 8:4473.

Moore, P. D., and Strauss, B. S., 1979, Sites of inhibition of in vitro DNA synthesis in carcinogen and UV-treated ØX174 DNA, Nature, 278:664.

Radman, M., 1974, Phenomenology of an inducible mutagenic DNA repair pathway in Escherichia coli: SOS repair hypothesis, in: "Molecular and Environmental Aspects of Mutagenesis," L. Prakash, F. Sherman, M. W. Miller, C. W. Lawrence, and H. W. Taber, eds., C. C. Thomas, Springfield, Illinois.

Rupp, W. D., Wilde, C. E., Reno, D. L., and Howard-Flanders, P., 1971, Exchanges between DNA strands in ultraviolet-irradiated Escherichia coli, J. Mol. Biol., 61:25.

Sanger, F., Nicklen, S., and Coulson, A. R., 1977, DNA sequencing with chain-terminating inhibitors, Proc. Natl. Acad. Sci. U.S.A., 74:5463.

Sedgwick, S. G., 1975, Inducible error-prone repair in Escherichia coli, Proc. Natl. Acad. Sci. U.S.A., 72:2753.

Sedgwick, S. G., 1976, Misrepair of overlapping daughter-strand gaps as a possible mechanism for UV-induced mutagenesis in uvr strains of Escherichia coli: A general model for induced mutagenesis by misrepair (SOS repair) of closely spaced DNA lesions, Mutat. Res., 41:185.

Sedliakova, M., Slezarikova, V., and Pirsel, M., 1978, UV-inducible repair. II. Its role in various defective mutants of Escherichia coli, Mol. Gen. Genet., 167:209.

Sirover, M., Duke, D. K., and Loeb, L. H., 1979, On the fidelity of DNA replication. VIII. Metal activation of Escherichia coli DNA polymerase I, J. Biol. Chem. 254:107.

Unrau, P., Wheatcroft, R., Cox, B., and Olive, T., 1973, The formation of pyrimidine dimers in the DNA of fungi and bacteria, Biochim. Biophys. Acta, 312:626.

Villani, G., Boiteaux, S., and Radman, M., 1978, Mechanism of ultraviolet-induced mutagenesis: Extent and fidelity of in vitro DNA synthesis on irradiated templates, Proc. Natl. Acad. Sci. U.S.A., 75:3037.

Volkert, M. R., George, D. L., and Witkin, E. M., 1976, Partial suppression of the lexA phenotype by mutations (rnm) which restore ultraviolet resistance but not ultraviolet mutability to Escherichia coli B/r uvrA lexA, Mutat. Res., 36:17.
Weymouth, L. A., and Loeb, L. A., 1978, Mutagenesis during in vitro DNA synthesis, Proc. Natl. Acad. Sci. U.S.A., 75:1924.
Witkin, E. M., 1976, Ultraviolet mutagenesis and inducible DNA repair in Escherichia coli, Bacteriol. Rev., 40:869.
Youngs, D. A., and Smith, K. C., 1976, Genetic control of multiple pathways of post-replicational repair in uvrB strains of Escherichia coli K-12, J. Bacteriol., 125:102.

CHAPTER 12

DEPURINATION OF DNA AS A POSSIBLE MUTAGENIC PATHWAY FOR CELLS

Roeland M. Schaaper, Thomas A. Kunkel, and
Lawrence A. Loeb

The Joseph Gottstein Memorial Cancer Research Laboratory
Department of Pathology SM-30, The University of
Washington, Seattle, Washington 98195

SUMMARY

The possible consequences of depurination for both spontaneous and induced mutagenesis were investigated using in vitro and in vivo assays. Depurination of synthetic polynucleotide templates such as poly[d(A-T))] or poly[d(G-C)] leads to increased misincorporation of noncomplementary nucleotides when these templates are copied by prokaryotic and eukaryotic DNA polymerases.

The ability of Escherichia coli DNA polymerase I to copy over apurinic sites was demonstrated using single-stranded circular DNA of bacteriophage ØX174 as a template and starting DNA synthesis at a fixed point. Analysis of the newly synthesized ØX174 restriction fragments on neutral and alkaline sucrose gradients shows that synthesis proceeded past apurinic sites. When using depurinated ØX174 DNA containing the am3 amber mutation as a template for copying by E. coli DNA polymerase I, an increased reversion to wild type is observed after transfection into E. coli spheroplasts. The enhancement in reversion frequency is proportional to the extent of depurination, suggesting that depurination is also mutagenic during copying natural DNA in vitro.

When noncopied depurinated ØX174 am3 DNA is transfected in E. coli spheroplasts, no increase in reversion frequency is observed above background level. However, when the spheroplasts are derived from bacteria in which the SOS response had been induced by UV irradiation, a substantial increase is observed for depurinated molecules, whereas no increase is observed for nondepurinated templates, suggesting in vivo mutagenesis at depurinated sites.

In each of the different assay systems investigated, the increase in misincorporation or reversion frequency is a linear function of the number of sites and is abolished by treatment of the depurinated templates with alkali, which rapidly induces strand breakage at apurinic sites.

INTRODUCTION

Depurination is the release of purine bases from DNA through breakage of the N-glycosylic bond that connects the purine base to the sugar-moiety of the sugar-phosphate backbone. Evidence suggests that this process occurs spontaneously at significant rates. From measurements on the rates of depurination in acid and at elevated temperatures, an in vivo rate constant has been estimated to be 3×10^{-11} per s (Lindahl and Nyberg, 1972). This corresponds to 20,000 apurinic sites per mammalian cell per 24-h period. The resulting apurinic site is relatively stable, with an estimated half-life of several hundred hours (Lindahl and Andersson, 1972). Thus, if unrepaired, depurination could represent a major and clearly intolerable loss of genetic information from the cell. A search for enzymes that might recognize and repair apurinic sites quickly led to the discovery of the ubiquitous and abundant apurinic endonucleases (Hanawalt, et al., 1979; Lindahl, 1979) that initiate excision repair of apurinc sites. In addition, activities have been described that insert purines back into the DNA (Deutsch and Linn, 1979) and have been designated as insertases. However, one still might expect a certain sub-fraction of the apurinic sites to remain unrepaired and be present at the time of DNA replication. Such lesions could result in misincorporations and contribute to spontaneous mutation frequencies. Furthermore, rates of depurination are enhanced several orders of magnitude by modification of purines at N-3 and N-7 positions (Lawley and Brookes, 1963). Since a large number of chemical carcinogens modify these positions, substantial depurination may be expected after exposure to these agents. Damage-specific N-glycosylases may also contribute in this process.

Early in vivo studies on heat, acid, and alkylation mutagenesis did focus on apurinic sites as premutagenic lesions, since apurinic sites seemed to be a common and relatively frequent product in all three treatments (Bautz and Freese, 1960; Freese, 1961). However, later, more detailed studies mainly with bacteriophage T4 (Baltz et al., 1976; Bingham et al., 1976; Drake and Baltz, 1976; Lawley and Martin, 1975) argued against the apurinic site and in favor of other simultaneously induced lesions as the major cause of mutations.

We have developed a number of in vitro systems to assess the fidelity with which purified polymerases synthesize DNA as a first step in trying to understand how cells can achieve such a highly accurate transfer of their genetic information content. We decided to investigate how this fidelity is affected by the presence of apurinic sites or other specific lesions in the template because this should provide us with information on the intrinsic mutagenic potential of such unrepaired lesions. From there on, it may be possible to make extrapolations to existing in vivo conditions.

In the following, we will present our evidence that such apurinic sites in the template lead to strongly increased error rates of the replicating enzyme. This is true on synthetic polynucleotide templates as well as on natural DNA. In at least one situation the presence of apurinic sites on DNA results in mutagenesis in vivo.

STUDIES WITH SYNTHETIC POLYNUCLEOTIDES

In order to assess the mutagenic consequences of depurination, we first made use of the homopolymer fidelity assay (Battula and Loeb, 1974). In these studies, poly[d(A-T)], poly[dA], or poly-[d(G-C)] templates were replicated with E. coli DNA polymerase I, avian myeloblastosis virus (AMV) DNA polymerase, or DNA polymerase β from human placenta (Shearman and Loeb, 1977, 1979). The fidelity assay measures the frequency by which these enzymes incorporate noncomplementary (incorrect) necleotides. Depurination of these templates and also of natural DNA (vide infra) can be done in a controlled manner by means of heat, a low pH, or a combination of both. Figure 1 shows the misincorporation of dGTP on a poly-[d(A-T)] template with increasing extent of depurination, indicating the increased error-proneness of DNA synthesis as a function of depurination. The response is linear with the number of depurinated sites introduced and is identical for E. coli Pol I, AMV DNA polymerase, and human DNA polymerase β (not shown) - enzymes with marked differences in their intrinsic accuracy. Extensive control experiments showed that this increased misincorporation occurs as single-base substitutions throughout the newly made DNA. Even though the rate of synthesis is reduced by depurination, the extent of synthesis after prolonged incubation is similar on depurinated and nondepurinated templates, suggesting that the DNA polymerases can copy over depurinated sites. From these data, the model emerged that errors arise as single-base substitutions opposite apurinic sites in the template strand. This model was in part confirmed by a nearest-neighbor analysis of the product DNA, which showed that all misincorporations on depurinated DNA occurred opposite the position of a template adenine in the case of poly[d(A-T)] and the position of a guanine in the case of poly[d(G-C)]. Secondly, treatment of the depurinated DNA with alkali before replication — thus breaking the DNA at positions of apurinic sites (Lindahl and Andersson, 1972) — abolished the mutagenic effect by 83 to 98% (Shearman and Loeb, 1979).

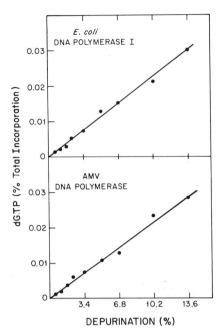

Fig. 1. Depurination of Poly d(A-T). Incorporation of dGTP due to depurination as a function of extent of depurination of the poly[d(A-T)] template using E. coli DNA polymerase I and AMV DNA polymerase. From Shearman and Loeb (1979) with permission. Incorporation was measured with [α-^{32}P]-dTTP, and dATP as complementary nucleotides and [^{3}H]dGTP as the noncomplementary nucleotide. Incubation was varied so that total incorporation was constant for each template. Incorporation due to depurination is defined as the difference between the noncomplementary nucleotide incorporated with the depurinated and nondepurinated templates.

COPYING EXPERIMENTS ON DEPURINATED ØX174 DNA

The single-stranded circular DNA of bacteriophage ØX174 can be copied by isolated DNA polymerases in the test tube if the molecule is primed to provide the enzyme with a 3'-OH terminus as a starting point for synthesis. This can be done conveniently with a ØX174 restriction endonuclease fragment. We used this system to test whether a DNA polymerase can proceed past an apurinic site on a natural DNA template. An experiment was performed (Kunkel et al., 1981) in which approximately 50 apurinic sites were introduced per ØX circle. After priming of the molecule, E. coli Pol I synthesized approximately 1,400 nucleotides per circle, whereas on a nondepurinated template approximately 4,800 nucleotides were incorporated. Since the entire ØX circle comprises 5,386 nucleotides, on an average basis, synthesis should have gone past approximately 13 apurinic sites. It was further demonstrated that synthesis has indeed proceeded past apurinic sites in the following way, as illustrated in Fig. 2. The idea is to restrict the double-stranded product DNA with a restriction endonuclease, to isolate a defined restriction fragment by gel electrophoresis, and then to demonstrate the presence of apurinic sites in the template strand of the fragment. On neutral sucrose gradients, both the template and the product strand of the restriction fragment should sediment coincidently. On alkaline gradients, the strands should separate and the template strand should be fragmented if containing apurinic sites.

Figure 2 shows how an isolated HaeIII restriction endonuclease fragment — ^{3}H-labeled in the template strand and ^{32}P-labeled in the product strand — sedimented on neutral and alkaline sucrose gradients. In all cases, the two strands show identical sedimentation behaviors, except in the case of a depurinated template on the alkaline gradient, in which the template strand sediments more slowly indicating the presence of alkali-labile apurinic sites. The change in sedimentation coefficient predicts two to three apurinic sites, which agrees well with estimates of two sites calculated from the length of the restriction endonuclease fragment and the extent of depurination.

MUTAGENESIS ON DEPURINATED ØX174 DNA

The accuracy of DNA polymerases has been measured using ØX174 am3 DNA as a template. In this assay, synthesis is initiated with a Z-5 fragment, 83 nucleotides distant from the amber mutation (Kunkel and Loeb, 1979, 1980; Weymouth and Loeb, 1978). Replication errors made at the middle position of the am3 TAG amber codon will revert the amber phage to wild type or pseudo-wild type. Thus, by transfecting the product DNA into E. coli spheroplasts and plating the progeny phage on permissive and nonpermissive indicator bacteria, one obtains a reversion frequency that relates in a direct way to the error rate of the enzyme. Table 1 shows the change in error rate for Pol I as a result of depurination. This experiment was performed with a tenfold dCTP pool bias. Pool biases are a convenient way to manipulate error rates of enzymes and can be used to increase the sensitivity of the assay (Kunkel and Loeb, 1979). It should be pointed out that am3 may be expected to be a good tool to study depurination mutagenesis since reversion occurs exclusively at the adenine of the amber codon, which should be susceptible to removal by depurination. Table I demonstrates a proportionality between the number of apurinic sites per ØX template and the

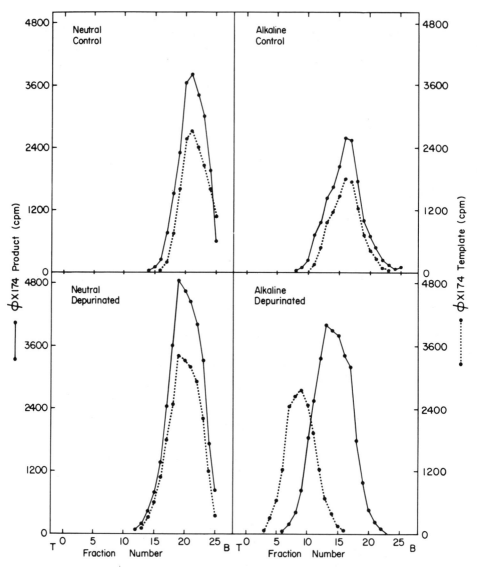

Fig. 2. Sucrose gradient centrifugation profile of a restriction fragment synthesized on a depurinated template. An average of 50 apurinic sites was introduced into ^{3}H-labeled single-stranded ØX174 DNA by treatment at pH 2.72 and 55°C for 25 min. Synthesis by Pol I using [α-^{32}P] nucleoside triphosphates substrates was initiated on a Z-1 primer. The product was digested with <u>HaeIII</u> restriction endonuclease and the resultant Z-7 (nondepurinated) and Z-8 depurinated fragments were separated and isolated by electroelution. Centrifugation was at 4°C in a SW 50.1 rotor for 16 h at 45,000 rpm [see Kunkel et al. (1981) for experimental details].

Table 1. Effect of Depurination on the Fidelity of Copying ØX174 am3 DNA by E. coli DNA Polymerase I

Apurinic Sites per Molecule	Nucleotides Added per Template	Permissive Titer (X 10^7)	Nonpermissive Titer (X 10^2)	Reversion Frequency (X 10^{-6})	Error Rate
0	0	25.7	4.80	1.87	—
0	520	25.9	10.4	4.02	1/90,700
2.2	425	5.15	14.0	27.2	1/7,700
2.2 (Alkali)	416	5.98	1.62	2.71	1/232,000
5.5	338	0.25	1.68	67.2	1/2,980

ØX174 DNA was depurinated and then used as a template in the fidelity assay. The method for calculating error rate is given in Kunkel and Loeb (1979). Pretreatment of the depurinated template with alkali is as described in Schaaper and Loeb (1981).

frequency of misincorporation by the polymerase. Also, treatment of the depurinated template with alkali before replication abolishes the enhancement in mutagenicity resulting from depurination.

IN VIVO MUTAGENICITY OF APURINIC SITES

If one transfects depurinated ØX174 am3 DNA into spheroplasts without copying, replication of the viral circle and production of the replicative form will occur in vivo by the E. coli Pol III holoenzyme complex. Extrapolation from the Pol I data as presented above would suggest that the holoenzyme complex might copy over apurinic sites and thus a mutagenic response might be observed. This is not the case. The major effect of depurination observed with normal spheroplasts is inactivation, to the extent that one apurinic site per circle constitutes a lethal event. No increase in reversion frequency above the background level of 10^{-6} is observed (Schaaper and Loeb, 1981). This puts an upper limit to the frequency of the mutagenic process of less than 1 in 1,400 per apurinic sites. This lack of mutagenicity can be interpreted in two ways: either depurinated molecules are destroyed in vivo by hydrolysis of the apurinic sites by one of the three apurinic endonucleases found to date in E. coli; or, alternatively, the Pol III holoenzyme complex is not able to copy over apurinic sites. The following arguments support the latter possibility. Transfection of depurinated single-stranded ØX DNA in the xth$^-$ mutants, which lack the exonuclease III/endonuclease VI enzyme that constitutes 90% of the apurinic endonuclease activity in a crude extract, does not change the survival (Loeb, unpublished data). This is in agreement with findings that all known E. coli apurinic endonucleases are specific for double-stranded DNA (Lindahl, 1979). Secondly, the E. coli Pol III holoenzyme is a much more faithful enzyme than Pol I (Loeb et al., 1980), and therefore might not incorporate a base as easily at an apurinic site.

We succeeded, however, in demonstrating mutagenesis of depurinated DNA in vivo by introducing the error-prone SOS response (Radman, 1974; Witkin, 1976) in the spheroplasts. This was done by UV irradiation and allowing expression of the SOS phenotype in the bacteria before converting them to spheroplasts. Under these conditions, a strong response (i.e., up to a 20- to 30-fold increase in reversion frequency above background) is observed using templates containing one to five apurinic sites per circle (Fig. 3). The increase is proportional to the number of apurinic sites introduced.

As demonstrated with synthetic polynucleotides (Shearman and Loeb, 1979) and natural DNA in vitro (Kunkel et al., 1981), exposure of depurinated DNA to alkali abolishes the mutagenic response. In Table 2 nondepurinated and depurinated DNA were incubated in alkali prior to transfection of spheroplasts. The mutagenic response observed in UV-irradiated spheroplasts was abolished, presumably by

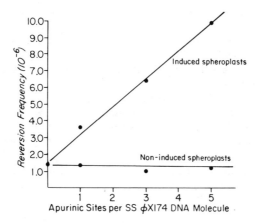

Fig. 3. Reversion frequency among am3 progeny phage as a function of the number of apurinic sites introduced. Single-stranded am3 DNA was depurinated and transfected into spheroplasts from normal or SOS-induced bacteria. The vertical bars represent the range of duplicate transfections. [Taken from Schaaper and Loeb (1981) with permission].

Table 2. Reversion Frequency Among Progeny Phage After Transfection of Depurinated and Alkali-Treated Single-Stranded ØX174 am3 DNA on Normal and Induced Spheroplasts

Apurinic Sites	Normal Spheroplasts		Induced Spheroplasts	
	No Alkali	Alkali	No Alkali	Alkali
0	1.2×10^{-6}	1.2×10^{-6}	1.2×10^{-6}	1.5×10^{-6}
4.5	1.5×10^{-6}	1.3×10^{-6}	20×10^{-6}	2.3×10^{-6}

After depuration, DNA samples were incubated with an equal volume of 0.2 N NaOH for 2.5 h at room temperature, neutralized, and transfected as normal. Control DNA samples were incubated at room temperature without alkali addition. Alkali treatment had no effect on the phage-producing ability of either normal or depurinated DNA. Adapted from Schaaper and Loeb (1981) with permission.

selective alkaline hydrolysis at apurinic sites. Also, the same reversal of mutagenicity can be obtained upon treatment of the depurinated DNA with highly purified HeLa cell apurinic endonuclease (to be published in detail elsewhere). Thus, the conclusion seems justified that apurinic sites can be the direct cause of mutations in vivo. In our case, we have shown this to be true on a single-stranded template. This may resemble rather closely the situation in a replication fork in which unrepaired apurinic sites from double-stranded DNA are faced by a DNA polymerase in a single-stranded region. The outcome would entirely depend on the ability of the enzyme to copy over such sites. If one assumes that the increased mutagenic response in SOS-induced cells is due to increased bypass of apurinic sites — which is at the heart of the current SOS hypothesis (although no definite proof for this has been presented) — then one can use the mutagenesis data to calculate bypass frequencies of the enzyme. In SOS-induced spheroplasts, it is approximately 1/50 as compared to a value of less than 1/1,400 in normal spheroplasts.

CONCLUSIONS

Little information is currently available on the ability of purified mammalian DNA polymerases to copy over apurinic sites. With synthetic polynucleotides, human placenta DNA polymerase β showed the same ability to synthesize and incorporate errors on a depurinated template as did E. coli DNA polymerase I (Shearman and Loeb, 1979). Further studies with natural DNA templates are required, especially in the light of the fact that none of the mammalian enzymes in purified form seem to possess an associated 3' → 5' exonuclease activity. In this respect, it is very stimulating to learn about the increasing evidence for SOS-type phenomena in eukaryotic cells as well (Cornelis et al., 1980; Sarasin and Benoit, 1980; Sarasin and Hanawalt, 1978) although such a response may not be absolutely required for spontaneous mutagenesis at apurinic sites.

If the above speculations pertain to mammalian cells, they provide a mechanism by which mutagenesis can result from modification of bases, particularly purines, by bulky chemical carcinogens. This mechanism is diagrammed in Fig. 4. Modification of purines by bulky carcinogens has been shown to terminate DNA in vitro (Moore and Strauss, 1979) and in vivo (Hsu et al., 1977). In such a model, halting the replication fork at the site of a modified base, or perhaps even at an apurinic site, would induce an SOS response that permits the replication to proceed past apurinic sites. Apurinic sites might result from spontaneous hydrolysis of glycosylic bonds or from enhanced hydrolysis of purines modified by chemical carcinogens, particularly at the N-3 and N-7 positions of guanine and adenine. Synthesis past apurinic sites would involve the nontemplate-directed insertion of nucleotides resulting in mutagenesis. This mechanism is notably attractive for mutagenesis by aflatoxin B_1, for which the principal adduct formed is on the N-7 position of guanine (Essigmann et al., 1977). Conceivably, the bulky aflatoxin could stall DNA replication and induce an alteration in the DNA replication

DEPURINATION OF DNA

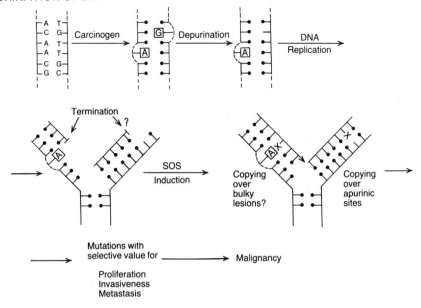

Fig. 4. Mutagenesis by Chemical Carcinogens Via Depurination.
Model for mutagenesis by depurination in cells expressing
SOS repair. This model does not distinguish between induction of SOS by blockage at bulky adducts or at apurinic
sites. It is possible that bypass occurs only after depurination.

complex. The altered replicating complex might proceed only after
removal of the aflatoxin adduct by depurination.

ACKNOWLEDGMENTS

This study was supported by grants from the National Institutes
of Health (CA-24845, CA-24498, AG-01751) and the National Science
Foundation (PCM 76-80439). R.M.S. is on leave from the Laboratory
of Molecular Genetics, University of Leiden, The Netherlands, and
performed segments of this study in partial fulfillment of the requirements for a Ph.D. degree.

REFERENCES

Baltz, R. H., Bingham, P. M., and Drake, J. W., 1976, Heat mutagenesis in bacteriophage T_4: The transition pathway, Proc.
    Natl. Acad. Sci. U.S.A., 73:1269.

Battula, N., and Loeb, L. A., 1974, The infidelity of avian myeloblastosis virus deoxyribonucleic acid polymerase in polynucleotide replication, J. Biol. Chem., 249:4086.

Bautz, E., and Freese, E., 1960, On the mutagenic effect of alkylating agents, Proc. Natl. Acad. Sci. U.S.A., 46:1585.

Bingham, P. M., Baltz, R. H., Ripley, L. S., and Drake, J. W., 1976, Heat mutagenesis in bacteriophage T_4: The transversion pathway, Proc. Natl. Acad. Sci. U.S.A., 73:4159.

Cornelis, J. J., Lupker, J. H., and Van der Eb, A. J., 1980, UV-reactivation, virus production and mutagenesis of SV-40 in UV-irradiated monkey kidney cells, Mutat. Res., 71:139.

Deutsch, W. A., and Linn, S., 1979, An apurinic DNA binding activity from cultured human fibroblasts that specifically inserts purines into depurinated DNA, Proc. Natl. Acad. Sci. U.S.A., 76:1089.

Drake, J. W., and Baltz, R. H., 1976, The biochemistry of mutagenesis, Annu. Rev. Biochem., 45:11.

Essigmann, J. M., Croy, R. G., Nadzan, A. M., Busby, W. F., Reinhold, V. N., Büchi, G., and Wogan, G. N., 1977, Structural identification of the major DNA adduct formed by aflatoxin B_1 in vitro, Proc. Natl. Acad. Sci. U.S.A., 74:1870.

Freese, E. B., 1961, Transitions and transversions induced by depurinating agents, Proc. Natl. Acad. Sci. U.S.A., 47:540.

Hanawalt, P. C., Cooper, P. K., Ganesan, A. K., and Smith, C. A., 1979, DNA repair in bacteria and mammalian cells, Annu. Rev. Biochem., 48:783.

Hsu, W. T., Lin, E. J., Harvey, R. G., and Weiss, S. B., 1977, Mechanism of phage ØX174 DNA inactivation by benzo[a]pyrene-7,8-dihydrodiol-9,10-epoxide, Proc. Natl. Acad. Sci. U.S.A., 74:3335.

Kunkel, T. A., and Loeb, L. A., 1979, On the fidelity of DNA replication: Effect of divalent metal ion activators and deoxyribonucleoside triphosphate pools on in vitro mutagenesis, J. Biol. Chem., 254:5718.

Kunkel, T. A., and Loeb, L. A., 1980, On the fidelity of DNA replication: The accuracy of Escherichia coli DNA polymerase I in copying natural DNA in vitro, J. Biol. Chem., 255:9961.

Kunkel, T. A., Shearman, C. W., and Loeb, L. A., 1981, Mutagenesis in vitro by depurination of ØX174 DNA, Nature, 291:349.

Lawley, P. D., and Brookes, P., 1963, Further studies on the alkylation of nucleic acids and their constituent nucleotides, Biochem. J., 89:127.

Lawley, P. D., and Martin, C. N., 1975, Molecular mechanisms in alkylation mutagenesis: Induced reversion of bacteriophage T4rII AP72 by ethyl methanesulphonate in relation to extent and mode of ethylation of purines in bacteriophage deoxyribonucleic acid, Biochem. J., 145:85.

Lindahl, T., 1979, DNA glycosylases, endonucleases for apurinic/apyrimidinic sites and base excision-repair, Prog. Nucleic Acid Res. Mol. Biol., 22:135.

Lindahl, T., and Andersson, A., 1972, Rate of chain breakage at apurinic sites in double-stranded deoxyribonucleic acid, Biochemistry., 11:3618.

Lindahl, T., and Nyberg, B., 1972, Rate of depurination of native deoxyribonucleic acid, Biochemistry, 11:3610.

Loeb, L. A., Kunkel, T. A., and Schaaper, R. M., 1980, Fidelity of copying natural DNA templates, in: "Mechanistic Studies on DNA Replication and Genetic Recombination," Vol. XIX, B. Alberts. C. F. Fox, and F. J. Stusser, eds., Academic Press, New York.

Moore, P., and Strauss, B. S., 1979, Sites of inhibition of in vitro DNA synthesis in carcinogen- and UV-treated ØX174 DNA, Nature, 278:664.

Radman, M., 1974, Phenomenology of an inducible mutagenic DNA repair pathway in Escherichia coli: SOS repair hypothesis, in: "Molecular and Environmental Aspects of Mutagenesis," L. Prakash, F. Sherman, M. W. Miller, C. M. Lawrence, and H. W. Taber, eds., C. A. Thomas, Springfield, Illinois.

Sarasin, A. R., and Benoit, A., 1980, Induction of an error-prone mode of DNA repair in UV-irradiated monkey kidney cells, Mutat. Res., 70:71.

Sarasin, A. R., and Hanawalt, P. C., 1978, Carcinogens enhance survival of UV-irradiated simian virus 40 in treated monkey kidney cells: Induction of a recovery pathway? Proc. Natl. Acad. Sci. U.S.A., 75:346.

Schaaper, R. M., and Loeb, L. A., 1981, Depurination causes mutations in SOS-induced cells. Proc. Natl. Acad. Sci. U.S.A., 78:1773.

Shearman, C. W., and Loeb, L. A., 1977, Depurination decreases fidelity of DNA synthesis in vitro, Nature, 270:537.

Shearman, C. W., and Loeb, L. A., 1979, Effects of depurination on the fidelity of DNA synthesis, J. Mol. Biol., 128:197.

Weymouth, L. A., and Loeb, L. A., 1978, Mutagenesis during in vitro DNA synthesis, Proc. Natl. Acad. Sci. U.S.A. 75:1924.

Witkin, E. M., 1976, Ultraviolet mutagenesis and inducible DNA repair in Escherichia coli, Bacteriol. Rev., 40:869.

CHAPTER 13

PASSIVE POLYMERASE CONTROL OF DNA REPLICATION FIDELITY:
EVIDENCE AGAINST UNFAVORED TAUTOMER INVOLVEMENT
IN 2-AMINOPURINE-INDUCED BASE-TRANSITION MUTATIONS

Myron F. Goodman, Susan M. Watanabe, and
Elbert W. Branscomb[1]

Department of Biological Sciences, Molecular Biology
Section, University of Southern California, University
Park, Los Angeles, California 90007, and the
[1]Biomedical Sciences Division, Lawrence Livermore
Laboratory, University of California, Livermore
California 94550

SUMMARY

We consider the role of unfavored tautomers in causing base-substitution transition mutations. Data obtained with the base analogue 2-aminopurine (AP) for the frequency of forming AP·T and AP·C base mispairs can be shown to be in probable conflict with tautomer model predictions. An alternative model, in which individual hydrogen bonds exhibit different bond strengths depending upon their ring position, is proposed to account for the frequencies of forming correct and incorrect base pairs. In this "differential H-bonding" model, disfavored tautomers of AP and those of common nucleotides play a generally insignificant role. A hydrogen-bonding free energy scale is derived in which free energy differences are obtained for all possible matched and mismatched base pairs. We also show that recent in vitro data for the formation of AP·C base pairs are consistent with a "passive polymerase" theoretical model in which base selection is governed not by the enzyme but by differences in base-pairing free energies.

INTRODUCTION

Nearly three decades have elapsed since Watson and Crick elucidated the structure of DNA and explained base-pairing rules in terms of stronger hydrogen bonding between correct vs incorrect base pairs.

At the same time they suggested that base mispairs could form during replication when unfavored tautomeric isomers of the bases formed stable hydrogen bonds in certain mispaired configurations (Watson and Crick, 1953a,b). Yet the physical-chemical mechanisms of forming base mispairs during DNA replication are still largely a mystery as are the mechanisms by which candidate nucleotides are selected or rejected by DNA polymerase during template-directed synthesis.

The base analogue 2-aminopurine (AP) is known to stimulate transition mutations bidirectionally, $A \cdot T \rightleftarrows G \cdot C$ (for a review, see Ronen, 1979). Based on structural considerations, Freese (1959) postulated that in the common amino form AP_{am} can form two Watson-Crick base pairs with T while in its imino form AP_{im} can form two base pairs with C. Indeed the ambiguous base-pairing properties of this analogue have provided some of the strongest putative arguments to date implicating unfavored tautomeric forms of the common nucleotides as the main participants in base mispairs.

In this paper we use recent data on the frequency of forming AP base mispairs to question the ideas that unfavored AP tautomers in particular and, by inference, unfavored tautomers in general play a role in causing base substitution transition mutations. We will present a counter proposal, which we call the "differential H-bonding" model, in which hydrogen bonds of differing strengths between only the common forms of the bases account for base mispairs. In this model, relative strengths are assigned to each hydrogen bond formed between various normal bases. This leads to a quantitative prediction of the relative hydrogen-bonding free energies for a variety of base pairs and mispairs. This in turn predicts the frequencies with which specific mispairs form during replication.

Finally, we will present some new data on the formation of $AP \cdot C$ base mispairs generated in vitro that offer further support for a "passive polymerase" model for the fidelity of DNA replication (Clayton et al., 1979; Galas and Branscomb, 1978). This model states that the fidelity of DNA-replicating enzymes is attributable to differences in base-pairing free energies between correct and incorrect base pairs and not to an intrinsic ability of a polymerase to distinguish between right and wrong nucleotides.

GENERAL PROCEDURES

Nucleotides

All nonradioactive deoxynucleotides were purchased from P-L Biochemicals. Radioactive nucleotides were purchased from Schwarz/Mann. 2-Amino-6-chloropurine, purchased from Sigma Chemical Co., was converted to [6-^{3}H]2-AP through catalytic reduction by

ICN Pharmaceuticals, Inc., Irvine, CA. [^{3}H]- and unlabeled-AP deoxynucleoside triphosphate were synthesized from the free base as described previously (Clayton et al., 1979). [^{3}H]dCTP was purified to remove [^{3}H]dUTP contaminant using high pressure liquid chromatography equipment purchased from Waters Associates, Inc.

Enzymes

T4 DNA polymerase was purified as described in Clayton et al. (1979). The highly purified enzymes calf thymus terminal deoxynucleotidyltransferase and DNA polymerase α from human KB cells were generous gifts from Drs. R. L. Ratliff (Los Alamos Scientific Laboratory) and P. Fisher and D. Korn (Stanford University), respectively.

Synthesis of Copolymer DNA Templates

Deoxynucleotide templates consisting either of dA and dAP primed with oligo(dT) or dC and dAP primed with oligo(dG) were constructed as described in Watanabe and Goodman (1981).

Assays

DNA polymerase deoxynucleotide incorporation and turnover assays and K_M, V_{max} enzyme kinetic measurements using natural DNA templates were carried out essentially as described in Clayton et al. (1979). Enzyme kinetic measurements using the copolymer templates have not been described previously and are as follows. Reaction mixtures of 0.1 ml contained 0.02 M Tris, pH 7.8, 2 mM β-mercaptoethanol, 1 mM MgCl$_2$, 0.2 mg/ml sterile gelatin, 3.0 units/ml DNA polymerase α (fx VIII), 20 μM p(dC·dAP)·oligo(dG), 70 μM dGTP, and either [^{3}H]dCTP (1.7 x 10^7 cpm/nmol for 4.5 μM to 9.0 μM measurements) and 4.7 x 10^6 cpm/nmol for 0.68 μM to 75 μM measurements). Incorporation for 10^6 cpm/nmol for 0.68 μM to 75 μM measurements). Incorporation for each nucleotide occurred in the absence of the other. Assays were incubated at 30°C for 60 min after which 2.0 ml of 15% TCA was used to precipitate acid-insoluble [^{3}H]cpm in the presence of 0.1 mg carrier DNA. The mixtures were centrifuged, resolubilized with 0.8 ml N/10 NaOH, and reprecipitated with 2.0 ml 15% TCA. The procedure was repeated two more times and acid-insoluble [^{3}H]cpm collected onto GF/C filters with 95% ethanol at 0°C.

The definition for the deoxynucleotide misinsertion ratio is I(W)/I(R) where I(W) and I(R) represent the insertion of the "wrong" and "right" nucleotides, respectively. For prokaryotic polymerases with 3'-exonuclease proofreading capability, I is measured as acid-precipitable radioactive counts per min incorporated into DNA plus acid-soluble counts per min (turnover) appearing as deoxynucleoside monophosphates. For mammalian α polymerase having no detectable exonuclease proofreading activity, I is measured as acid-precipitable counts per min incorporated into DNA.

AP mutagenesis and the measurement of AP·T base pairs in T4 bacteriophage in vivo is described in Goodman et al. (1977); the measurement of AP·T base pairs in vitro is in Clayton et al. (1979); the assay for AP·C heteroduplex-heterozygotes in T4 is described in Hopkins and Goodman (1979), and AP·C measurements in vitro are contained in Watanabe and Goodman (1981).

MISPAIRING OF 2-AMINOPURINE

We begin by summarizing what has recently been learned about the mispairing behavior of AP. Consider the nucleotide pair competition experiment illustrated in Fig. 1 where dATP and AP deoxyribonucleoside triphosphate (dAPTP) are competing simultaneously for insertion opposite a template T site. It has been shown for the case of five polymerases studied to data, T4 mutator (L56), wild type (43^+), antimutator (L141), Escherichia coli Pol I, and mammalian α, that the ratio of AP/A insertion is the same for all of the above enzymes in the range of 13 to 16% for equimolar concentrations of deoxyribonucleotide substrates (Bessman et al., 1974; Clayton et al., 1979). A similar AP insertion frequency (13%) has recently been observed in a HeLa cell nuclei system (Wang et al., 1981), and it has been shown (Liu et al., 1978) that the insertion specificity for AP by the bacteriophage T4 wild-type DNA polymerase is not altered in the presence of the T4 replication complex composed of 44/62, 45, and 32 proteins.

Next consider nucleotide pair competition where dCTP and dTTP are competing for insertion opposite AP (Fig. 1). We recently observed that dCMP is inserted about 5% (4.6 ± 2%) as frequently as dTMP at template AP sites (Watanabe and Goodman, 1981). This 5% value is in excellent agreement with earlier predictions (Goodman et al., 1977) and with subsequent in vivo measurements of AP·C/AP·T base-pair misincorporation ratios at T4 rII marker loci (Hopkins and Goodman, 1979; Ripley, 1981). These in vitro data are summarized in Fig. 1.

Based on the data in Fig. 1, namely, (1) When dATP and dAPTP compete for insertion by DNA polymerase opposite template T sites, AP is inserted at a frequency of 14%, and (2) when dTTP and dCTP compete opposite template AP sites, C is inserted at a frequency of 5%, we propose that tautomer models cannot account for AP·T, AP·C-mispairing frequencies but instead appear to be contradicted by the data.

INSERTION

Template-Substrates	Mispaired Bases	I(W)/I(R)	Implication
dAPTP, dATP / T (3')	AP·T / A·T	$\frac{I(AP)}{I(A)} = 14\%$	$AP_{im} \geq 86\%$
dCTP, dTTP / AP (3')	AP·C / AP·T	$\frac{I(C)}{I(T)} = 5\%$	$AP_{im} \leq 5\%$

Fig. 1. Evidence against AP tautomeric base mispairs. The first column shows two deoxynucleotide substrates competing for insertion at a template site: dAPTP and dATP competing opposite thymine (upper diagram) and dCTP and dTTP competing opposite template AP. The second column indicates a ratio of the base pairs formed in each competition. The DNA polymerase misinsertion ratio is given in the third column, where the data for I (AP)/I(A) = 13.9 ± 2.7% is taken from Clayton et al. (1979) and the ratio I(C)/I(T) = 4.6 ± 2.0% is taken from Watanabe and Goodman (1981). The fourth column indicates the contradictory bounds on the "unfavored" tautomer frequencies based on assuming that unfavored tautomers can account for AP base-mispairing frequencies.

First consider the case where dATP and dAPTP compete opposite template T sites. As amino tautomers, AP_{am} and A_{am} can form two hydrogen bonds with T; as imino tautomers, AP_{im} and A_{im} can form a single bond with T (Fig. 2a). In terms of the types and numbers of bonds formed, the pairing of A and AP with T appear sterically identical. How can we then explain the fact that A is inserted opposite T 86% of the time, i.e., sevenfold more readily than is AP?

It has generally been supposed that the mispairing behavior of AP is due to its being more often in the imino configuration than is A (Freese, 1959). But to explain the AP·T mispairing data in tautomer terms requires that AP be in the imino form 86% of the time; 86% is actually a minimum frequency for the presence of the unfavored AP_{im} tautomer since it is arrived at by assuming that AP_{im}·T base-pairing frequencies are negligible compared with AP_{am}·T frequencies.

However, the conclusion that AP is predominantly imino conflicts with AP·C base-pairing data showing that T is strongly preferred over C for pairing with template AP. AP_{am} forms one hydrogen bond with C, whereas AP_{im} forms two (Fig. 2b). If AP_{am} were 14% of the total, then only that fraction could bind significantly to T, while the remaining 86% AP_{im} should bind strongly to C. But the data using mammalian α polymerases (Watanabe and Goodman, 1981) (lacking any

discernible proofreading activity) clearly show that AP·C base pairs are formed at a frequency of only 5% compared with AP·T base pairs.

Suppose we now reverse the argument and attempt to fit, using a tautomer rationale, the AP·C data first. We assume as before that AP_{am} pairs strongly with T but negligibly with C, and AP_{im} pairs strongly with C but negligibly with T. The observed AP·C frequencies imply that the frequency of AP_{im} = 5%; even 5% seems an implausibly large imino fraction (Wolfenden, 1969). The clear implication is that now 95% of the AP is in the amino form, which should compete equally with A for pairing opposite T. Yet AP·T base pairs are formed only 14% of the time in obvious disagreement with the value AP_{am} = 95%.

Before continuing, we need to call attention to the fact that the experimental data for AP·T base pairs has dAPTP as a substrate whereas the measurement for AP·C base pairs has AP located in the template (Fig. 1). Would it then be reasonable to suppose that AP changes form from predominantly AP_{im} (86%) to predominantly AP_{am} (95%) depending upon whether AP is present as a substrate dAPTP or as a part of the DNA template?

There are several lines of evidence which argue against this possibility, but before discussing these we would first like to address a possible alternative explanation of the data that supposes that the tautomeric equilibria are influenced significantly by stacking interactions with neighboring bases. Fresco et al. (1980) recently reported a study in which A·C and I·U mispairs were imbedded in poly(dA·dU) and poly(dI·dC) double helices. They found that in annealed helices one base in each mispair existed in its unfavored tautomer; which base was affected depended on the neighboring bases. In terms of known differences in base-stacking energies for different pairs of bases, it appeared that the tautomer configuration adopted was the one that maximized the sum of hydrogen-bonding and base-stacking energies.

As interesting as the Fresco et al. (1980) findings are, the effects observed cannot account for the AP-mispairing data. First, the 20-fold preference for T over C when pairing with template AP

Fig. 2. Base-pairing diagrams for (a) adenine and AP pairing with thymine and (b) with cytosine. Hydrogen-bonding patterns are shown for adenine and AP in their amino (A_{am} and AP_{am}) and in their imino (A_{im} and AP_{im}) tautomers. The ring atoms involved in hydrogen-bonding are assigned their conventional numberings. Hydrogen bonds are indicated with dotted lines (·····).

was observed with AP imbedded in three markedly different base-stacking environments: AP in poly(dA) and in poly(dC) in vitro (Watanabe and Goodman, 1981) and AP in T4 DNA in vivo (Hopkins and Goodman, 1979). Further evidence suggesting that AP has not substantially altered its tautomeric character when acting either as substrate (dAPTP) or template comes from the 3'-exonuclease proofreading behavior of wild-type T4 DNA polymerase. We find that when AP serves as a template base, the polymerase removes an inserted T at a frequency similar to the removal of an inserted AP substrate opposite a template T (Watanabe and Goodman, unpublished data). Evidence that AP ribopolymer templates exist predominantly in an amino form comes from studies of Janion and Shugar (1973) who found that AP ribopolymers formed double-stranded complexes with poly(rU) that exhibited sharp melting profiles typical of helix coil transitions, while poly(rAP) failed to complex with poly(rC). Our most compelling evidence that AP substrates (dAPTP) are in amino form is based on deoxynucleotide competition data from Clayton et al. (1979) that clearly demonstrate that AP is inserted predominantly opposite template T compared with template C sites.

Until now, the ambiguous base-pairing behavior of AP has provided perhaps the strongest rationale for postulating tautomer involvement in causing transition mutations. But, based on the data in Fig. 1 and the arguments given above, the failure of tautomers to offer a consistent explanation for the measured frequencies of forming AP·T and AP·C base pairs leads us to propose a counter hypothesis in which tautomers play no essential role in causing base-transition mispairs to occur. We will now state this hypothesis and pursue its consequences to arrive finally at a semiempirical but predictive base-pairing free energy ladder. This free energy ladder (Fig. 3) predicts free energy differences for a series of different base-pairing partners beginning with the most stable base pair G·C and proceeding up the free energy scale to less stable A·T, AP·T, AP·C, ..., including G·T and A·C mismatched base pairs.

DIFFERENTIAL H-BONDING MODEL

We suggest first that base-pairing specificities observed with AP (Bessman et al., 1974; Clayton et al., 1979; Watanabe and Goodman, 1981) can be interpreted plausibly in terms of differences in hydrogen-bond strengths between bases in their favored tautomeric configurations, i.e., A, AP, and C are in amino rather than imino forms, and G and T in keto rather than enol forms. In the differential H-bonding model, the six- to sevenfold preference for A over AP in pairing with T could be explained if the base-pairing bonds formed between T and A are stronger than those formed between T and AP by about 1.1 kcal/mol. Because the N3-N1 bond is common to both pairs,

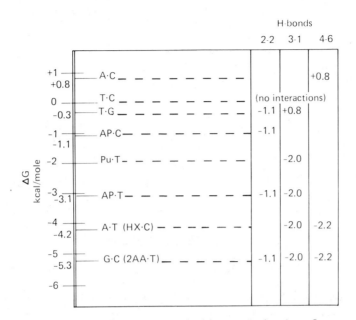

Fig. 3. Base-pairing free energy ladder. Relative free energy assignments proposed for the hydrogen bond base-pairing strengths. The free energy scale runs from +1 to -6 kcal/mol with the base mispair T·C assigned the value zero. The columns to the right indicate which hydrogen bonds, 2-2, 3-1, or 4-6 strengths of -1.1, -2.0, and -2.2 kcal/mol, respectively, are involved. Also indicated are the repulsive forces (of approximately +0.8) due to H-H charge conflict which we presume are involved in the A·C and T·G base pairs. Besides the conventional A, T, G, C notations, we have used the abbreviations AP, 2-aminopurine; Pu, purine; HX, hypoxanthine; and 2AA, 2-6 diaminopurine.

a plausible interpretation of this difference is that the N3-N1 bond formed in the AP·T base pair is 1.1 kcal/mol weaker than the similar O4-N6 bond in the A·T pair (Fig. 2a).

The three bonds, O2-N2, N3-N1, and O4-N6, involved in A·T and AP·T base pairs mimic in atomic character and position all of the hydrogen bonds formed in most base pairs and mispairs of importance. We now hypothesize that they mimic these other bonds in strength as well. Indeed, in situations where the geometry and atomic character of two bonds are the same and where the electronegativity of the ring atoms involved in the bonds is arguably the same, then the hydrogen bonds should have the same strength.

Based on such considerations, we think it reasonable to suppose (1) that an N-O hydrogen bond between the PuC6 and PyC4 positions is about 1 kcal stronger than a similar bond between the PuC2 and PyC2 positions (Fig. 2a); (2) that the PuN6-PyO4 bond in A·T has roughly the same strength as the PuO6-PyN4 bond in the G·C pair; these bonds have similar NHO atomic character but a reversed orientation with respect to whether the pur or pyr contains the amino group; and (3) that the N3-N1 bond strength is independent of whether the pur or pyr nitrogen carries the hydrogen.

When AP_{am} pairs with C, the only hydrogen bond available is the PuN2-PyO2 bond (Fig. 2b), also formed in AP·T and G·C pairs. Since AP·T has the additional N1-N3 bond, we would expect AP·T to be preferred over AP·C by the strength of the additional bond. Indeed, as we discussed above, AP·T is preferred over AP·C by approximately 20:1, or about 1.8 kcal/mol. For various reasons we believe these data may slightly underestimate the preference. Therefore, in keeping with the above assumptions, we will take 2.0 kcal/mol as a first estimation of the strength of the N3-N1 bond.

We have so far used the two discrimination ratios for which adequate data presently exist to derive the value of one of the standard base-pair hydrogen bonds, N3-N1, and a relation between the other two that states that the O4-N6 (or N4-O6) bond is 1.1 kcal/mol stronger than the O2-N2 bond. At present no further base-competition data exist that could be confidently used to determine the missing number. However, there are several reasons, mentioned below, for believing that the 2-2 bond contributes approximately 1.1 kcal/mol. We have then the following tentative assignments for the three "normal" hydrogen bonds: (2-2) = 1.1 kcal/mol, (3-1) = 2.0 kcal/mol, and (4-6) = 2.2 kcal/mol (relative to bases that have no enthalpic base-pair interaction in normal helix geometry). As a result, the normal base pairs have been assigned hydrogen bond free energies (again, relative to a presumed null interaction configuration like T·C) of -5.3 kcal for G·C and -4.2 kcal for A·T. The difference in strength of these two base-pair bonds agrees with the observed difference in their melting temperatures of approximately 40°C (Marmur and Doty, 1962) and provides additional support for the 1.1 kcal assignment for the 2-2 bond. This value is also roughly supported by measurements (Borer et al., 1974) where an average difference between G·C and U·A pairs of approximately 1.0 kcal was observed. We have provisionally assigned the T·C mispair a hydrogen-binding energy of zero on the assumption that there are no significant enthalpic interactions between these bases when they oppose each other in the normal double helix. The energy difference between this base pair and the AP·C pair must then equal the strength assigned to the 2-2 bond which, in turn, according to our assumptions, must equal the energy difference between A·T and G·C.

These arguments may be taken a step further by considering such mispairs as A·C and T·G. These two pairs, if arranged in the normal geometry, may generate repulsive forces through hydrogen-hydrogen charge conflicts. The pair T·G interacts at two of the normal hydrogen bond sites: the normal O-N hydrogen bond at (2-2), plus an H-H charge conflict at (3-1). On the other hand the only interaction in the A·C base pair is the H-H charge conflict at the (4-6) site. The interface energy of this pair should then be more positive than that of the T·G pair, by roughly the energy (assumed to be 1.1 kcal/mol) of the normal (2-2) H bond. In fact, Hibner and Alberts (1980) found the T·G mispair to be formed more readily than the A·C mispair by about a factor of ten. These data also give some support to assigning a value of roughly 1.1 kcal to the (2-2) H bond.

Finally, the data discussed in Watanabe and Goodman (1981) on the incorporation of C in place of T opposite A indicate that the free energy difference in this discrimination is approximately 5 kcal/mol. If this range is correct, then the charge conflict interactions present in T·G and A·C must make a destabilizing contribution of approximately +0.8 kcal/mol. Based on these conjectures, we arrive at the base-pairing free energy ladder given in Fig. 3.

We have also placed on the diagram in Fig. 3 the base pairs Pu-T (purine-thymine), HX-C (hypoxanthine-cytosine) and 2AA-T (2-6 diamino-purine-thymine). This is based on the fact that, in normal geometry, Pu-T should form only the (1-3) bond, and HX-C and 2AA-T should form bonds nearly identical to those of A·T and G·C, respectively.

While we do not regard the above speculation as a theory of hydrogen-bond strengths, it serves to summarize, perhaps suggestively, some recently obtained data. It also makes a number of predictions that can be compared with base-competition data yet to be obtained.

Perhaps the most questionable aspect of these bond-strength assignments (Fig. 3) is that the bond on the minor groove side of the bases is about half as strong as the other two. There are at least two reasons, however, why this might be the case (Petruska, personal communication). First, the minor groove is the more hydrophilic of the two. Water molecules may therefore compete somewhat more effectively for the hydrogen bond on that side of the helix and lower its free energy accordingly. Second, both of the atoms involved in the O2-N2 hydrogen bond are attached to carbons that have relatively more positive charge than do the corresponding carbons involved in the O4-N6 bond. This is because the C2 carbons are between two ring nitrogens whereas the Py4 and Pu6 carbons are between a nitrogen and a carbon. As a result of this asymmetry, the

N2 and O2 atoms will be less negative. This will decrease the hydrogen-bonding charge on the oxygen but increase it on the hydrogen; these changes should have opposing effects on the strength of the hydrogen bond. However, since there is an additional intervening bond on the nitrogen side, these inductive effects should be the greatest for the oxygen and thus have the net effect of decreasing the strength of the bond.

A number of potentially important issues have been disregarded in deriving the above scheme. One of these is the possible protection from competition with water afforded the middle hydrogen bond (Py3-Pu1) by the presence of the other bonds. Thus, the strength of the middle bond could depend on the presence or absence of the other bonds. We have also tacitly assumed that entropic factors are the same for all competing bases and specifically that the hydrogen bonds formed between various unpaired bases and water are not significantly different.

These considerations notwithstanding, the most important complicating factor not treated in the preceding arguments is the effect of base-stacking interactions (see e.g. Topal et al., 1980). Stacking energies, at least in the free helix, are now known to be comparable to base-pair hydrogen-bonding energies and strongly dependent on the bases involved in the stacking. The most complete description of these effects presently available is found in Tinoco et al. (1973) and Borer et al. (1974). In addition, as noted earlier, the recent experiments of Fresco et al. (1980) show that stacking interactions can cause one of the bases in a mispair to assume a normally disfavored tautomeric form, at least for pairs imbedded in double helical DNA. In particular stacking and mispairing environments, the rare tautomer has sufficiently increased stacking and base-pairing energies to overcome the estimated 6 kcal/mol energy requirement for establishing the rare tautomer (Fresco et al., 1980).

However, as Fresco et al. (1980) note, these effects cannot apply without modification to the problem of DNA replication since they would imply mispairing frequencies vastly higher than are observed. Indeed, because base-stacking energies depend on "irrelevant" base context information, they can only act to degrade the fidelity of DNA replication, whether they act through indirect tautomeric effects or directly by adding extraneous base-pair stabilizing energy. We might therefore expect that the polymerase in conjunction with its accessory replication protein complex is designed to minimize stacking interactions, perhaps by stabilizing the helix in a slightly unwound configuration. Specifically with regard to possible effects of base stacking on tautomer equilibria, it is important to reemphasize that the frequency of forming AP-base mispairs does not seem to depend upon whether AP is acting as a substrate (dAPTP) or as a template base.

AP·C BASE-PAIR FORMATION AND THE PASSIVE POLYMERASE MODEL

The previous discussion was directed towards establishing two main points. The first point was to demonstrate a probable incompatability between the data for AP·T and AP·C base pairs and tautomer predictions for base-mispairing frequencies (Fig. 1). The second was to propose a simple model in which H-bonds between favored tautomers played the central role in determining base-mispairing frequencies. In this final section we consider a third point that has to do with the question of base selection by DNA polymerases: Do polymerases exert an active site discrimination to select the right base and reject the wrong base for incorporation into DNA? We will use the data from Fig. 1 and some new (Watanabe and Goodman, unpublished data) and earlier (Clayton et al., 1979) results to show that polymerase-active site discrimination need not be invoked to account for the fidelity of DNA synthesis.

In a series of papers (Clayton et al., 1979; Galas and Branscomb, 1978; Goodman et al., 1980) a "passive polymerase" model has been developed and undergone partial experimental evaluation. We will not attempt to provide a complete summary of the model here but instead limit ourselves to recounting its most important features and stating one fundamental prediction of the model for which additional data is now available.

The simplest assumption we can make regarding the mechanism responsible for fidelity during DNA synthesis is that differences in base-pairing strengths account entirely for the discrimination against mispairs. In this model the DNA polymerase activity and associated 3'-exonuclease, if present, are viewed as passive elements in the sense that they cannot distinguish between correct and incorrect nucleotides per se. Suppose that two deoxyribonucleoside triphosphate substrates compete for access to a polymerase triphosphate binding site. The association rates are assumed to be diffusion-limited and therefore the same for both right (properly paired) and wrong (mispaired) substrates when the two are at equal concentration. However, the substrate dissociation rate can be markedly different. A "wrong" substrate can simply diffuse off the enzyme-template complex at a more rapid rate than a "correct" substrate.

For a passive polymerase, the deoxyribonucleoside triphosphate dissociation rate is controlled by the sum of three independent binding terms: (1) nucleotide nonspecific binding at the polymerase triphosphate binding site, (2) hydrogen bonds between substrate and template bases, and (3) nearest neighbor base-stacking interactions. Assuming that nonspecific binding at the triphosphate binding site is of roughly equal magnitude both for right and wrong substrates, then the dNTP dissociation rates are given by an exponential of the difference in total substrate-binding energy between right and wrong bases and are therefore a function of the difference in base-pairing stabilities (H-bonding plus base-stacking) at the template site. Thus the misinsertion ratio for two competing dNTPs is given by

$$\frac{I(W)}{I(R)} = \frac{[dWTP]}{[dRTP]} e^{-\frac{\Delta G}{RT}} = \frac{[dWTP]}{[dRTP]} \frac{K_{eq}^R}{K_{eq}^W} \qquad (1)$$

where $I(W)$ and $I(R)$ are the rates of inserting wrong and right nucleotides, [dWTP] and [dRTP] are the concentrations of the competing wrong and right nucleotides, ΔG is the difference in base-pairing free energies for the two nucleotides, and K_{eq}^R/K_{eq}^W is the ratio of the equilibrium dissociation constants (assuming, as is certainly true, that on-off rates for dNTP-polymerase binding are rapid compared with forward polymerization rates).

Simply stated, the misinsertion frequency for an enzyme that does not utilize some special active-site property to accept right or reject wrong nucleotides, i.e., a passive polymerase, will insert any nucleotide that happens to be resident at its active site. The difference in residence times for right and wrong nucleotides is governed by differences in base-pairing free energies. In order to demonstrate that the passive model is wrong would only require that we show that Eq. (1) is grossly violated, that is, $I(W)/I(R) \neq \{[dWTP]/[dRTP]\} (K_{eq}^R/K_{eq}^W)$. A related model prediction is that the maximum velocities for inserting wrong and right dNTPs are approximately the same, $V_{max}^W \approx V_{max}^R$.

How are passive polymerase model predictions borne out for the case of AP·C base-pair formation? When dCTP and dTTP, at equimolar concentrations, compete for insertion opposite a template AP, $I(C)/I(T) \approx 5\%$ (Fig. 1). These measurements were made with highly purified α-DNA polymerase from KB cells (Fisher and Korn, 1977) containing no measurable 3'-exonuclease activity. In Fig. 4, we show individual K_M and V_{max} determinations for the insertion of C and T opposite AP. Note that the ratio $V_{max}^C \approx V_{max}^T$ and $K_M^T/K_M^C = 3.4 \pm 1\%$ which is quite similar to $I(C)/I(T) = 4.6\%$. It is important to emphasize that the two measurements (Figs. 1 and 4) are completely independent. The $I(C)/I(T)$ determination is obtained by allowing dCTP and dTTP to compete simultaneously for insertion opposite a template T while the K_M, V_{max} measurements are made separately for each nucleotide in the absence of its competitor.

From Eq. 1 we obtain a free-energy difference between AP·C and AP·T base pairs of $\Delta G \approx 1.8$ kcal/mol. This value serves, in a sense, as a partial confirmation of the value contained in Fig. 3 since originally the information used to obtain the 1.8 kcal/mol value in Fig. 3 came from an in vivo measurement of AP·C base-pairing

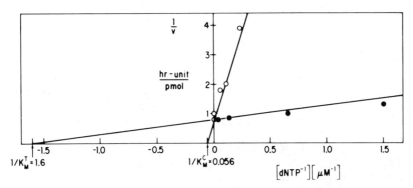

Fig. 4. Double-reciprocal plot of v^{-1} vs $[dNTP]^{-1}$ for dTTP (●) and dCTP (○) (◐ represents coinciding v^{-1} for dCTP and dTTP). Reaction mixtures of 0.1 ml were as described in General Procedures Assays with 0.3 units DNA polymerase α (fx VIII), 70 μM dGTP, 20 μM p(dC·dAP)·oligo(dG), and either [^{3}H]dCTP (1.7 x 10^7 cpm/nmol or 4.7 x 10^6 cpm/nmol) or [^{3}H]TTP (5.8 x 10^6 cpm/nmol). Incorporation of each nucleotide was measured in the absence of the other. Assays were incubated at 30°C. The straight-line fit to the data was generated by the weighted and nonlinear regression method (Wilkinson, 1961) specifically designed to obtain a best estimate for the kinetic parameters K_M and V_{max} of the Michaelis-Menten equation.

frequencies using T4 (43$^+$) bacteriophage (Hopkins and Goodman, 1979). Moreover, in an earlier study, Clayton et al. (1979) showed that in the formation of AP·T base mispairs (dATP and dAPTP competing opposite template T sites, Fig. 1) I(AP)/I(A) ≃ 14%, K_M^A/K_M^{AP} ≃ 14% (ΔG = 1.1 kcal/mol), and V_{max}^{AP} ≃ V_{max}^A. Thus, there are now two independent in vitro tests of passive polymerase model predictions that involve the two mispaired intermediates in the AP-induced A·T → G·C transition mutation pathways.

ACKNOWLEDGEMENTS

We thank Drs. J. J. Hopfield and I. Tinoco, Jr, for valuable discussions concerning this work. We are especially indebted to Dr. John Petruska for his open generosity in sharing with us his creative insights concerning tautomers and base-stacking interactions. We thank Ms. Sarah E. Wright for her expert preparation of the manuscript. Work reported here was supported by Grants GM21422 and CA17358 from the National Institutes of Health, by Grant 595 from the American Heart Association, and by the U.S. Department of

Energy, Contract W-7405-ENG-48 and Interagency Agreement 91-D-X0533 with the Environmental Protection Agency.

REFERENCES:

Bessman, M. J., Muzyczka, N., Goodman, M. F., and Schnaar, R. L., 1974, Studies on the biochemical basis of spontaneous mutation. II. The incorporation of a base and its analogue into DNA by wild-type mutator and antimutator DNA polymerases, J. Mol. Biol., 88:409.

Borer, P. N., Dengler, B., Tinoco, Jr., I., and Uhlenbeck, O. C., 1974, Stability of ribonucleic acid double-stranded helices, J. Mol. Biol., 86:843.

Clayton, L. K., Goodman, M. F., Branscomb, E. W., and Galas, D. J., 1979, Error induction and correction by mutant and wild-type T4 DNA polymerases, J. Biol. Chem., 254:1902.

Fisher, P. A., and Korn, D., 1977, DNA polymerase α purification and structural characterization of the near homogeneous enzyme from human KB cells, J. Biol. Chem., 252:6528.

Freese, E., 1959, The specific mutagenic effect of base analogues on phage T4, J. Mol. Biol., 1:87.

Fresco, J. R., Broitman, S., and Lane, A-E., 1980, Base mispairing and nearest-neighbor effects in transition mutations, in: "Mechanistic Studies of DNA Replication and Genetic Recombination," B. Alberts, and C. F. Fox, eds., ICN-UCLA Symposia on Molecular and Cellular Biology, Vol. XIX, Academic Press, New York.

Galas, D. J., and Branscomb, E. W., 1978, Enzymatic determinants of DNA polymerase accuracy theory of coliphage T4 polymerase mechanisms, J. Mol. Biol., 124:653.

Goodman, M. F., Hopkins, R., and Gore, W. C., 1977, 2-Aminopurine-induced mutagenesis in T4 bacteriophage: A model relating mutation frequency to 2-aminopurine incorporation in DNA, Proc. Natl. Acad. Sci. U.S.A., 74:4806.

Goodman, M. F., Hopkins, R. L., Watanabe, S. M., Clayton, L. K., and Guidotti, S., 1980, On the molecular basis of mutagenesis: Enzymological and genetic studies with the bacteriophage T4 system, in: "Mechanistic Studies of DNA Replication," B. Alberts, and C. F. Fox, eds., ICN-UCLA Symposia on Molecular and Cellular Biology, Vol. XIX, Academic Press, New York.

Hibner, U., and Alberts, B. M., 1980, Fidelity of DNA replication catalyzed in vitro on a natural DNA template by the T4 bacteriophage multi-enzyme complex, Nature, 285:300.

Hopkins, R., and Goodman, M. F., 1979, Asymmetry in forming 2-aminopurine·hydroxymethylcytosine heteroduplexes: A model giving misincorporation frequencies and rounds of DNA replication from replication from base-pair populations in vivo, J. Mol. Biol., 135:1.

Janion, C., and Shugar, D., 1973, Preparation and properties of poly 2-aminopurine ribotidylic acid, Acta Biochim. Pol., 20:271.

Liu, C. C., Burke, R. L., Hibner, U., Barry, J., and Alberts, B. M., 1978, Probing DNA replication mechanisms with the T4 bacteriophage in vitro system, Cold Spring Harbor Symp. Quant. Biol., 43:469.

Marmur, J., and Doty, P., 1962, Determination of the base composition of deoxyribonucleic acid from its thermal denaturation temperature, J. Mol. Biol., 5:109.

Ripley, L. S., 1981, The influence of diverse gene 43 DNA polymerases on the in vivo incorporation and replication of 2-aminopurine at A·T base-pairs in bacteriophage T4, J. Mol. Biol., in press.

Ronen, A., 1979, 2-Aminopurine, Mutat. Res., 69:1.

Tinoco, I., Jr., Borer, P. N., Dengler, B., Levine, M. D., Uhlenbeck, O. C., Crothers, D. M., and Gralla, J., 1973, Improved estimation of secondary structure in ribonucleic acids, Nature (New Biol.), 246:40.

Topal, M. D., DiGuiseppi, R., and Sinha, N. K., 1980, Molecular basis for substitution mutations, J. Biol. Chem., 255:11717.

Wang, M-L. J., Stellwagen, R. H., and Goodman, M. F., 1981, Evidence for the absence of DNA proofreading in HeLa cell nuclei, J. Biol. Chem., 256:7097.

Watanabe, S. M., and Goodman, M. F., 1981, On the molecular basis of transition mutations: The frequencies of forming 2-aminopurine·cytosine and adenine·cytosine base mispairs in vitro, Proc. Natl. Acad. Sci. U.S.A., 78:2864.

Watson, J. D., and Crick, F. H. C., 1953a, Genetical implications of the structure and deoxyribonucleic acid, Nature, 171:946.

Watson, J. D., and Crick, F. H. C., 1953b, The structure of DNA, Cold Spring Harbor Symp. Quant. Biol., 18:123.

Wilkinson, G. N., 1961, Statistical estimations in enzyme kinetics, Biochem. J., 80:324.

Wolfenden, R. V., 1969, Tautomeric equilibria in inosine and adenosine, J. Mol. Biol., 40:307.

CHAPTER 14

MUTATORS, ANTIMUTATORS, AND DNA REPLICATION ERRORS:

A SUMMARY AND PERSPECTIVE

Maurice J. Bessman

Department of Biology and McCollum-Pratt Institute
The Johns Hopkins University, Baltimore, Maryland
21218

Since the discovery of Escherichia coli DNA polymerase in the late 1950s, numerous papers have been written on the role of this enzyme and of several related enzymes in assuring that the nucleotide sequence of the parental DNA is transmitted with the high degree of fidelity to the progeny DNA as is seen during normal replication. The simplest view is that the nucleotide sequence of the daughter strand is determined uniquely by the complementary sequence in the parental or template strand according to the rules laid down by Chargaff and conceptualized by Watson and Crick. The polymerase molecule merely catalyzes the condensation of deoxynucleoside triphosphates according to the predetermined sequence in the template and, itself, does not contribute significantly to the accuracy of replication. This view had to be reevaluated in the light of the experiments of Speyer and of Drake, who showed, respectively, that various mutants in the DNA polymerase gene of phage T4 have markedly increased or decreased spontaneous mutation rates. It was these genetic experiments which focused the attention of enzymologists on the mechanism of action of DNA polymerases. Although we have learned a great deal about various aspects of these reactions, the four papers in this Section clearly point out that the interaction of polymerases with the template DNA is a complicated process and the accuracy of DNA synthesis is intimately linked to properties of the polymerases themselves, as well as to other factors in the replication complex.

The paper by Ripley and Shoemaker addressed the role of DNA polymerase in frameshift mutagenesis. They developed a clever assay based on suppression of known frameshifts in the rIIB cistron. Several interesting observations emerged. First, out of 19 recombinationally distinct alleles in gene 43, 9 showed more than threefold effects (some 50- to 60-fold) over the wild-type spontaneous revertants. This is rather convincing evidence that T4 DNA polymerase is involved in assuring frameshift fidelity. It is interesting that, in general, those gene 43 alleles which were known to have large effects on base-substitution mutations were also most active in frameshift mutagenesis. In a comparison of different frameshifts, it was observed that the particular "target" also plays a role. For example, one of the alleles (A58) that had comparatively low activity against the first frameshift tested had one of the highest activities against the second frameshift in a different locus of the rIIB gene. Also, frameshifts located in different regions of the gene showed different susceptibilities to the mutant genes. An interesting project for the near future will be to try to correlate the specific effects of the polymerase alleles with specific nucleotide sequences in the particular frameshifts.

Moore and Strauss discussed their studies on the template properties of DNA which had been damaged by known mutagens. Purified preparations of DNA polymerases (T4D, Pol I, Pol III holoenzyme, α from human lymphoma, and avian myeloblastosis virus (AMV) reverse transcriptase) were compared in respect to their ability to synthesize past specific blocks in ØX DNA. These blocks were produced by treatment with acetylaminofluorene (AAF) or UV light. In no case were these lesions bypassed, although specific differences were noted in the ability of certain of the polymerases to fill in a nucleotide opposite AAF. Most stop synthesizing DNA one nucleotide prior to the lesion. It would appear that the interruption of synthesis is not related to the proofreading exonuclease because α polymerase contains none. Also, we cannot generalize that the inability to move on is due to the lack of insertion of a correct nucleotide opposite the lesion. AMV transcriptase puts a cytosine in opposite AAF-derivatized guanine, but it stops at this point anyhow. Since these studies clearly demonstrate a very strong block in the synthesis of DNA off these specifically derivatized DNA preparations, it would seem that this type of DNA would be useful in searching for enzymes or factors induced under SOS conditions, which allow these lesions to be bypassed.

The paper by Schaaper, Kunkel, and Loeb also concerned itself with the effect of DNA lesions on DNA synthesis. In this case, the lesions are not produced by external mutagens such as AAF or ultraviolet light but by depurination of the DNA. The spontaneous depurination of DNA at neutral pH is no minor reaction, for it has been estimated by Lindahl to be in the range of 20,000 depurinations per day per mammalian cell. Although several mechanisms are known which would repair most of the depurinated regions, it would be expected that some sites would escape repair. Are these unrepaired sites mutagenic? Several approaches were used to investigate this question. First, synthetic polynucleotides were depurinated by mild heat or acid treatment and the incorporation of noncomplementary nucleotides was measured. The results suggested that single-base substitutions occurred opposite the depurinated sites and that the number of errors was independent of the source of the enzyme (E. coli Pol I, AMV DNA polymerase, or human DNA polymerase α). These observations should not be over-interpreted because of the special nature of these templates. Both poly[d(A-T)] and poly[d(G-C)] are structurally flexible molecules with less constraints than natural DNA, and they have been shown to behave quite differently in a number of enzymatic reactions. ØX174 DNA was used to advantage. It was depurinated and annealed with a complementary restriction fragment which served as a primer. From the length of DNA synthesized off this depurinated template, it could be shown that E. coli Pol I could synthesize past the depurinated sites. In a related series of experiments, depurinated ØX174 carrying an amber mutation was used as a template in vitro and then transfected into E. coli spheroplasts. The higher reversion frequency in the depurinated vs the nondepurinated control indicated that depurination leads to a higher error rate during DNA synthesis in vitro. An interesting offshoot of these experiments was the observation that depurinated ØX174 DNA was more mutagenic in a host previously irradiated with ultraviolet light. This suggests that the SOS system can also bypass depurinated templates but in so doing is error prone. It would have been interesting to see whether this increased mutagenicity were dependent on the recA protein.

The last paper in the Session, by Goodman, Watanabe, and Branscomb, addressed the mechanism of base-transition mutations. They pointed out that, although conceptually attractive, there was no strong experimental evidence that base substitutions were substantially caused by mispairings of unfavored tautomeric forms of the purines and pyrimidines. The authors focused their studies on 2-aminopurine (AP) mutagenesis but suggested that their observations may well be applicable to other base analogues as well as to the normal bases themselves. On the basis of measured insertion frequencies of dATP and dAPTP during DNA synthesis in vitro and on the misincorporation of dCTP using AP-containing templates, they argue that there is a strong inconsistency with a model based on tautomerism of AP. Instead, they offer a different model to account for the experimental observations which is based on the differential H-bonding capacities of adenine vs AP. In this model, they look on DNA polymerase as a "passive" enzyme that will insert any nucleotide which happens to be resident in its active site. Whether or not this nucleotide is incorporated into DNA in respect to a competing nucleotide will depend on the relative energies of the hydrogen bonds between the competing nucleotides and the template nucleotide for which they are competing. Support for this idea is the demonstration that the maximum velocities (obtained from Lineweaver-Burke plots) of the incorporation of C and T opposite template AP are equal, whereas the K_M value for C is 29-fold higher than the K_M for T. This is in good agreement with the relative insertion of T or C opposite AP, which in independent measurements was shown to be 21-fold higher for T. These observations have been generalized into the "passive polymerase" model which implicates the relative hydrogen-bond energies as the principal determinant of insertion frequencies and relegates tautomerism to a lesser role in base mispairings. The authors have generated a table listing possible base pairs in the order of their hydrogen-bond energies. It summarizes some of their recently obtained data and they suggest it has predictive value for base-competition data yet to be obtained. It will certainly be of interest to evaluate future data on the incorporation of base analogues in terms of the passive polymerase model.

CHAPTER 15

LOW LEVEL AND HIGH LEVEL DNA REARRANGEMENTS IN

Escherichia coli

        Ahmad I. Bukhari and Hajra Khatoon

        Cold Spring Harbor Laboratory, Cold Spring Harbor
        New York   11794

SUMMARY

    It can be argued that all organisms exhibit two levels of DNA rearrangements. At a low level they may occur sporadically in cells, perhaps largely because of spontaneous activity of transposable genetic elements. A high level may be induced in special circumstances if functions that cause rearrangements are hyperactive. As an example of low level genetic rearrangements, we have studied the occurrence of spontaneous polar mutations in the early regions of prophage Mu. We isolated 49 independent prophage mutants, which are defective in replication and expression of late genes; 44 were in the B region and 5 were in the A region. In the B region, 68% were IS1 insertions, 9% were IS5 insertions and 9% were IS2 insertions; 14% showed no insertion. In the A region, all 5 were IS5 insertions. Thus most spontaneous polar mutations in Escherichia coli appear to be insertions. IS1 is the most common insertion; however, certain DNA regions may show preference for a specific element. High level DNA rearrangements are exemplified by DNA fusion and DNA dissociation that occur when replication-transposition functions of Mu are induced.

INTRODUCTION

    The major emphasis in this symposium has been on base-pair changes in DNA. The production and repair of these lesions in DNA have been studied extensively. However, it is becoming clear that complex changes in DNA, that is, mutations other than single base-pair changes, are also biologically important and are perhaps the primary determinants in genetic diversification and evolution.

These complex changes include a whole set of different DNA rearrangements - transpositions, deletions, inversions, and duplications.

It can be argued that all organisms exhibit two levels of DNA rearrangements. There is always the low level of DNA rearrangements occurring in the cells, perhaps largely because of spontaneous activity of transposable genetic elements. A high level may be induced in special circumstances if functions that cause rearrangements become hyperactive for some reason. In Escherichia coli, insertions and deletions occur spontaneously at a certain frequency. When a powerful transposable element such as bacteriophage Mu is present in the cell, a potentially explosive situation exists. If the element becomes fully active, a high level of DNA rearrangement occurs. The low levels of DNA rearrangements are an important form of mutagenesis since it seems that most detectable spontaneous mutations in E. coli are insertions rather than base-pair changes. In some cases a DNA sequence may be particularly susceptible to deletion formation. In that case most of the mutations at that locus may be deletions, and this high mutation rate will then not be related to the activity of transposable elements.

LOW LEVEL REARRANGEMENTS

Insertion Mutations in E. coli

Although many diverse DNA rearrangements occur at a low frequency, the insertion mutations are perhaps the most important in terms of mutagenesis. There are presumably two factors which regulate the frequency of insertion mutations: (1) The number of insertion sequence (IS) elements. In E. coli there are several known IS elements that can transpose independently. (2) The number of sites available for insertion at the target locus or the target chromosomes. Different elements have different specificities; some integrate at only a few places, whereas others can be less specific.

In E. coli there are many copies of IS elements. For example, there are several copies of IS1 and IS5, and five copies of IS2 (see Bukhari et al., 1977; Calos and Miller, 1980; Cold Spring Harbor Symposium on Quantitative Biology, 1981; Kleckner, 1977; Nyman et al., 1981). Most of these elements can integrate at multiple sites within genes. Thus, insertion mutations in E. coli occur at a frequency that is easily detectable. This frequency can be enhanced manyfold if a new transposable element or a transposon is introduced into the cells. Cells in natural populations carrying such an element are likely to accumulate insertion mutations at a higher frequency.

We have made some observations on the occurrence of spontaneous insertion mutations in prophage Mu in E. coli. When E. coli cells carrying an Mucts prophage (carrying a temperature-sensitive repressor owing to a mutation in the c gene) are plated at 42°C, the cells are killed because of prophage induction. The survivors are mutants of the prophage in which the killing functions of Mu have been eliminated. The location of the genes responsible for cell death, the A, B, and kil genes of Mu, is shown in Fig. 1. This defect in the killing functions of Mu can occur (1) by a deletion of the entire left end of Mu, covering A, B, and the kil genes of Mu; (2) by nonsense polar mutations in the B gene; or (3) by insertions in A or B genes that are also polar for the downstream genes (Bukhari, 1975; van de Putte et al., 1981). All three groups of mutants have been isolated (Bukhari, 1975). However, the last group is the most significant; from 5 to 90% of the survivors can fall into this group. The wide variation in numbers of survivors containing insertions may result from sudden bursts of activity of transposable elements at different times during the growth of the culture. The second group containing nonsense mutations forms the minority class. It would seem that, at least in E. coli, spontaneous polar mutations primarily result from insertions.

Mucts prophages which are A^+ B^- (that is, are defective in DNA replication and killing) can be excised from the host chromosome at a low frequency. These defective prophages have been called the X mutants (Bukhari, 1975). Almost all the X mutants have been found to have insertions in the B gene (Bukhari, 1975; Bukhari and Taylor, 1975; Bukhari et al., 1976; Khatoon and Bukhari, 1981). We isolated 44 independent X mutants from Mucts prophages located at different sites in the lacZ gene of E. coli carried on a F' pro^+lac episome. We also examined 5 A^- mutants, which cannot be excised. Each of the prophages was induced in the presence of a helper Mucts. Phage particles, mixtures of wild-type and mutant phages, were purified. DNA was extracted with phenol and cut with restriction endonuclease EcoRI to see if the left end of Mu DNA carried any insertions in the mutants as has been described earlier (Bukhari, 1975; Bukhari and Taylor, 1975). As shown in Fig. 2, if there is an insertion in the B gene and if it does not carry an EcoRI cleavage site, a left-end fragment larger than the wild-type fragment will be obtained. If the insertion carries an EcoRI site, then two smaller fragments will be obtained. An example of this screening is shown in Fig. 3. Among the 18 mutants examined, 15 showed an insertion of about 800 base pairs in the left end, two others showed insertions of about 1400 base pairs carrying an EcoRI cleavage site, and one showed an insertion of about 1400 base pairs carrying no EcoRI restriction site. We have identified these insertions by DNA-DNA hybridization. The small insertions are IS1, insertions with an RI site are IS5, and insertions with no RI site are IS2. Table 1 shows the results of analysis of 44 X mutants and 5 A^- mutants of Mu. The insertions in the B gene were predominantly IS1, whereas IS5 and IS2 insertions were equal in frequency. Six mutants did not show any detectable insertion or deletion and have not been studied.

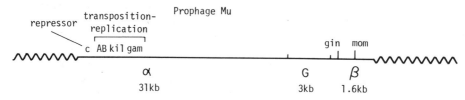

Fig. 1. Genetic map of prophage Mu showing functions responsible for replication-transposition and host cell killing. Repressor controls the synthesis of A and B proteins and other functions through a complicated regulatory circuit. Full expression of the transposition-replication region kills the host cells. Mutants with insertions in A or B genes are not killed. For a description of the kil and gam functions and the whole early region, see van de Putte et al. (1981). α, G, and β refer to different segments of Mu DNA. The sizes are given in kilobases. The gin function is involved in the inversion of the G segment and the mom function brings about modification of DNA.

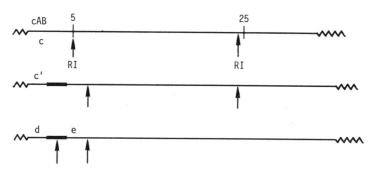

Fig. 2. The left end EcoRI fragment of Mu. The left end fragment (c) is 5 kilobases. An insertion in this region will give rise to a fragment (c') larger than c. If the insertion has an EcoRI cleavage site, then two fragments, d and e, will be generated, the combined lengths of which will exceed the length of the c fragment by the size of the insertion. The two EcoRI cleavage sites in Mu are indicated by arrows. The numbers indicate kilobases from the left end.

Surprisingly, all insertions in A gene were found to be IS5. Based on the size of the proteins, A being 70,000 molecular weight and B 32,000 (van de Putte et al., 1981), it can be estimated that the B gene is about 1000 base pairs in length, whereas A is at least twice as long. It would seem from the analysis of the Mu mutants that

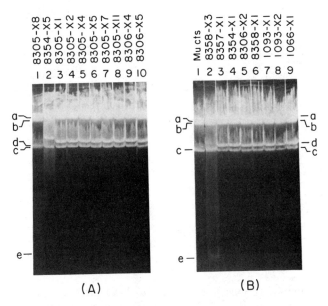

Fig. 3. Screening of Mucts X mutants for insertions near the left end of Mu. Mucts X mutants were induced in the presence of a helper Mucts prophage. The phage preparation thus yielded a mixture of wild-type and mutant DNAs. An EcoRI digest of Mucts is shown in column 1 in panel B. The left end fragment is labeled as c (a is the middle fragment and b is the right fragment; these fragments are irrelevant for the purpose of the present discussion). Most of the mutants yield two fragments, one is larger than c by 800 base pairs. In one case, (column 2, panel A) the fragment is larger by about 1300 base pairs. In two cases (column 1, panel A, and column 2, panel B) there are two additional fragments, d and e. The sum total of d and e is about 6400 base pairs (i.e., 1400 base pairs are added to the fragment c). The numbers at the top refer to the insertions in the lacZ gene, from which the X mutants were derived (see Bukhari, 1975 and Khatoon and Bukhari, 1981).

IS1 is the most frequently transposed element followed by IS5 and IS2. However, if a DNA region is strongly preferred by a transposable element, then that element may be the most frequent cause of mutations in that region. This may be the case for IS5 in the A gene of Mu.

Table 1. Spontaneous Polar Mutations in Genes A and B of Bacteriophage Mu

Region	Mutations	
	Insertion Type	Number (% of Total Examined)
A	IS5	5 (100)
B	IS1	30 (68)
	IS5	4 (9)
	IS2	4 (9)
	None	6 (14)

Localized Mutagenesis by Transposable Elements

Transposable elements generate deletions or inversions in their vicinity (Bukhari et al., 1977; Calos and Miller, 1980; Cold Spring Harbor Symposium on Quantitative Biology, 1981; Kleckner, 1977). Thus, a sequence containing a transposable element may be particularly susceptible to DNA rearrangements. In one class of rearrangements the element is retained at the site of deletions. This property is exhibited by IS elements, transposons, and Mu. In another class of rearrangements, the element is removed (imprecise excision), resulting in deletions that go either on one or the other side, or possibly on both sides, of the element. This is shown in Fig. 4. These deletions have been seen as a consequence of imprecise excision of the MuX mutants (Khatoon and Bukhari, 1981). IS4 is perhaps another element that can generate deletions spanning both sides of the element (Klaer et al., 1981). Thus, the imprecise excision of transposable elements can cause fusion of host sequences. The frequency of such events is low, 10^{-5} or 10^{-8} per cell in a culture. The elements can also be excised perfectly (precise excision), resulting in reversion of mutations caused by the insertion of the element. The frequency of precise excision is at least 10- to 100-fold lower than imprecise excision.

HIGH LEVEL DNA REARRANGEMENTS

The activities of transposable elements are generally regulated in such a manner that they do not cause DNA rearrangements that would kill the host cells. Transposons such as Tn3 and Tn5 encode functions that repress the synthesis of their transposases (see Cold Spring Harbor Symposium on Quantitative Biology, 1981, in general; Berg et al., 1981; Biek and Roth, 1981; Casadaban et al., 1981;

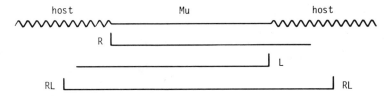

Fig. 4. Deletions as a consequence of Mu excision. Most deletions begin from one end of the prophage, remove the whole prophage and extend to the right (R) or left (L). Some deletions span the whole prophage (RL). The frequency overall of the three types of deletions ranges between 10^{-5} to 10^{-8} per cell (Khatoon and Bukhari, 1981).

Heffron et al., 1981). Transposons such as Tn10 appear to keep their transposition functions under check by mechanisms that interfere with the transcription of the transposition functions (N. Kleckner, personal communication). Bacteriophage Mu encodes a repressor (see Fig. 1) that keeps the A and B genes repressed in a lysogen allowing Mu prophage to be maintained stably. Mu thus behaves like a classical temperate phage.

Whenever transposition functions of an element are induced by inactivation of the repressor or by some other mechanisms of escape synthesis, there is a burst of transposition activity which is accompanied by various DNA rearrangements. For a discussion of these rearrangements see Harshey and Bukhari (1981), Shapiro (1979), and Toussaint et al. (1977). Common rearrangements are (1) transposition, in which an element moves to a new site by replication and leaves a copy behind; (2) replicon fusion or cointegrate formation, in which a molecule carrying a transposable element may fuse with another DNA molecule in such a manner that the transposable element is present at each junction of the fused molecules (Fig. 5); (3) DNA dissociation and deletion formation, in which segments of DNA next to the transposable element are pinched off as circles which contain a copy of the transposable element [the original host DNA molecule thus carries a deletion and still retains a copy of the element (Fig. 5).]; (4) inversion, in which transposition within a replicon generates an inversion of the sequences between the target site and the donor site.

These rearrangements clearly have dramatic genetic consequences for the cell. For transposons such as Tn3, the frequency of transposition is about 10^{-2} to 10^{-3} per cell. Under fully derepressed conditions this frequency goes up by about 100-fold such that essentially at least one event takes place in each cell. In bacteriophage Mu, however, the frequency of transposition accompanied by various rearrangements is extraordinarily high, about 100 transposition

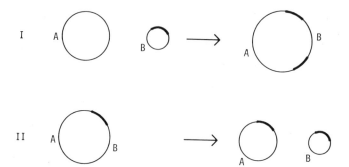

Fig. 5. Replicon fusion and DNA dissociation engineered by transposable elements. (I) Fusion: DNA molecules A and B are fused as a consequence of replication-integration of the transposable element (thick lines) forming structures known as cointegrates. The element serves as a linker and is present at each junction of the fused DNA molecules. (II) Dissociation: Replication-integration leads to a dissociation of a segment of DNA located next to the transposable element. What is left behind is a deletion. Both DNA products of this reaction carry a copy of the element.

events per cell. Mu has acquired functions that enhance the basal level of transposition by apparently increasing the efficiency of Mu DNA replication. Thus, $A^+ B^-$ Mu prophage behaves as a transposon, but $A^+ B^+$ prophage (carrying other necessary functions also) behaves as a virus as well as a transposon.

CONCLUSIONS

Thus, it would appear that in natural populations transposable elements are repressed and function only sporadically. This would result from spontaneous induction of their functions, just as is the case of spontaneous induction of prophages in lysogens. It is quite reasonable to suppose, however, that certain environmental or physiological conditions favor the induction of these functions. A clear example is a transposon for mercury resistance whose functions are induced in the presence of mercury (Sherratt et al., 1981). In bacteriophage Mu a full induction of the functions kills the cells. The survivors of such a cataclysmic event would probably have various degrees of changes in their genomes.

Do such events operate in cells higher than bacteria? Transposable elements have been clearly found in yeast (Fink et al., 1981) and in Drosophila (Rubin et al., 1981). Whether their functions are induced or regulated, as is the case with the transposons in bacteria, is not clear. McClintock (1975) in her classical studies on controlling elements in maize described excision and transposition activities of elements that are sporadic in nature but are controlled in time. In mammalian cells where transposition-like activities have so far been associated only with viruses (see Cold Spring Harbor Symposium on Quantitative Biology, 1981), indigenous transposable elements, which are normal residents of the genome, have not been detected so far. It is possible that these elements are normally quiescent and to detect them a cataclysmic event analogous to the induction of bacteriophage Mu has to be triggered.

ACKNOWLEDGMENTS

This work was supported by grants from the National Science Foundation and the National Institutes of Health. H.K. was on leave from the Department of Microbiology, University of Karachi, Pakistan (present address).

REFERENCES

Berg, D. E., Egner, C., Hirschel, B. J., Howard, J., Johnsrud, L., Jorgensen, R. A., and Tisty, T. D., 1981, Insertion, excision and inversion of Tn5, Cold Spring Harbor Symp. Quant. Biol., XLV:115.

Biek, D., and Roth, J. R., 1981, Regulation of Tn5 transposition, Cold Spring Harbor Symp. Quant. Biol., XLV:189.

Bukhari, A. I., 1975, Reversal of mutator phage Mu integration, J. Mol. Biol., 96:87.

Bukhari, A. I., Froshauer, S., and Botchan, M., 1976, The ends of bacteriophage Mu DNA, Nature, 264:580.

Bukhari, A. I., Shapiro, J., and Adhya, S., 1977, "DNA Insertion Elements, Plasmids, and Episomes," Cold Spring Harbor Laboratory, New York.

Bukhari, A. I., and Taylor, A. L., 1975, Influence of insertions on packaging of host sequences covalently linked to bacteriophage Mu DNA, Proc. Natl. Acad. Sci. U.S.A., 72:4399.

Calos, M. P., and Miller, J. H., 1980, Transposable elements, Cell, 20:579.

Casadaban, M. J., Chou, J., Lemaux, P., Tu, C.-P. D., and Cohen, S. N., 1981, Tn3: Transposition and control, Cold Spring Harbor Symp. Quant. Biol., XLV:269.

Cold Spring Harbor Symposium on Quantitative Biology, 1981, "Movable Genetic Elements," Vol. XLV, Cold Spring Harbor Laboratory, New York.

Fink, G., Farabaugh, P., Roder, G., and Chaleff, D., 1981, Transposable elements (ty) in yeast, Cold Spring Harbor Symp. Quant. Biol., XLV:575.

Harshey, R. M., and Bukhari, A. I., 1981, A mechanism of DNA transposition, Proc. Natl. Acad. Sci. U.S.A., 78:1090.

Heffron, F., Kostriken, R., Morita, C., and Parker, R., 1981, Tn$\underline{3}$ encodes a site-specific recombination system: Identification of essential sequences, genes and the actual site of recombination, Cold Spring Harbor Symp. Quant. Biol., XLV:259.

Khatoon, H., and Bukhari, A. I., 1981, DNA rearrangements associated with reversion of bacteriophage Mu induced mutations, Genetics, 98:1.

Klaer, R., Kühn, S., Fritz, J.-J., Tillmann, E., Saint-Girons, I., Habermann, P., Pfeifer, D., and Starlinger, P., 1981, Studies on transposition mechanisms and specificity of IS$\underline{4}$, Cold Spring Harbor Symp. Quant. Biol., XLV:215.

Kleckner, N., 1977, Translocatable elements in procaryotes, Cell, 11:11.

McClintock, B., 1975, The control of gene action in maize, Brookhaven Symp. Biol., 18:162.

Nyman, K., Nakamura, K., Ohtsubo, H., and Ohtsubo, E., 1981, Distribution of the insertion sequence IS$\underline{1}$ in gram-negative bacteria, Nature, 289:609.

Rubin, G. M., Brorein, W. J., Jr., Dunsmuir, P., Flavell, A. J., Levis, R., Strobel, E., Toole, J. J., and Young, E., 1981, Copia-like transposable elements in the Drosophila genome, Cold Spring Harbor Symp. Quant. Biol., XLV:619.

Shapiro, J. A., 1979, Molecular model for the transposition and replication of bacteriophage Mu and other transposable elements, Proc. Natl. Acad. Sci. U.S.A., 76:1933.

Sherratt, D., Arthur, A., and Burke, M., 1981, Transposon-specified, site-specific recombination systems, Cold Spring Harbor Symp. Quant. Biol., XLV:275.

Toussaint, A., Faelen, M., and Bukhari, A. I., 1977, Mu-mediated illegitimate recombination as an integral part of the Mu life cycle, in: "DNA Insertion Elements, Plasmids, and Episomes," (A. I. Bukhari, J. Shapiro, and S. Adhya, eds., Cold Spring Harbor Laboratory, New York.

van de Putte, P., Giphart-Gassler, M., Goosen, N., Goosen, T., and van Leerdam, E., 1981, Regulation of integration and replication functions of bacteriophage Mu, Cold Spring Harbor Symp. Quant. Biol., XLV:347.

CHAPTER 16

MUTANTS OF Escherichia coli K12 WHICH AFFECT EXCISION OF TRANSPOSON TN10

Victoria Lundblad and Nancy Kleckner

Department of Biochemistry and Molecular Biology
The Biological Laboratories, Harvard University
Cambridge, Massachusetts   02138

SUMMARY

We have described three illegitimate recombination events associated with, but not promoted by, transposon Tn10: precise excision, nearly precise excision, and precise excision of a nearly precise excision remnant. All three are structurally analogous: excision occurs between two short direct repeat sequences, removing all intervening material plus one copy of the direct repeat. In each case, the direct repeats border a larger inverted repeat. We report here the isolation of host mutants of Escherichia coli K12 which exhibit increased frequencies of precise excision of Tn10. Nineteen of the 39 mutants have been mapped to five distinct loci on the E. coli genetic map and have been designated texA through texE (for Tn10 excision). Mapping and genetic characterization indicate that each tex gene corresponds to a previously identified gene involved in cellular DNA metabolism: recB and/or recC, uvrD, mutH, mutS, and dam. The role of these various DNA repair and recombination genes in an illegitimate recombination process such as Tn10 excision will be discussed. In addition to an increase in precise excision frequency, all 39 tex mutants display an increased frequency for nearly precise excision. However, none of the mutants are increased for the third excision event, precise excision of a nearly precise excision remnant, supporting the idea that precise and nearly precise excision occur by closely related pathways which are distinct from those pathways which promote the third type of excision event.

INTRODUCTION

Illegitimate recombination can be defined as chromosomal alterations which are independent of normal recombination genes (such as recA), occur in the absence of gross sequence homology, and are not promoted by specific elements such as transposons. No extensive characterization of the host genes that mediate these DNA rearrangements has been reported, largely because these processes are rare and often difficult to assay. We have described three DNA excision events involving Tn10 which are examples of illegitimate recombination; these events are recA independent and are not promoted by Tn10. Because Tn10 excision is simple to assay and involves potentially interesting interactions between inverted and direct repeats of DNA, we have chosen it as a model system for illegitimate recombination. We present in this report an analysis of Escherichia coli K12 host mutants which affect excision of transposon Tn10.

THREE Tn10 EXCISION EVENTS

The transposable tetracycline-resistance element Tn10 is 9300 base pairs (bp) in length. The ends of the element are 1400 bp inverted repeats, with the 2500 bp tetracycline-resistance determinant located asymmetrically within the nonrepeating 6500 bp material. Transposition of Tn10 into a new site results in a 9 bp duplication of the target DNA sequence and insertion of the transposon between 9 bp direct repeats. Precise excision of Tn10 can occur by deletion of material between these 9 bp direct repeats in such a manner that the interrupted gene is restored to its original wild-type sequence (see Fig. 1).

Nearly precise excision utilizes other short repeat sequences which occur within and very near the ends of Tn10. Deletion of Tn10 material between these direct repeats, in a manner analogous to precise excision, results in excision of all but 50 bp of Tn10 DNA, with one 24 bp direct repeat still remaining.

The identification of nearly precise excision derivatives as a particular class of high-reverting polarity-relief revertants has also revealed a third Tn10 excision event, precise excision of the 50 bp nearly precise excision remnant. Nearly precise excision derivatives still retain the 9 bp direct repeats of the original insertion (see Fig. 1). Deletion of the remaining Tn10 material plus one of the two 9 bp repeats can occur, resulting in restoration of the wild-type target DNA sequence.

All three of these Tn10 excision events are structurally analogous: they all involve deletion between two short direct repeats, removing the intervening material and one copy of the direct repeat. In addition, the direct repeats always border a larger inverted

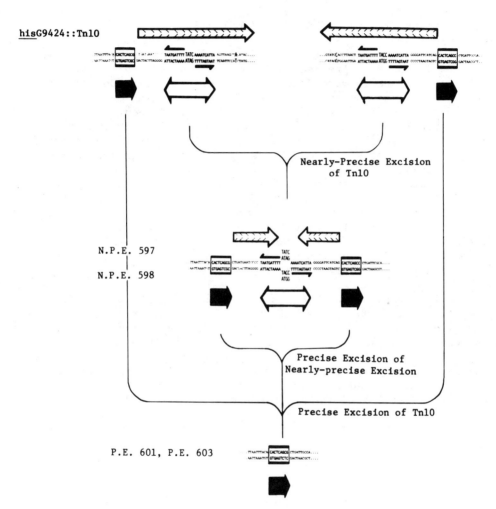

Fig. 1. Three Tn10-associated excision events: precise excision of Tn10, nearly-precise excision of Tn10, and precise excision of the nearly-precise excision "remnant" (from Foster et al., 1981).

repeat, which is itself deleted in the process of excision. A variety of genetic observations (Botstein and Kleckner, 1977; Egner and Berg, 1981; Foster et al., 1981) have suggested that these excision events are not promoted by the transposon itself but are, instead, mediated by the host cell. To understand what types of cellular DNA-handling processes might be involved in Tn10 excision, we have isolated and characterized bacterial host mutants which are affected for Tn10 excision.

ISOLATION OF PRECISE EXCISION HOST MUTANTS

Mutants of E. coli K12 were isolated that displayed increased frequencies of Tn10 precise excision for a particular lacZ::Tn10 insertion. Mutants were identified by plating on MacConkey lactose indicator plates and screening for individual colonies which displayed increased numbers of Lac$^+$ papillae, a procedure first developed by Hopkins et al. (1980) for Tn5. A total of 39 independent mutants were identified from 60,000 colonies of ethyl methanesulfonate- or N-methyl-N'-nitro-N-nitrosoguanidine-mutagenized bacteria. All 39 mutants had acquired one or more lesions in the bacterial genome, rather than in the Tn10 element itself, as shown by transduction of the lacZ::Tn10 insertion from the unmutagenized strain back into Lac$^+$ revertants of each mutant; each reconstructed strain exhibited a papillation phenotype identical to that of the original mutant. In addition to the papillation phenotype, all 39 mutants demonstrated increased frequencies of Lac$^+$ revertants in liquid reversion tests with increases of 10-to 10,000-fold.

All 39 mutants exhibit a single phenotype with regard to the three Tn10 excision events. In addition to an increase in precise excision frequency, all of the mutants display an increased frequency for nearly precise excision (ranging from 5- to 500-fold), with an approximate correlation between the degree of increase for the frequency of precise excision and that for nearly precise excision. However, none of the mutants are increased more than fivefold for the third excision event, precise excision of the nearly precise excision remnant. This suggests that precise excision and nearly precise excision occur by closely related pathways and that these pathways are distinct from the pathway(s) for the third type of excision event, despite the structural features common to all three events.

Each of the mutants was also tested for three additional phenotypes which could be sensitive to changes in cellular DNA metabolism: sensitivity to ultraviolet light, spontaneous mutation rate, and altered growth of several bacteriophage λ derivatives mutated in the phage recombination genes red and gam. The viability of these particular phage mutants is known to be altered by mutations in recA, polA, lig, recB, and recC (Zissler et al., 1971). Twelve of the 39 precise excision host mutants are wild type for all three phenotypes. None of the mutants display a marked increase in sensitivity

to UV, although some show a slight increase. Three mutants fail to
plate λ red⁻gam⁻ phages, eight mutants exhibit increases or decreases
of 5-to 500-fold in spontaneous mutation frequency, and 16 mutants
are altered in both of these properties [see Foster et al. (1981)
for further details].

GENETIC LOCATION AND CHARACTERIZATION OF PRECISE EXCISION MUTANTS

Twenty of the 39 precise excision host mutants have been mapped
by Hfr matings and subsequent P1 transduction to five distinct loci
on the genetic map of E. coli (see Table 1). These five genes have
been designated texA through texE (for Tn10 excision). On the basis
of genetic phenotypes and frequency of co-transduction with nearby
markers, each of these five tex genes appears to correspond to a
previously identified gene known to be involved in cellular DNA
metabolism: texA with recB/recC, texB with uvrD, texC with mutH,
texD with mutS, and texE with dam. None of the 12 tex mutants which
are normal for the DNA metabolism phenotypes tested have yet been
mapped.

texA

Four observations suggest that the three mutations in the texA
locus are alleles of either recB or recC (Lundblad, Taylor, Smith,
and Kleckner, manuscript in preparation): (1) all three map with
recB and recC between thyA and argA on the E. coli K12 map (approximately minute 60.3); (2) λ gam⁻ mutants (λ gam210) grow very poorly
on texA mutants, and the only known function of the gamma gene product is its ability to inactivate the recBC nuclease (Karu et al.,
1974); (3) Andrew Taylor and Gerald Smith (personal communication)
have shown that recombination among lambda phages carrying the
genetic element Chi, which is a recBC-dependent process (Stahl and
Stahl, 1977), is altered in texA mutants; and (4) Andrew Taylor
(personal communication) has shown that texA343 is complemented by
F'15recB21 but not by F'15recC22, suggesting that texA343 at least
maps in recC (however, since recC22 is an amber mutation, texA343
could map in a gene downstream from recC).

None of the three texA mutants exhibit recombination phenotypes
typical of classical recB and recC alleles (see Table 1). Not only
are texA mutations defective for plating λ gam⁻ phages (while recB21
and recC22 strains plate such phages better than a Rec⁺ strain), but
texA mutants give normal frequencies of recombinants when used as
recipients in Hfr- and P1-mediated crosses, while recB21 and recC22
mutants exhibit decreases of up to 50-fold in such recombination
tests. In addition, while texA strains show increases in precise
and nearly precise frequencies of 5-to 80-fold (Table 1), the

Table 1. Classification of 20 Host Mutants by Map Position and Phenotypes

Tex Locus	Total of mutants	Probable gene assignment	Increases of Tn10 Excision[a]			Recombination Phenotypes			Other Phenotypes		Map Position
			PE of Tn10	NPE of Tn10	PE of NPE	Plating of λ gam⁻red⁻	Hyper-rec phenotype[b]	Recombination (Hfr crosses)	UV^{SC}	Increase in Frequency of Spontaneous Mutations	(P1 co-transduction)
texA	3[d]	recC	5-80x	5-40x	normal	−	normal	normal	+/−	normal	60% w. argA; 70% w. thyA
texB	14[e]	uvrD	15-100x	25-80x	normal	−	increased	nt	−/+	15-80x	38-75% w. metE
texC	1	mutH	15x	7.5x	normal	+	nt	nt	normal	150x	76% w. thyA
texD	1	mutS	5x	nt	normal	+	nt	nt	normal	35x	18% w. srl
texE	1	dam	200x	370x	normal	+	nt	nt	normal	300x	12% w. rpsL

[a]PE, precise excision; NPE, nearly precise excision; PE of NPE, precise excision of the nearly precise excision "remnant"; x, fold increase;
[b]nt, not tested.
[c]UV^S, sensitivity to ultraviolet light.
[d]From three independent mutagenized cultures.
[e]From five independent mutagenized cultures.

standard alleles recB21 and recC22 affect Tn10 excision no more than
two- to threefold (Table 2). Finally, texA mutations do not confer
any reduced viability on a growing cell culture, whereas recB21 and
recC22 reduce cell viability to about 40% that of Rec$^+$ strains (data
not shown). The unusual phenotypes of texA mutants could reflect
either overproduction of normal recBC nuclease or some qualitative
alteration in recBC action. Recent evidence from Gerald Smith's
laboratory (Taylor and Smith, personal communication) favors the
latter possibility, and this issue is under active investigation.

texB

More than one-third of all tex mutants appear to map in a single
locus, texB, at about minute 84. The correspondence between the map
positions and phenotypes of known uvrD mutants and of texB mutants,
shown in Tables 1 and 2, strongly suggests that texB mutants are
alleles of uvrD. Alleles of uvrD have previously been isolated on
the basis of several different phenotypes: increased frequency of
spontaneous mutations (mutU; Siegel, 1973), increased sensitivity
to ultraviolet light (uvrD, uvrE; Ogawa et al., 1968; van de Putte
et al., 1965), decreased recombination in the recF pathway (recL152;
Horii and Clark, 1973), or a "hyper-rec" phenotype (uvrD210; Lloyd
and Scott, 1978). All of these mutations appear to map in a single
gene, designated uvrD (Kushner et al., 1978) and all have been
demonstrated to confer slight UV sensitivity and a modest mutator
phenotype (although for some uvrD alleles, UV sensitivity and mutator
activity can vary in different genetic backgrounds; Arthur and
Lloyd, 1980) and to display an increased frequency of genetic exchanges (hyper-rec phenotype). The effects of these alleles of uvrD
on precise excision and on plating of λ red$^-$ gam$^-$ phages are shown
in Table 2. Based on these two phenotypes, the four alleles of uvrD
fall into two categories. MutU4 and uvrD3 have the same phenotype
as texB mutants; they do not plate the lambda mutants and are increased for Tn10 precise excision by 15- to 20-fold. By contrast,
uvrE502 and recL152 are wild type for both phenotypes.

texC, texD, and texE

Eight of the 39 tex mutants comprise another distinct category
with regard to spontaneous mutation rates and plating of λ red$^-$gam$^-$
mutants: they have increased mutator activity by 10- to 300-fold,
but plate λ red$^-$gam$^-$ variants normally. Three of these mutants have
been mapped to three different loci, designated texC, texD, and
texE. Co-transduction experiments suggest that texC, texD, and texE
correspond to the known mutator loci mutH, mutS, and dam, respectively (Table 1). Consistent with this assignment, standard alleles
of these three genes all increase precise excision by 5- to 15-fold
(see Table 2). These three genes, in addition to mutL, have been

Table 2. Effect of Known DNA Metabolism Mutants on Tn10 Excision[a]

Mutation	Plating of λγ⁻red⁻	PE of Tn10	NPE of Tn10	PE of NPE	References
recA	−	normal	nt	nt	Howard-Flanders and Theriot (1966)
recB21	++	normal	normal	normal	Emmerson (1968)
recC22	++	normal	normal	normal	
uvrE502	+	normal	nt	nt	Smirnov and Skavronskaya (1971)
recL152	+	normal	nt	nt	Horii and Clark (1973)
mutU4	−	20x↑	nt	nt	Siegel (1973)
uvrD3	−	10x↑	nt	nt	Ogawa et al. (1968)
mutH3	+	6x↑	nt	nt	Hill (1970)
mutS3	+	10x↑	nt	nt	Siegel and Bryson (1964)
dam-3	+	15x↑	nt	nt	Marinus and Morris (1973)
polA1	−	normal	normal	70x↑	De Lucia and Cairns (1969)
polA214	−	normal	normal	40x↑	Kingsbury and Helinski (1973)
top10	+	6x↑	6x↑	2.5x↑	Sternglanz et al. (1981)
arl-1	+/−	11x↑	nt	nt	Hays and Korba (1979)
arl-2	nt	2x↑	nt	nt	Hays and Korba (1979)
arl-4	+/−	75x↑	nt	nt	Hays and Korba (1979)
arl-5	nt	4x↑	nt	nt	Hays and Korba (1979)

[a] Abbreviations defined in Table 1 footnotes; ↑, increase.

postulated to interact in a methylation-instructed mismatch repair pathway (see Conclusions). Mapping of the remaining five tex mutants is in progress to determine if all eight mutants in this particular category lie in the four genes that comprise this repair pathway.

Other Mutants of E. coli K12 with Altered Tn10 Excision Frequencies

Mutations in a number of other genes known to be involved in DNA metabolism have been tested for their effect on Tn10 excision, as shown in Table 2. These include two alleles of polA, shown to map in the two different domains of DNA polymerase I (Kelley, 1979), top10, believed to be a nonsense mutation in the structural gene for topoisomerase I (Sternglanz et al., 1981), and four alleles of arl, a subclass of E. coli hyper-rec mutants which accumulate recombinogenic lesions in DNA (Hayes and Korba, 1979). Unlike the 39 tex mutants described in this report, polA had no effect on precise or nearly precise excision; however, precise excision of a nearly precise excision is increased 40-to 70-fold in both polA mutants. The top10 mutation increased all three Tn10 excision events by three- to eightfold, and the arl mutations increase precise excision (the only Tn10 excision event tested so far) by 2-to 75-fold. Investigation of the unmapped tex mutants is in progress to determine if any tex mutants map in either top or arl.

Several other mutants of E. coli have previously been shown to affect precise excision of transposons. HimA mutations, isolated because of their failure to support site-specific integration of bacteriophage lambda, also reduce the frequency of precise excision of Tn5 and Tn10 by nearly 1000-fold (Miller and Friedman, 1980). Hopkins et al. (1980) have isolated two classes of mutants of the E. coli sex factor F which affect precise excision of transposons. Mutations in one of these two loci, ferB, also stimulate homologous recombination between IS3 sequences residing on F.

CONCLUSIONS

The three Tn10 excision events described in Fig. 1 are particularly intriguing types of illegitimate recombination events because of the structural involvement of direct and inverted repeats. As the actual site of the recombination event, the direct repeats in all three cases must be explicitly involved in excision, possibly because of "re-pairing" or "slippage" between the two repeat sequences. A role for the inverted repeats in excision comes from genetic evidence suggesting that a reduction in the length of the inverted repeats reduces precise excision of Tn5 and both precise and nearly precise excision of Tn10 (Egner and Berg, 1981; Foster et al., 1981). The inverted repeats may interact with one another either as two intact duplex segments or, more likely, by intrastrand pairing to form "snap-back" or "cruciform" structures. Presumably the interaction between inverted repeats helps to bring together in space the two short direct repeat sequences which must directly interact.

The texA mutants appear to be unique alleles of recC or (less likely) recB, two genes whose products specify an ATP-dependent DNase which plays an important role in general recombination (Clark, 1973). A number of different activities have been described for this enzyme, including exonucleolytic degradation of DNA, DNA unwinding, and endonucleolytic cleavage (Goldmark and Linn, 1972; Rosamund et al., 1979; Taylor and Smith, 1980). The action of the RecBC enzyme in general recombination is also known to involve specific DNA sequences called Chi sites (Stahl and Stahl, 1977). The texA mutants might increase or decrease the level of one of the RecBC activities and/or they might involve a qualitative alteration in the way this enzyme recognizes Chi sites. The in vitro activities of wild-type RecBC enzyme normally require either a double-stranded DNA end or a single-stranded gap, which suggests that Tn10 excision might involve one or both of these structures.

Four of the five identified tex loci correspond to functions involved in DNA repair and correction of base-pair mismatches (uvrD, mutS, mutH, and dam; Nevers and Spatz, 1975; Rydberg, 1977, 1978). The identification of mutations in these four genes provides further support for their interrelated action in vivo and suggests that we may be able to identify some particular feature of Tn10 excision which is sensitive to the action of these functions. Mutations in mutS and mutH are known to suppress certain phenotypes of dam mutants (Glickman and Radman, 1980; Marinus, 1980; McGraw and Marinus, 1980). However, mutations in all three of these genes stimulate Tn10 excision. In light of the postulated roles of mutS, mutH, and mutL in excision of base-pair mismatches, one might imagine that these functions normally recognize and destroy some structure(s) which are important for Tn10 excision, while lesions in the dam function might stimulate the formation of such structures. The product of uvrD might also be postulated to destroy some important intermediate in the excision pathway. If Tn10 excision does involve snapback or cruciform structures, such structures might themselves be recognized by repair enzymes as abnormal forms of DNA. Also, the inverted repeat segments of Tn10 are not identical; they differ at seven positions within 200 bp of the ends of the element (Foster et al., 1981). Thus, formation of intrastrand snapbacks during excision would generate standard base-pair mismatches, and these mismatches could be the targets of mutH and mutS action. On the other hand, it may be that mutations in all of these DNA repair genes exert an indirect effect on precise excision. For example, they might cause the production or accumulation of DNA damage which would in turn stimulate the activity of those functions which are directly involved in the excision process.

A number of the mutations shown thus far to increase Tn10 excision also stimulate genetic exchanges (the hyper-rec phenotype). Mutations in uvrD, mutH, mutL, mutS, dam, and arl all have a hyper-rec phenotype. On the other hand, polA mutations, which also confer a hyper-rec phenotype, do not increase Tn10 precise excision (Table 2). The relationship between increased Tn10 excision and the hyper-rec phenotype remains to be understood.

We cannot yet say exactly how tex mutants stimulate precise excision and nearly precise excision of Tn10. Experiments are under way to map the remaining tex mutations, to probe the relationships among dam, uvrD, mutH, mutS, and mutL in the excision pathway, to understand how unusual alleles of recBC can influence Tn10 excision, to examine the effect of tex mutations on spontaneous deletion formation, and to determine whether these mutations are specifically influencing interactions involving direct and/or inverted repeats. It is already clear that analysis of these mutants will shed light on the action of previously identified DNA-handling functions; it is also likely that new functions may be represented among the mutations which have not yet been mapped.

ACKNOWLEDGMENTS

The experiments reported here were supported by research grants to N.K. from the National Science Foundation (PCM 79-23508) and the National Institutes of Health (5 ROI GM 25,326-03).

We gratefully acknowledge the assistance of Anna Ferri in preparation of this manuscript.

REFERENCES

Arthur, M., and Lloyd, R. G., 1980, Hyperrecombination in uvrD mutants of Escherichia coli K-12, Mol. Gen. Genet., 180:185.
Botstein, D., and Kleckner, N., 1977, Translocation and illegitimate recombination by the tetracycline-resistance element Tn10, in: "DNA Insertion Elements, Plasmids and Episomes," A. I. Bukhari, J. A. Shapiro, and S. Adhya, eds., Cold Spring Harbor, New York.
Clark, A. J., 1973, Recombination deficient mutants of Escherichia coli and other bacteria, Annu. Rev. Genet., 7:67.
De Lucia, P., and Cairns, J., 1969, Isolation of an E. coli strain with a mutation affecting DNA polymerase, Nature, 224:1164.
Egner, C., and Berg, D., 1981, Excision of transposon Tn5 is dependent on the inverted repeats but not on the transposase function of Tn5, Proc. Natl. Acad. Sci. U.S.A., 78:459.

Emmerson, P. T., 1968, Recombination deficient mutants of Escherichia coli K12 that map between thyA and argA, Genetics, 60:19.

Foster, T. J., Lundblad, V., Hanley-Way, S., Halling, S. M., and Kleckner, N., 1981, Three Tn10-associated excision events: relationship to transposition and role of direct and inverted repeats, Cell, 23:215.

Glickman, B. W., and Radman, M., 1980, Escherichia coli mutator mutants deficient in methylation-instructed DNA mismatch correction, Proc. Natl. Acad. Sci. U.S.A., 77:1063.

Goldmark, P. J., and Linn, S., 1972, Purification and properties of the recBC DNase of Escherichia coli K-12, J. Biol. Chem., 247:1849.

Hays, J. B., and Korba, B. E., 1979, DNA from recombinogenic λ bacteriophages generated by arl mutants of Escherichia coli is cleaved by single-strand-specific endonuclease S1, Proc. Natl. Acad. Sci. U.S.A., 76:6066.

Hill, R. F., 1970, Location of genes controlling excision repair of UV damage and mutator activity in Escherichia coli WP2, Mutat. Res., 9:341.

Hopkins, J. D., Clements, M. B., Liang, T., Isberg, T., and Syvanen, M., 1980, Recombination genes on the Escherichia coli sex factor specific for transposable elements, Proc. Natl. Acad. Sci. U.S.A., 77:2814.

Horii, Z. I., and Clark, A. J., 1973, Genetic analysis of the recF pathway to genetic recombination in Escherichia coli K-12: Isolation and characterization of mutants, J. Mol. Biol., 80:327.

Howard-Flanders, P., and Theriot, L., 1966, Mutants of Escherichia coli K-12 defective in DNA repair and in genetic recombination, Genetics, 53:1137.

Karu, A., Sakaki, Y., Echols, H., and Linn, S., 1974, In vitro studies of the gam gene product of bacteriophage λ, in: "Mechanisms in Recombination," R. F. Grell, ed., Plenum Press, New York.

Kelley, William S., 1980, Mapping of the polA locus of Escherichia coli K12: genetic fine structure of the cistron, Genetics, 95:15.

Kingsbury, D. T., and Helinski, D. R., 1975, Temperature-sensitive mutants for the replication of plasmids in E. coli. I. Isolation and specificity of host and plasmid mutations, Genetics, 74:17.

Konrad, E. B., 1977, Method for the isolation of Escherichia coli mutants with enhanced recombination between chromosomal duplications, J. Bacteriol., 130:162.

Kushner, S. R., Shepard, J., Edwards, G., and Marples, V. R., 1978, uvrD, uvrE and recL represent a single gene, J. Supramol. Struct. Suppl. 2:59.

Lloyd, R. G., and Scott, H. M., 1978, A non-selective assay of recombination in Escherichia coli K-12, Soc. Gen. Microbiol. Q., 6:42.

Marinus, M. G., 1980, Influence of uvrD3, uvrE502, and recL152 mutations on the phenotypes of Escherichia coli K-12 dam mutants, J. Bacteriol., 141:223.

Marinus, M. G., and Morris, N. R., 1973, Isolation of deoxyribonucleic acid methylase mutants of Escherichia coli K12, J. Bacteriol., 114:1143.

McGraw, B. R., and Marinus, M. G., 1980, Isolation and characterization of Dam$^+$ revertants and suppressor mutations that modify secondary phenotypes of dam-3 strains of Escherichia coli K-12, Mol. Gen. Genet., 178:309.

Miller, H. I., and Friedman, D. I., 1980, An E. coli gene product required for lambda site-specific recombination, Cell, 20:711.

Nevers, P., and Spatz, H-C., 1975, Escherichia coli mutants uvrD and uvrE deficient in gene conversion of λ-heteroduplexes, Mol. Gen. Genet., 139:233.

Ogawa, H., Shimada, K., and Tomizawa, J., 1968, Studies on radiation-sensitive mutants of E. coli. I. Mutants defective in the repair synthesis, Mol. Gen. Genet., 101:227.

Rosamund, J., Teleander, K. M., and Linn, S., 1979, Modulation of the action of the recBC enzyme of Escherichia coli K-12 by Ca^{++}, J. Biol. Chem., 254:8646.

Rydberg, B., 1977, Bromouracil mutagenesis in Escherichia coli: Evidence for involvement of mismatch repair, Mol. Gen. Genet. 152:19.

Rydberg, B., 1978, Bromouracil mutagenesis and mismatch repair in mutator strains of Escherichia coli, Mutat. Res., 52:11.

Siegel, E. C., 1973, Ultraviolet-sensitive mutator strain of Escherichia coli K-12, J. Bacteriol., 113:145.

Siegel, E. C., and Bryson, V., 1964, Selection of resistant strains of Escherichia coli by antibiotics and antibacterial agents: Role of normal and mutator strains, Antimicrob. Agents Chemother., 1963:629.

Smirnow, G. B., and Skavronskaya, A. G., 1971, Location of uvr502 mutation on the chromosome of Escherichia coli K-12, Mol. Gen. Genet., 113:217.

Stahl, F. W., and Stahl, M. M., 1977, Recombination pathway specificity of Chi, Genetics, 86:715.

Sternglanz, R., DiNardo, S., Voelkel, K. S., Nidhimuts, Y., Hirota, Y., Becherer, K., Zumstein, L., and Wang, J., 1981, Mutations in the gene coding for Escherichia coli DNA topoisomerase I affect transcription and translation, Proc. Natl. Acad. Sci. U.S.A., in press.

Taylor, A., and Smith, G. R., 1980, Unwinding and rewinding of DNA by the recBC enzyme, Cell, 22:447.

Van de Putte, P., van Sluis, C. A., van Dillewijn, R., and Rorsch, A., 1965, The location of genes controlling radiation sensitivity in Escherichia coli, Mutat. Res., 2:97.

Zissler, J., Signer, E., and Schaefer, F., 1971, The role of recombination in growth of bacteriophage lambda. I. The gamma gene, in: "The Bacteriophage Lambda," A. D. Hershey, ed., Cold Spring Harbor Press, Cold Spring Harbor.

CHAPTER 17

GENE CONVERSION: A POSSIBLE MECHANISM FOR ELIMINATING

SELFISH DNA

Robin Holliday

National Institute for Medical Research, The Ridgeway
Mill Hill, London NW7 1AA, England

A strong case has recently been made for the existence of selfish DNA in higher organisms (Doolittle and Sapienza, 1980; Orgel and Crick, 1980). Given that genetic elements exist which have the means of reproducing themselves by transposition to new locations, they would be expected to gradually accumulate in the genome, especially in diploid organisms. The essence of the argument is that each insertion of a nonfunctional DNA sequence, which only very slightly increases the size of the genome, would have a neglible phenotypic effect on the individual and therefore would not be selected against. For the amount of DNA to increase, all that is required is that the probability of a DNA sequence being added to the genome is greater than the probability of it being eliminated by natural selection or by random deletion. There will also be positive selection for genetic elements which increase by mutation their efficiency of transposition. Not only will genomes increase in size, but also most of the DNA will become "junk," with no function for the organism. This conclusion may help solve the C-value paradox, namely, that higher organisms appear to have much more DNA than is required for the synthesis and control of gene products, and that there is enormous variation of DNA content between different taxonomic groups or even between quite closely related species of higher organisms.

The argument critically depends on the assumption that the probability of insertion of a sequence of selfish DNA is greater per generation than the probability of its removal or the removal of an equivalent sequence elsewhere in the genome. I question this assumption and suggest that organisms may have evolved an efficient mechanism for removing, or at least keeping in check, selfish DNA. This mechanism depends on gene conversion during meiosis. After homologous pairing of chromosomes at pachytene of meiosis, maternal and paternal DNA sequences interact at one or more positions along the length of the chromosome. This interaction is necessary for recombination, and it is very likely that it involves the formation of hybrid or heteroduplex DNA, which subsequently leads to gene conversion and/or crossing over (for reviews, see Holliday, 1977; Pukkila, 1977; Stahl, 1979). The means by which hybrid DNA is formed is not known in detail, but it may require DNA unwinding proteins or topoisomerases. In any event, if the hybrid DNA spans a region containing a sequence of selfish DNA present in one parental chromosome but not the other, a looped-out single-stranded structure will be formed, with normal hydrogen-bonded double-stranded DNA on either side (Fig. 1). Such a major distortion in DNA is likely to be a substrate for repair enzymes; in particular, active single-stranded endonucleases are known which could efficiently degrade the loop (Ando, 1966; Pukkila, 1978; Rabin et al., 1972; Sutton, 1971). Ligation of the resulting single-stranded nick will restore normal double-stranded DNA. The upshot of this conversion event will be the elimination of the insert of selfish DNA. It is assumed that the presence of the structural heterozygosity (i.e., an addition) will not prevent hybrid DNA formation. This appears to be the case in yeast (Fink and Styles, 1974), but in Ascobolus the frequency of conversion is reduced by large deletions (Girard and Rossignol, 1974). It also assumed that the conversion event is polarised, that is, the tendency is always to remove the loop. As yet, no evidence on this point is available from higher organisms. In yeast, conversion of deletions can occur in either direction (see Radding, 1978), but in bacteriophage T4 of Escherichia coli (Benz and Berger, 1973) and bacteriophage SPP1 of Bacillus subtilis (Trautner et al., 1972), single-stranded loops in heteroduplex DNA are preferentially removed, as suggested here. [Yeast may be a special case because it may have a mechanism for the transposition of genetic elements by gene conversion (see Leupold, 1980)].

The frequency of conversion per gene in fungi is in the range 0.1 to 5%, with a few loci showing as much as 20% conversion (Fogel et al., 1978; MacDonald and Whitehouse, 1979; Rossignol et al., 1978). In higher eukaryotic organisms a direct measurement cannot be made since tetrad analysis is not possible. The existence of gene conversion can be inferred from crosses between heteroalleles, which generate wild-type recombinants with parental outside markers, and in Drosophila by the use of the attached X chromosome (half-tetrads). In higher eukaryotes, intragenic recombination is commonly 0.1 to 0.2%, but in Drosophila it can be one or two orders of magnitude less frequent (see Thuriaux, 1977). These are minimum estimates since co-conversion of heteroalleles in hybrid DNA is not detected genetically. In E. coli the frequency of transposition is fairly low: the probability that a transposon will be inserted in a structural gene is about 10^{-7} to 10^{-8} per cell generation (Kleckner, 1977). If the frequency is similar in eukaryotes, then it is possible that for

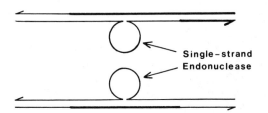

Fig. 1. The formation of hybrid DNA structures during meiotic recombination of DNA molecules which are homologous, with the exception of a DNA insert in one parent. Single strands from each parental molecule (thick and thin lines) form duplex hydrogen-bonded structures containing a single-stranded loop in the region of nonhomology. This may be a substrate for single-strand-specific endonucleases and result in gene conversion by elimination of the insert. Above: hybrid DNA with crossing over of parental DNA molecules. Below: hybrid DNA without crossing over (nonreciprocal recombination).

any given gene the equilibrium will be against the accumulation of selfish DNA. However, the probability of transposition of any inserted sequence to another location in the genome has also to be taken into account. There are, therefore, three types of events which may occur, as the following scheme shows.

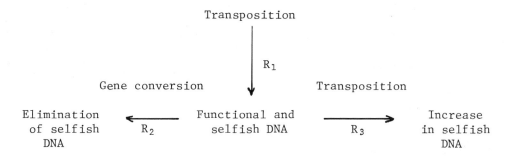

R_1 is the rate on insertion of a transposable element at any one locus and R_3 its rate of transposition elsewhere in the genome. [In the transposition of genetic elements, the parental sequence remains in situ and a replicated sequence is inserted elsewhere (see Kleckner, 1977)]. According to the hypothesis, rate R_2 will be greater than R_1 for any one genetic locus, and if R_2 is also greater than R_3, then selfish DNA will not accumulate. If, however, R_3 is greater than R_2, then selfish DNA will accumulate in the genome, but at a much lower rate than would be the case in the absence of meiosis and gene conversion.

The conventional explanation for the almost universal existence of sexual reproduction in higher organisms is that meiosis and fertilisation result in the re-assortment of genetic material, thus generating variability on which natural selection can act. In other words, beneficial mutations in different parents can, as a result of sexual reproduction, be combined into one individual that is fitter in Darwinian terms than either parent. It is maintained that this process is essential if species are to adapt themselves to new environments. The argument depends on the acceptance of "group selection," that is, species or populations which have sexual reproduction will be favoured in evolution over those which do not. There are, however, strong grounds for doubting the importance of group selection, which have been reviewed by several authors (Crow and Kimura, 1970; Emlen, 1973; Maynard-Smith, 1976a). With regard to sexual reproduction, the main problem has been discussed in most detail by Maynard-Smith (1976b). It is that any female which dispensed with this method of reproduction and used parthenogenesis instead, would be fitter in Darwinian terms than a sexually reproducing female, since she would be contributing more of her own genes to the next generation. Therefore, it would not be expected that sexual reproduction would be maintained in most plant and animal species.

I suggest, however, that sexual reproduction plays a positive role in benefiting individuals by helping them to eliminate selfish DNA. Let us suppose that an organism requires a mutation which makes it parthenogenetic, that is, it circumvents sexual reproduction and meiosis. Since R_2 will now be zero, there will immediately be a more rapid accumulation of selfish DNA, which may well be phenotypically disastrous. There may be no strong selection against the progeny of a parthenogenetic individual in the first few generations or so, but if accumulation of selfish DNA is exponential (like a cancerous growth), fitness will deteriorate progressively, leading eventually to the elimination by natural selection of the whole clone of asexual individuals.

The hypothesis suggested is compatible with the evidence that recombination may be largely confined to structural genes in higher organisms. This was deduced by Thuriaux (1977) from a comparison of recombination frequencies within such genes and the DNA content of the genome of several eukaryotic species. It is not clear how such a pattern of recombination might evolve, but it could certainly have the effect of protecting essential genes from inactivation by DNA inserts. It is also compatible with the general observation that recombination does not occur in heterochromatin, whether in the Y chromosome, in B chromosomes, or in the heterochromatic regions of autosomes. The absence of recombination correlates with the accumulation of noncoding, possibly functionless, DNA in heterochromatin. It would also be predicted that those species which do reproduce parthenogenetically may contain more DNA, especially certain classes of repetitive DNA, than their sexually reproducing relatives. Correlations might also be found between the frequency and direction of conversion of additions or deletions and the genome size of higher eukaryotes.

ACKNOWLEDGMENTS

I thank F. Doolittle, T. B. L. Kirkwood, S. G. Sedgwick, L. E. Orgel, F. H. C. Crick, J. Maynard-Smith, J. J. Bull, D. Charlesworth, and B. Charlesworth for helpful discussion and comments.

REFERENCES

Ando, T., 1966, A nuclease specific for heat-denatured DNA isolated from a product of Aspergillus oryzae, Biochem. Biophys. Acta, 114:158.
Benz, W. C., and Berger, H., 1973, Selective allele loss in mixed infections with T4 bacteriophage, Genetics, 73:1.
Crow, J. F., and Kimura, M., 1970, "Introduction to Population Genetics Theory," Harper and Row, New York.
Doolittle, W. F., and Sapienza, C., 1980, Selfish genes, the phenotype paradigm and genome evolution, Nature (London), 284:601.
Emlen, J. M., 1973, "Ecology: An Evolution Approach," Addison Wesley, Reading, Massachusetts.
Fink, G. R., and Styles, C. A., 1974, Gene conversion of deletions in the HIS4 region of yeast, Genetics, 77:231.
Fogel, S., Mortimer, R., Lusnak, K., and Tavares, F., 1978, Meiotic gene conversion: A signal of the basic recombination event in yeast, Cold Spring Harbor Symp. Quant. Biol., 43:1325.
Girard, J., and Rossignol, J. L., 1974, The suppression of gene conversion and intragenic crossing over in Ascobolus immersus: Evidence for modifiers acting in the heterozygous state, Genetics, 76:221.
Holliday, R., 1977, Recombination and meiosis, Philos. Trans. R. Soc. London, Ser. B., 277:359.
Kleckner, N., 1977, Translocatable elements in procaryotes, Cell, 11:12.
Leupold, U., 1980, Transposable mating-type genes in yeast, Nature (London), 283:811.
MacDonald, M. V., and Whitehouse, H. L. K., 1979, A buff spore colour mutant in Sordaria brevicollis showing high-frequency conversion. 1. Characteristics of the mutant, Genet. Res., 34:87.
Maynard-Smith, J., 1976a, Group selection, Q. Rev. Biol., 51:277.
Maynard-Smith, J., 1976b, "The Evolution of Sex," Cambridge University Press, Cambridge.

Orgel, L. E., and Crick, F. H. C., 1980, Selfish DNA: The ultimate parasite, Nature (London), 284:604.

Pukkila, P. J., 1977, Biochemical analysis of genetic recombination in eukaryotes, Heredity, 39:193.

Pukkila, P. J., 1978, The recognition of mismatched based pairs in DNA by DNase I from Ustilago maydis, Mol. Gen. Genet., 161:245.

Rabin, E. Z., Tenenhouse, H., and Fraser, M. J., 1972, An exonuclease of Neurospora crassa specific for single-stranded nucleic acids, Biochem. Biophys. Acta, 259:50.

Radding, C. M., 1978, On the mechanism of conversion of deletions and insertions, Cold Spring Harbor Symp Quant. Biol., 43:1315.

Rossignol, J. L., Paquette, N., and Nicholas, A., 1978, Aberrant 4:4 asci, disparity in the direction of conversion, and frequencies of conversion in Ascobolus immersus, Cold Spring Harbor Symp. Quant. Biol., 43:1343.

Stahl, F. W., 1979, "Genetic Recombination," W. H. Freeman, San Francisco.

Sutton, W. D., 1971, A crude nuclease preparation suitable for use in DNA reassociation experiments, Biochem. Biophys. Acta, 240:522.

Thuriaux, P., 1977, Is recombination confined to structural genes on the eukaryotic genome? Nature (London), 268:460.

Trautner, T. A., Spatz, H. Ch., Behrens, B., Pawlek, B., and Behuke, M., 1972, Exchange between complementary strands of DNA?, Adv. Biosci., 8:79.

CHAPTER 18

TRANSPOSONS AND ILLEGITIMATE RECOMBINATION IN PROKARYOTES:

A SUMMARY AND PERSPECTIVE

Nancy Kleckner

Department of Biochemistry and Molecular Biology
Harvard University, Cambridge, Massachusetts 02138

MECHANISMS FOR GENOME REARRANGEMENT

Many of the DNA rearrangements in prokaryotic organisms can be accounted for by one of three relatively well-defined, mechanistically distinct processes: general (homology-dependent) recombination, site-specific recombination, and transposon-promoted replicative recombination.

General Recombination

General recombination is strongly dependent upon pairing between relatively longer, i.e. greater than 100 base-pair (bp) stretches of DNA sequence homology on the two interacting DNA segments. It may be reciprocal or nonreciprocal, may involve greater or lesser amounts of DNA synthesis, and is promoted by specific functions encoded by bacteria and/or bacteriophages. If the two interacting homologous segments occur at different positions in a single genome or on different genomes, homologous recombination can generate deletions, inversions, duplications, and genome fusions. Amplification of antibiotic resistance and biosynthetic genes in response to selective pressure and integration/excision of plasmid genomes into and out of the bacterial chromosome are two particularly important examples of recA-promoted rearrangements. (See, for example, Anderson and Roth, 1978; Guyer et al., 1981; Schmidt et al., 1981; Yagi and Clewell, 1977.)

Site-specific Recombination

Site-specific recombination involves the action of particular proteins upon a pair of specific sites to generate reciprocal double-strand break/join events at those sites. Such events are "conservative": they do not require DNA synthesis or the action of DNA nucleases or ligases. Depending on the positions of the interacting sites and the details of the reaction, such events can result in deletion, inversion, or genome fusion. Examples of site-specific recombination are integration of bacteriophage lambda, Mu G-loop inversion, Salmonella flagellar antigen phase variation, TnpR-promoted cointegrate reduction in transposons gamma-delta and Tn3, and P1 site-specific recombination (Howe, 1980; Iino and Kutsukake, 1981; Kostriken et al., 1981; Nash, 1981; Reed, 1981; Silverman et al., 1981; Sternberg et al., 1981). Because of its highly site-specific nature, this type of recombination has been applied by prokaryotic organisms to highly specialized tasks.

Transposon-promoted Rearrangements

Transposon-promoted rearrangements include transpositions, deletions, inversions, and replicon fusion cointegrates. A summary of these events is shown in Fig. 1. Current models for transposition suggest that all of these events occur by similar or identical mechanisms which are characterized by: (a) separation of a _single_ strand at each end of the element from adjacent donor molecule sequences; (b) religation of two such single-stranded ends to a target DNA molecule that has been broken by a pair of staggered nicks; and (c) replication across the transposing element in such a way that transposon sequences are duplicated _without_ duplication of adjacent donor or target molecule sequences. The duplicative nature of transposon-promoted recombination is seen directly in the structure of replicon-fusion cointegrates (Fig. 1C). One circular genome containing a single copy of the transposon interacts with a second circular genome lacking the transposon to generate a structure containing one copy of each circular replicon but _two_ copies of the transposon. [See Bukhari (1981) and Kleckner (1981) for further discussion of transposition models and mechanisms.]

Transposons interact efficiently with many different target DNA sites and are not limited to highly specific sites nor to regions of homology. As a result they have great potential as agents for promoting genetic rearrangements. On the other hand, as seen in Fig. 1 events promoted by transposons always leave a copy of the transposable element at each newly created fusion junction. Thus transposon cannot promote direct joining of two unrelated target DNA sequences.

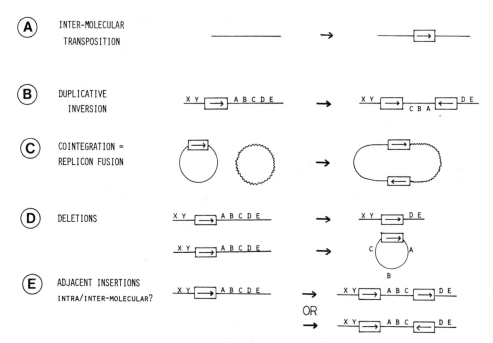

Fig. 1. Types of transposon-promoted events. See Kleckner (1981) for further discussion of the nature of and relationships among these vents.

We can suggest several ways in which transposons make particularly important or unique contributions to genome rearrangements. (1) Transposon insertions often contribute most or all of the spontaneous mutations obtained in selections which demand total abolition of gene function or specific turn-on of a gene lacking its own functional promoter. [See Kleckner (1981) for discussion of how transposons turn on genes.] Presumably, single-base changes having these consequences are rare. (2) In many situations transposons account for a large proportion of spontaneous deletions. (3) Transposable elements are uniquely able to effect the transposition of other, unrelated genetic determinants. The most notable example is the incorporation of antibiotic resistance genes into transposons and the resulting spread of such genes among many different plasmid genomes and bacterial hosts. (4) Transposons can directly mediate fusions between and subsequent dissociations of unrelated replicons. When coupled with systems for physical transfer of DNA from one bacterium to another, such events play a major role in reassorting genetic information among widely differing organisms which may share little or no direct DNA sequence homology. (5) Transposons can serve as portable regions of homology. Action of general recombination functions on duplicate copies of an element at two different locations can generate all of the rearrangements discussed under homologous recombination above.

ILLEGITIMATE RECOMBINATION

Many genome rearrangements in prokaryotes do not occur by one of these above pathways. Such events are collectively referred to as "illegitimate recombination," and their mechanisms are for the most part unknown. Certain types of deletions, duplications, and aberrant excisions of integrated prophages are classical examples of illegitimate recombination [see Franklin (1971) and Weisberg and Adhya (1977) for reviews]. More recently, other types of illegitimate recombination events associated with (but not necessarily promoted by) transposons have been described. For convenience these events can be grouped into three major classes.

Precise Excision of Transposons

Insertion of all transposons is accompanied by duplication of a short (3-12 bp) target DNA sequence, with the transposon integrated between the duplicated segments. The length of the duplication is characteristic of the particular element involved. All transposons are capable of excising precisely from DNA. Deletion between the flanking direct repeats regenerates the target DNA to its original wild-type sequence.

Precise excision can occur in the absence of transposon-encoded functions, and must, therefore, be mediated by host DNA-handling processes. Despite the involvement of direct repeats, excision does not require host homologous recombination functions. For transposons Tn10 and Tn5, excision is also stimulated by structural interaction between long (greater than 1000 bp) inverted repeats at the ends of the elements (Egner and Berg, 1981; Foster et al., 1981). For bacteriophage Mu, transposition functions specifically promote or stimulate precise excision above a background level of host-promoted events (Khatoon and Bukhari, 1981). Mu-dependent excision could reflect aberrant resolution of a nearly completed transposition event (Harshey and Bukhari, 1981). Alternatively, Mu functions may stimulate excision simply by making nicks in the DNA near the ends of the element or by facilitating the "coming together" of the two ends of Mu without further participation in excision itself. It is possible that other transposable elements also stimulate their own excision but at levels below that of host-mediated events.

Abberant Transposon-promoted Deletions

Classical transposon-promoted recombination events involve the joining of transposon ends to new target DNA sequences (Fig. 1). Precise excision of Mu is one instance in which transposon functions stimulate or promote another type of joining event. More generally, it is possible that proteins which normally see the ends of transposons could sometimes act nonspecifically at other DNA sequences or that replication in situ (like in situ replication of a non-excising prophage) could stimulate the occurrence of host-mediated illegitimate recombination events at nonspecific sites. Mu promotes several types of aberrant prophage excision (Khatoon and Bukhari, 1981). IS4 stimulates the formation of deletions which do not retain a copy of the element intact at the end of the deletion and instead have random end points extending in both directions from the site of the element (Klaer et al., 1981). Interesting deletions between one of Tn3's termini and a similar sequence nearby have also been described (McCormick et al., 1981).

Local Sequence Rearrangements

Analysis of IS2 and Tn10 polarity-relief revertants has uncovered several rearrangements in which DNA replication and/or repair synthesis has acted upon short direct and/or inverted repeats to generate small local alterations of DNA sequence. There is no indication that transposon functions are involved. Some of these events are complex rearrangements which can be economically accounted for by a replication slippage mechanism which depends upon an elongating single strand folding back on itself at an inverted repeat and then serving as a further template for synthesis (Ghosal and Saedler, 1978). Others are excision events between very close direct repeats; these excisions are probably stimulated by the presence of inverted repeats between the excision sites (Ahmed et al., 1981; Foster et al., 1981; Ghosal and Saedler, 1979).

Two of the papers in this volume (Bukhari and Khatoon, 1981; Lundblad and Kleckner, 1981) discuss illegitimate recombination events involving transposable elements: transposon Tn10 and bacteriophage Mu respectively.

REFERENCES

Ahmed, A., Bidwell, K., and Musso, R., 1981, Internal rearrangements of IS2 in *Escherichia coli*, Cold Spring Harbor Symp. Quant. Biol., 45:141.

Anderson, R. P., and Roth, J. R., 1978, Tandem genetic duplications in *Salmonella typhimurium*: Amplification of the histidine operon, J. Mol. Biol., 126:33.

Bukhari, A. I., 1981, Models of DNA transposition, Trends Biol. Sci., 6:56.
Bukhari, A. I., and Khatoon, H., 1981, Low level and high level DNA rearrangements in Escherichia coli, this volume.
Egner, C., and Berg, D. E., 1981, Excision of Tn5 is dependent on the inverted repeats but not on the transposase function of Tn5, Proc. Natl. Acad. Sci. U.S.A., 78:459.
Foster, T. J., Lundblad, V., Hanley-Way, S., Halling, S., and Kleckner, N., 1981, Three Tn10-associated excision events: Relationship to transposition and role of direct and inverted repeats, Cell, 23:215.
Franklin, N., 1971, The N operon of lambda: Extent and regulation as observed in fusions to the tryp operon of E. coli, in "The Bacteriophage Lambda," A. D. Hershey, ed., Cold Spring Harbor Laboratory, New York.
Ghosal, D., and Saedler, H., 1978, DNA sequence of the mini insertion sequence 2-6 and its relation to the sequence of insertion sequence 2, Nature, 275:611.
Ghosal, D., and Saedler, H., 1979, IS2-61 and IS2-611 arise by illegitimate recombination from IS2-6, Mol. Gen. Genet., 176:233.
Guyer, M. S., Reed, R. R., Steitz, J. A., and Low, K. B., 1981, Identification of a sex factor affinity site in E. coli as gamma delta, Cold Spring Harbor Symp. Quant. Biol., 45:135.
Harshey, R. M., and Bukhari, A. I., 1981, A mechanism of DNA transposition, Proc. Natl. Acad. Sci. U.S.A., 78:1090.
Howe, M. W., 1980, The invertible G segment of phage Mu, Cell, 21:605.
Iino, T., and Kutsukake, K., 1981, Trans-acting genes of bacteriophages P1 and Mu mediate inversion of a specific DNA segment involved in flagellar phase variation of Salmonella, Cold Spring Harbor Symp. Quant. Biol., 45:11.
Khatoon, H., and Bukhari, A. I., 1981, DNA rearrangements associated with reversion of bacteriophage Mu-induced mutations, Genetics, 98:1.
Klaer, R., Kuhn, S., Fritz, H. J., Tillman, E., Saint-Girons, I., Habermann, P., Pfeifer, D., and Starlinger, P., 1981, Studies on transposition mechanisms and specificity of IS4, Cold Spring Harbor Symp. Quant. Biol., 45:215.
Kleckner, N., 1981, Transposable elements in prokaryotes, Annu. Rev. Genet., 15, in press.
Kostriken, R., Morita, C., and Heffron, F., 1981, The transposon Tn3 encodes a site-specific recombination system: identification of essential sequences, genes and the actual site of recombination, Proc. Natl. Acad. Sci. U.S.A., in press.
Lundblad, V., and Kleckner, N., 1981, Mutants of E. coli K12 which affect excision of transposon Tn10, this volume.
McCormick, M., Wishart, W., Ontsubo, H., Heffron, F., and Ohtsubo, E., 1981, The structure of recombinant plasmids mediated by

the transposable DNA element Tn3 and Tn3 mutants, Gene, in press.
Nash, H. A., 1981, Integration and excision of bacteriophage λ: The mechanism of conservative site-specific recombination, Annu. Rev. Genet., 15, in press.
Reed, R. R., 1981, Resolution of cointegrates between transposons gamma-delta and Tn3 defines the recombination site, Proc. Natl. Acad. Sci. U.S.A., 78:3428.
Schmidt, F. J., Jorgensen, R. A., de Wilde, M., and Davies, J. E., 1981, A specific tetracycline resistance determinant of transposon Tn1771 in E. coli, Genet. Res., 33:253.
Silverman, M., Zieg, J., Mandel, G., and Simon, M., 1981, Analysis of the functional components of the phase variation system, Cold Spring Harbor Symp. Quant. Biol., 45:17.
Sternberg, N., Hamilton, D., Austin, S., Yarmolinsky, M., and Hoess, R., 1981, Site-specific recombination and its role in the life cycle of bacteriophage P1, Cold Spring Harbor Symp. Quant. Biol., 45:297.
Weisberg, R. A., and Adhya, S., 1977, Illegitimate recombination in bacteria and bacteriophage, Annu. Rev. Genet., 11:451.
Yagi, Y., and Clewell, D. B., 1977, Identification and characterization of a small sequence located at two sites on the amplifiable tetracycline resistance plasmid pAMα1 in Streptococcus faecalis, J. Bacteriol., 129:400.

CHAPTER 19

MUTAGENESIS AND REPAIR IN YEAST MITOCHONDRIAL DNA

E. Moustacchi and M. Heude

Institut Curie-Biologie, Centre Universitaire
Bâtiment 110, 91405 Orsay, France

SUMMARY

The mitochondrial DNA (mtDNA) is a dispensable genome in the facultative aerobe Saccharomyces cerevisiae. Since cell viability is maintained even if mtDNA is nonfunctional, mitochondrial mutagenesis can be studied. Marked dissimilarities exist between nuclear and mtDNA with respect to base composition, structure, multiplicity, and heterogeneity of molecules; differences in mutagenic processes are consequently expected. Mitochondrial mutants belong to two classes that both exhibit non-Mendelian inheritance; the rho^- (or "petite" mutation) results from massive deletions with a preferential loss of a specific segment, accompanied by repetitions of retained sequences; the other type is due to point mutations or small deletions and includes antibiotic resistant mutants and mutants defective in the synthesis of one or more components of the mitochondrial membrane complex (mit^-) or in the mitochondrial protein machinery (syn^-). The rho^- mutation is extremely frequent even spontaneously (from 1 to a few percent), whereas mitochondrial point mutations are relatively rare even with induced mutagenesis. Nuclear mutations due to molecular modifications similar to those seen in rho^- are rather infrequent. The genetic control of mitochondrial mutagenesis requires gene products involved either (1) in both nuclear and mitochondrial mutagenesis, the genes being located in the nucleus; or (2) specifically in mitochondrial mutagenic processes, the genes being located either in the nuclear or in the mitochondrial genomes.

Since the frequency of rho^- mutants and the loss of mitochondrial genetic markers by ultraviolet light can be modulated according to growth phases, the genetic background (rad, uvsρ, gam genes, etc.), dark holding in presence or absence of protein synthesis inhibitors, etc., it was hypothesized that mtDNA-induced lesions could be repaired in certain conditions. Indeed, pyrimidine dimers in mtDNA can be photoreactivated but not removed in a controlled process, as in the nucleus, by excision repair. When growing cells are UV-treated, a degradation of the mtDNA is observed and this is accompanied by a reduction in size of mtDNA parental molecules. However, this is followed upon transfer in growth medium by restitution of high molecular weight mtDNA molecules. Consequently, it is proposed that the recovery of the rho$^+$ phenotype is due to a nonexcision dark repair process possibly depending upon replication and/or recombination acting on damaged mtDNA.

The comparative susceptibility of the nuclear and mitochondrial genomes to different families of chemicals will be presented.

INTRODUCTION

Yeast has played a central role in the study of mitochondrial genetics beginning with the discovery of a cytoplasmically inherited respiratory deficiency or "petite" (also designated rho$^-$) mutation (Ephrussi et al., 1949). Recent advances have increased our understanding of the organization and expression of the mitochondrial genome (for review see Borst and Grivell, 1978). It is generally agreed that in Saccharomyces cerevisiae the mitochondrial genome consists of a closed circular duplex DNA of about 75 kilobase pairs that codes for a part of the mitochondrial protein synthesizing machinery and for enzyme components of the inner mitochondrial membrane. The order of the different genes on the mitochondrial genome has been established and the genetic map correlates with the physical map obtained by restriction endonuclease enzyme analysis of the mitochondrial DNA (mtDNA). Sequence data of some mitochondrial genes have been published in the last 2 years and soon the complete sequence is likely to be established.

COMPARATIVE FEATURES OF NUCLEAR AND MITOCHONDRIAL GENOMES

In various strains the mtDNA represents around 5 to 20% of the total DNA in yeast cells. Marked dissimilarities exist between nuclear and mtDNA. Indeed, the mole percentage of A+T is 82% in mtDNA vs 60% in nuclear DNA. The base composition along the mtDNA molecule is heterogeneous, with AT-rich regions making up about 50% of this DNA and GC-rich clusters (Faugeron-Fonty et al., 1979; Prunell and Bernardi, 1974). Nuclear DNA is much less heterogeneous. The mitochondrial genome does not seem to have the nucleosomal organization characteristic of the nuclear genome (Caron et al., 1979). Approximately 50 copies of mtDNA molecules are present in each haploid cell (Williamson, 1970). Moreover, the mitochondrion is an extremely dynamic system. The number of individual mitochondria and their distribution undergo extensive modifications according to physiological state and changes in life cycle. For instance, rapidly growing cells contain less than 10 mitochondria per cell, whereas stationary phase cells contain up to 50 per cell (Stevens, 1981). Such continual fusion and fragmentation of organelles, probably accompanied by recombination and segregation of the mitochondrial genome, together with the dissimilarities between nuclear and mitochondrial genomes in base composition, intramolecular homogeneity, structural organization and multiplicity of the molecules are expected to lead to differences between these two organelles in mutagenic processes.

Two classes of mitochondrial mutations occur in the facultative aerobe S. cerevisiae. The first, called petite or rho$^-$, consists of more or less extended deletions of mtDNA accompanied by tandem or palindromic repetitions of the retained sequences (Faye et al., 1973). This is a nonreverting mutation with pleiotropic effects, the major one being the loss of respiratory competence that leads to inability to utilize nonfermentable carbon sources. The rho$^-$ mutation occurs spontaneously at a high frequency ($\sim 10^{-3}$ to 10^{-1}) and is induced by a variety of unrelated mutagens. It should be recalled that certain nuclear mutants were also found to exhibit the petite phenotype (Chen et al., 1950). To date approximately 40 nuclear pet$^-$ genes, each exhibiting the petite phenotype, most of which are unlinked, have been isolated (Sherman and Slonimski, 1964). Proper genetic tests allow us to distinguish the nuclear from the mitochondrial petites. In spontaneous and induced petite populations, the frequency of pet$^-$ compared to that of rho$^-$ is so low that it is negligible.

Large deletions and rearrangements like those responsible for the mitochondrial rho$^-$ mutation are rarely recovered from the nuclear genome even after mutagenic treatments (Sherman et al., 1975). Such gross alterations are certainly more easily detected in the mitochondrial genome, which is nonessential, than in the nuclear genome where they would probably lead to cell lethality. This difference in susceptibility to deletion events is, however, so large that it cannot be excluded that it results from structural dissimilarities.

The second class of mitochondrial mutants consists of the syn$^-$, the mit$^-$, and the antibiotic-resistant mutants. Lesions in such mutants are restricted to single sites and may involve point mutations, which are frequently revertible, or small deletions. The syn$^-$ mutations are conditional mutants and affect the activity of components of the mitochondrial protein-synthesizing system. The mit$^-$ mutants are affected in components of the respiratory chain and oxidative phosphorylation. Spontaneous mutations of this class are produced at low rates (in the range 10^{-4} to 10^{-8} per generation) and have, in fact, been induced by growing yeast in the presence of Mn^{2+} (Putrament et al., 1973). Other agents are also capable of inducing these point mutations and, in general, the response of nuclear and mitochondrial genes is qualitatively similar.

Although it is accepted that mtDNA replicates in a semiconservative manner, its precise mode of replication is still a matter of controversy. It has been reported that in wild-type yeast the replication of mtDNA, measured by ^{15}N density labeling, appears to be dispersive (Sena et al., 1975; Williamson and Fennell, 1974), probably as a consequence of extensive and rapid recombination between replicated and unreplicated mtDNA molecules. More recently, non-dispersive replication was found (Leff and Eccleshall, 1978) in a strain auxotrophic for thymidine monophosphate (mtDNA was density-labeled by brominated precursors of DNA). In these contradictory reports different strains and different methods of labeling were used. However, even if recombination is not as extensive and rapid as initially assumed, the extreme flexibility of the mitochondrial genome (which is able to genetically recombine even in crosses between rho⁻ mutants) and the results of density tracings in the latter report suggest that some recombination takes place.

Neither the site(s) of replication origin(s) nor their number on wild-type mtDNA is yet known. The maintenance of mtDNA in rho⁻ mutants presumably requires the presence of an origin. Making use of suppressiveness, a property of rho⁻ strains involving suppression of respiratory competence in diploids resulting from crosses between rho⁻ and wild-type rho⁺ strains, it was possible to demonstrate that supersuppressive rho⁻ mutants (> 95% suppressivity) conserved a common region of about 260 to 300 base pairs (Blanc and Dujon, 1980; De Zamaroczy et al., 1979). These retained sequences are considered to be origins of replication and have been mapped in six different regions of the mitochondrial genome (Bernardi et al., personal communication). However, this does not tell us yet if one or more than one of these segments function as origin(s) of replication in wild-type mtDNA molecules.

GENETIC CONTROL OF MITOCHONDRIAL MUTAGENESIS

Spontaneous Mutability

Yeast strains in laboratory collections exhibit a wide range of spontaneous rho⁻ frequencies. Strains producing high proportions of petites were already genetically analyzed in the early days of this field (Ephrussi and Hottinguer, 1951). However, it is only relatively recently that mutants that alter the frequency of spontaneous rho⁻ mutants were purposely isolated and characterized.

Two types of genetic defects may alter the stability of the wild-type mtDNA. First, disturbance of mtDNA transcription/translation or structural modifications of the mitochondrial components resulting from indirect metabolic alterations may lead to mtDNA instability. This type of effect is exemplified by the nuclear pet mutants, which are known to exhibit high rates of rho$^-$ mutants (Sherman and Slonimski, 1964). Second, mutations in the enzymes responsible for mtDNA replication and repair or genetic alterations modifying the mechanism of segregation of the mitochondrial multicopy system or the way in which mtDNA molecules interact with each other are likely to result in increased frequencies of rho$^-$. These mechanisms are difficult to distinguish from each other experimentally.

Among all the mutants affecting spontaneous rho$^-$ production, only a few were also analyzed for mutator activity toward induction of point mutations such as resistance to antibiotics (e.g., erythromycin and chloramphenicol).

Nuclear mutations affecting spontaneous mitochondrial mutability. Certain DNA metabolism mutants affect rho$^-$ production. This is the case with cdc8 and cdc21 mutants, which are defective in continued replication during the S phase of the cell cycle at nonpermissive temperature (Newlon and Fangman, 1975; Newlon et al., 1979); cdc21 is allelic to tmp1 and the strain is deficient in thymidylate synthetase. The precise function of CDC8 is unknown. Both gene products are required for nuclear and mtDNA synthesis. Mutations in these agents result in 6- to 11-fold increase in the rate of production of rho$^-$ even at the permissive temperature. These rho$^-$ mutants contain mtDNA; therefore, they do not result from underreplication of mtDNA. The mutation rates for three nuclear and one mitochondrial gene tested are almost the same in cdc8, cdc21, and the original normal strain.

The mmc (mitochondrial mutability control) and the pet-ts1, pet-ts2, pet-ts10, pet-ts52, and pet-ts53 genes constitute nuclear determinants involved in the control of spontaneous rho$^-$ production (Marmiroli et al., 1980b). The mutation rates for two nuclear genes (ade2 → ADE and CAN1 → can1) did not display any mmc-dependent variation, whereas they were significantly higher for two mitochondrial genes (ERY^S → ERY^R and CAP^S → CAP^R) than in the MMC normal strain. On the basis of indirect arguments (Marmiroli et al., 1980a), it was suggested that these genes, although not directly involved in mtDNA metabolism, provide structures that contribute to ensuring the proper conformation of a "replicative complex" for mtDNA. Growth of three other conditional nuclear mutants, tsp-20, tsp-25, and tsp-30 (Backaus et al., 1978), at high temperature (35°C) resulted in the rapid production of rho$^-$ cells. A concomitant decrease in the ability to transmit mitochondrial genetic information to the rho$^+$ progeny of crosses was observed.

In one case it has been shown that the presence of the nuclear mutation mum, which enhances the spontaneous rate of rho^- production by a factor of 10, is associated with increased in vivo digestion of AT-rich regions in mtDNA (Lusena and James, 1976). It was proposed that this gene is concerned with the regulation of nuclease activity. More recently, 14 ethyl methanesulfonate (EMS)-induced mutants were selected for increased production of spontaneous erythromycin-, oligomycin-, and spiramycin-resistant mitochondrial gene mutants (Johnston, 1979). All of these mutants, which fell into two nuclear complementation groups, showed changes in the frequency of rho^- production even though they did not display mutator activity for nuclear genes. The molecular nature of the defect in these mutants is not yet known.

Other nuclear mutations that exhibit enhanced spontaneous levels of rho^- and/or mitochondrial gene mutations were initially selected, after ultraviolet light (UV) or EMS mutagenesis, for their sensitivity to radiation-induced mitochondrial mutagenesis. This is the case of uvsρ5, uvsρ6, and uvsρ34 (Moustacchi et al., 1975) in which the spontaneous frequency of rho^- is increased by a factor of 5, 4, and 14, respectively, with no concomitant effect on nuclear or mitochondrial gene mutability. Such mutations are recessive and mutant cells resemble their wild-type parent with respect to transmission and recombination of mitochondrial markers (Moustacchi et al., 1976). The amount of mtDNA per cell is elevated in uvsρ5 strain, relative to wild type. Therefore, it is unlikely that the rho^- mutants produced by uvsρ5, for instance, result from a reduction of the input of mtDNA in daughter cells or from a defective interaction of mtDNA molecules. Foury and Goffeau (1979) screened γ-ray sensitive mutants for enhanced levels of spontaneous mutation in mitochondrial genes. Five nonallelic nuclear mutations with this property were isolated. Three mutants, gam1, gam2, and gam4, only weakly sensitive to γ rays, showed increased levels of spontaneous mutation to antibiotic resistance (erythromycin, oligomycin, diuron) and to the rho^- condition. Two mutants, gam3 and gam5, which are very sensitive to γ rays, increased specifically the mutation rate of particular alleles of the mtDNA; gam3, which does not affect rho^- production, showed enhanced levels of spontaneous mutation in two of the three nuclear genes tested.

Among 10 mut strains selected for high spontaneous nuclear gene mutability (Hastings et al., 1976; von Borstel et al., 1971), we found that four of them (mut5, mut6, mut8, and mut10) show a three- to fivefold increase in rho^- production (Heude, unpublished data).

In summary, all these mutations demonstrate Mendelian patterns of segregation and consequently are likely to depend on single nuclear genes. They are generally recessive, indicating the loss of an enzymatic activity. They can affect (a) only rho^- production,

(b) rho⁻ production and mitochondrial gene mutation, (c) rho⁻ production and both nuclear and mitochondrial gene mutation, or (d) nuclear and mitochondrial gene mutation with no effect on rho⁻ production (gam3). Although proper complementation tests have not been performed between the different mutants, it is likely, from the differences in phenotypes, that several nuclear genes govern specifically the spontaneous mutability of mtDNA and that several steps, not involved in nuclear mutagenesis, are concerned. The existence of certain mut and gam3 mutations indicates, however, that both nuclear and mitochondrial mutagenic responses share a few common steps.

Cytoplasmic mutations affecting spontaneous mitochondrial mutability. A number of cases of nonnuclear mutations affecting the maintenance of the rho⁺ character have been reported (Bolotin-Fukuhara et al., 1977; Handwerker et al., 1973; Moustacchi et al., 1975; Slonimski and Tzagoloff, 1976; Storm and Marmur, 1975; Trembath et al., 1975). The mitochondrially encoded tsm-8 mutation, for instance, renders 90% of the cell population rho⁻ within five to six generations of growth on fermentable substrate at the nonpermissive temperature (35°C) (Handwerker et al., 1973; Schweyen et al., 1976). Similarly, when crosses are performed between newly arisen spontaneous rho⁻, respiratory competent colonies can form. Some of such restored clones are highly unstable and yield a high proportion of rho⁻ cells. The high-frequency rho⁻ formation is inherited in a non-Mendelian manner (Oakley and Clark-Walker, 1978). Sequence changes, absence of bands, and duplications are detected by restriction endonuclease enzyme analysis in the mtDNA of such hfp strains (Oakley, 1979).

Apart from these mutations, two uvsρ mutants, uvsρ72 and uvsρ13, which were selected for their UV sensitivity to rho⁻ induction, gave a non-Mendelian pattern of segregation of the high spontaneous rho⁻ mutability trait. Elimination of mtDNA from these strains by prolonged treatment with a high concentration of ethidium bromide (EB) resulted in the loss of this feature in subsequent crosses (Moustacchi et al., 1975). When compared to the original normal strain, the uvsρ72 mutant grown in nonfermentable substrate for selection of the rho⁺ phenotype did not demonstrate any difference in the mtDNA restriction enzyme pattern (Moustacchi and Fukuhara, unpublished data). However, we found morphological evidence for an unstable chondriome and defective mitochondrial biogenesis in uvsρ72 strains (Stevens and Moustacchi, 1976). In cells growing on nonfermentable media, "doublet" mitochondria in which a transverse septum produces figure-eight shapes were unusually common. When grown in glucose to allow the survival of rho⁻ cells, a fraction of the cells appeared to have a "mixed" mitochondrial population. They contained abnormal rho⁻-type mitochondria in the same cytoplasm with normal mitochondria. The observed rearrangement of the cristae suggests that the $UVS_\rho 72$ product plays an important role in the structural organization of the mitochondrial membrane and hence on the interaction between the mtDNA and the inner membrane.

The existence of the tsm-8, hfp, $uvs_\rho 72$ mutations, among other extrachromosomal mutations, indicates that the maintenance of the wild-type mtDNA is not only governed by nuclear genes but is also partly under the control of mitochondrial determinants. The interlock between persistence of the mitochondrial genetic information and the mitochondrial protein machinery encoded by both nuclear and mitochondrial genes is evidenced by the different classes of pet, pet-ts, tsm, cold-sensitive, etc. mutants that intervene in rho^- production.

Induced Mutagenesis

Inducing agents. The rho^- mutation can be efficiently induced by a large number of physical and chemical agents. It was shown that while UV light was an effective inducer of rho^-, having its maximum effect at 260 nm which is the wavelength of maximum absorption by nucleic acids, X-irradiation had relatively little effect (Moustacchi 1964; Raut and Simpson, 1955). UV also induces mitochondrial mutations conferring resistance to erythromycin with an overall efficiency that is comparable to that observed for induction of nuclear gene reversion; the dose-response kinetics, however, appear to differ (Ejchart and Putrament, 1979). Growth at elevated temperature and heat shocks (Schenberg-Frascino and Moustacchi, 1972; Sherman, 1958) are also effective in rho^- induction.

In general, rho^- mutants are induced by intercalating compounds. This is the case of various polycyclic aromatic molecules such as acridines, acriflavine, and euflavine (Ephrussi, 1953). More recently the phenanthridine dye EB (Slonimski et al., 1968), and berenil, a benzamidine derivative (Mahler and Perlman, 1973), have been used as powerful rho^- inducers. Drugs of this type are either weakly or nonmutagenic at the nuclear level. A direct effect of EB on mtDNA structure was suggested by the fact that it acts in the absence of cell multiplication and transforms all the mother cells into rho^- with first-order kinetics (Slonimski et al., 1968). Further studies demonstrated that a complex process is, in fact, involved since EB mutagenesis depends on intramitochondrial energy supply (Mahler and Bastos, 1974). Glucose repression (Hollenberg and Borst, 1971) and anaerobiosis (Pinto et al., 1975) strongly inhibit this process.

Intercalation of psoralen derivatives in DNA followed by covalent binding by photoaddition (365 nm irradiation) on pyrimidines also induces the rho^- mutation in yeast (Averbeck and Moustacchi,

1975; Swandbeck, 1976). The frequency of induced rho⁻ as a function of dose or survival is about 2.5 times higher after photoaddition of the monofunctional derivative 3-carbethoxypsoralen (3-CPs) than after treatment with the bifunctional compound 8-methoxypsoralen, which forms both monoadducts and DNA interstrand cross-links. The reverse is true for nuclear mutagenesis and recombination (Averbeck and Moustacchi, 1979, 1980). This observation has been extended to a number of other mono- and bifunctional psoralen derivatives (Averbeck et al., 1978). This appears to directly reflect the photoreactivity of the molecules with DNA (Magana-Schwencke et al., 1980), since, as we will see below, psoralen-mediated photo-induced damage is very inefficiently repaired at the mtDNA level under the conditions used. On the contrary, an error-prone component of cross-link repair leads to high frequencies of nuclear events. A number of factors, including ploidy, stage in the cell cycle, and genetic background, influence the dose-effect relationship for rho⁻ induction by psoralen photoaddition (Henriques and Chanet, personal communication; Henriques et al., 1977).

A group of miscellaneous chemicals, including heavy metals such as copper, nickel, cobalt, and manganese, some antifungal antibiotics, and a number of carcinogens, are also more or less efficient rho⁻-inducing agents (for review see Egilsson et al., 1979; Nagai et al., 1961). Regardless of the initial mutagenic step of rho⁻ induction by this variety of agents, the outcome is the loss of sequences of the wild-type mtDNA or even complete elimination of the mtDNA under prolonged drug treatment (Goldring et al., 1970; Nagley and Linnane, 1970). Strains lacking mtDNA have been designated as rho°.

As far as point mutations are concerned, it appears that the nuclear and mitochondrial genomes respond in a qualitative similar manner to mutagens. UV (as seen above), methyl methanesulfonate, diepoxybutane, nitrous acid, Mn^{2+} ions, all of which act on nuclear DNA, have been shown to produce mitochondrial point mutations as well as rho⁻ mutations (Baranowska and Putrament, 1979; Ejchart and Putrament, 1979; Polakowska and Putrament, 1979). The quantitative comparison in the response of the two genomes to these mutagens is difficult to make in spite of qualitative similarities. The limitations in interpretation of the data are essentially due to the differences in multiplicity of DNA molecules, in repair abilities, and in physiological factors known to modulate mutagenic responses.

Besides the agents cited above, which are likely to directly induce lesions in mtDNA (intercalation, pyrimidine dimers, monoadditions on purine or pyrimidines, strand breaks, interstrand cross-links, depurination, etc.), a wide range of inducers appear to principally act indirectly. This can occur by inhibition of the mitochondrial protein synthesis, as in the case of erythromycin (Williamson et al., 1971), or by modification of mitochondrial products such as depletion of the ATP intramitochondrial pool (Subik et al., 1978) or by fluorinated pyrimidines, which introduce anomalies in mitochondrial RNA (Oliver and Williamson, 1976). Agents or conditions that interfere with DNA metabolism, such as the folic acid analogue methotrexate (Wintersberger and Hirsch, 1973) or thymidilate deprivation of a dTMP auxotroph (Barclay and Little, 1978), will also lead to induction of rho⁻ and also, in the latter case, to high frequencies of mitochondrial drug-resistance mutations.

<u>Genes controlling induced mitochondrial mutagenesis</u>. Nuclear mutations induced by physical or chemical agents are thought to arise as a result either of a subtle alteration in base sequence promoting directly base-pairing errors (misreplication) or of gross DNA lesions leading indirectly through genetically controlled enzymatic repair processes to a nonspecific spectrum of base changes (misrepair) (Radman et al., 1979). However, in yeast at least, these two types of mutagenic processes are probably not as distinctly different since misreplication occurs more commonly in cells undergoing repair and the majority of agents appear to act as indirect mutagens (for discussion see Lawrence, 1981).

Since induced mutagenesis of nuclear genes is controlled by genes involved in repair, it was asked whether the same is true for induced mutagenesis of mitochondrial genes:

(a) When compared to the normal wild-type (<u>RAD</u>) strains, excision-deficient mutants (<u>rad3</u> type) show enhanced nuclear mutability to UV, 4NQO, and photoaddition of psoralens (for reviews see Lawrence, 1981; Lemontt, 1980). This has been interpreted as indicating that the excision-repair process is essentially error free, the premutational lesions being directed into error-prone repair in excision-defective mutants. In contrast to the nuclear response, it was shown that most of the <u>rad3</u>-type mutants led to the same UV response as the isogenic wild type for rho⁻ induction (Moustacchi, 1969, 1972). This indicated that the products of the <u>RAD3⁺</u>-type genes are not involved in the expression of mitochondrial premutagenic UV-induced lesions.

(b) A number of mutants belonging to the <u>rad6</u> epistasis group are deficient in mutagenesis of nuclear genes induced by UV, γ rays, and chemical mutagens (for review see Lawrence, 1981). However, all mutants within this group do not share this phenotype. Similarly, the frequency of UV-induced rho⁻ mutants is lower in <u>rad6-1</u> (Moustacchi, 1972) and in <u>rev2-1</u> (Lemontt, 1970) strains in comparison to the wild type. However, <u>rad18</u> (Moustacchi, 1972), <u>rev3</u> (Lemontt, 1970; Tuite and Cox, 1980), <u>umr1</u> (Tuite and Cox, 1980) or <u>pso1-1</u> and <u>pso2-1</u> (Henriques and Chanet, personal communication) strains, which are also mutationally defective at the nuclear level, exhibit higher frequencies of UV-induced rho⁻ mutants than does the wild type. These results show that the fate of UV-induced damage in the mitochondrial genome is, at least partly, under the genetic control of RAD6-type genes. Variations in the rho⁻ induction response have been encountered (Moustacchi, 1972; Tuite and Cox, 1980). This is likely to be due to differences in genetic background between strains.

(c) Mutants with defective radiation-induced intragenic recombination such as rec4 and rec5 (Rodarte-Ramon and Mortimer, 1972), an effect which appears to be localized to certain genes such as arg4 (Mortimer, personal communication), also exhibit a reduction in the frequency of UV-induced rho⁻ mutants (Moustacchi, 1973). This suggests a correlation between recombination ability and rho⁻ production.

(d) The rad52-type mutants, which are sensitive to ionizing radiations (for review see Lawrence, 1981) and to photoaddition of psoralen derivatives (Henriques and Moustacchi, 1980a) but normally resistant to UV, exhibit a normal response to γ-ray-induced mutagenesis of nuclear genes (McKee and Lawrence, 1979). Moreover, radiation-induced mitotic recombination is reduced or absent in strains carrying mutations at the RAD51 through RAD57 loci (Prakash et al., 1979; Saeki et al., 1980). We have recently shown that induction of rho⁻ mutants by psoralen derivative photoaddition is strongly reduced in such mutants concerned with a type of repair that depends on genetic recombination (Moustacchi, unpublished).

A series of mutants called ebi, less inducible by EB than the parental strain for the rho⁺ to rho⁻ mutation, has been isolated after EMS mutagenesis (Dujardin and Dujon, 1979). These mutations are inherited as nuclear genes. Certain ebi mutants did not affect transmission or recombination of mitochondrial markers while others led to noncoordinated transmission of the oli1 marker with respect to other mitochondrial markers. However, in all cases no significant decrease in the frequency of recombinants has been detected, and in some mutants this frequency was clearly enhanced. Genetic recombination of mitochondrial markers, as well as rho⁻ production, is likely to require transient interruptions in mtDNA molecules. The possible coupling between these events is still unclear.

Another class of mutants, called uvsρ, was selected for normal sensitivity to the lethal effect of UV and enhanced sensitivity to rho⁻ induction. We have already mentioned that spontaneous rho⁻ mutability is generally enhanced in such mutants, whereas both spontaneous and induced nuclear mutagenesis are not affected. The existence of three complementing uvsρ mutations (uvsρ5, uvsρ6, and uvsρ34) that segregate like Mendelian determinants indicates that nuclear genes are specifically involved in induced rho⁻ mutagenesis. The uvsρ5 mutant has been the object of detailed analysis

(Moustacchi et al., 1976). It differs from the wild type in its extreme susceptibility to rho⁻ induction by UV, while induction by EB or 3-CPs photoaddition is greatly diminished. On the other hand, uvsρ5 does not demonstrate the characteristic degradation of mtDNA after UV or EB treatment. Therefore, it is likely that the uvsρ5 mutant is blocked in one of the steps leading to the reduction in size of mtDNA subsequent to introduction of the initial damage produced by exposure to UV, EB, or 3-CPs plus 365 nm.

Two other uvsρ mutants (uvsρ72 and uvsρ13) demonstrate a non-Mendelian pattern of meiotic segregation for the phenotype of high frequency rho⁻ production (spontaneous, UV- or EB-induced). As seen above, this trait is associated with mtDNA.

Finally, strains carrying gam3 and gam5 (Foury and Goffeau, 1979), which are sensitive to γ rays (but are not allelic to rad52-type genes) and exhibit selective effects on spontaneous mitochondrial mutability, show the same response as the wild type for rho⁻ induction by γ ray (Cassier, personal communication). This indicates that the products of these genes are not involved in γ-ray-induced rho⁻ mutagenesis.

From these comparisons between wild-type and repair-defective mutants a complex network of interactions between the nuclear and the mitochondrial genomes emerges. A number of nuclear genes involved in repair of damage to nuclear DNA do not interfere with the mitochondrial response (rad3-type), whereas certain genes that play an important role in nuclear error-prone (rad6-type) or in recombination (rad52-type) types of repair modulate mitochondrial mutagenesis. Finally, some nuclear (uvsρ5-type) and mitochondrial (uvsρ72-type) genetic determinants appear to govern specifically the expression of mitochondrial genetic damage. Table 1 recapitulates the main features of some of the nuclear mutations that interact with the mitochondrial response.

THE FATE OF LESIONS INDUCED IN mtDNA

Photorepair of UV-Induced Mutational Damage Leading to rho⁻ Mutations

More than 20 years ago it was shown that photoreactivation would reduce the frequency of rho⁻ mutants in surviving populations of UV-irradiated S. cerevisiae cells (Pittman and Pedigo, 1959; Sarachek 1959) with a concomitant enhancement of survival. Since the specific function of the photoreactivating enzyme is the splitting of UV-induced pyrimidine dimers, the photoreversal of rho⁻ mutagenesis suggests that dimers are split by this photorepair process in mitochondria. The direct demonstration that dimer content was indeed reduced in mtDNA from yeast cells exposed to UV irradiation and then to visible light (Waters and Moustacchi, 1974) brought support to

Table 1. Nuclear Mutations That Affect the Frequencies of Spontaneous and Induced rho⁻ or Frequencies of Mitochondrial Mutations (Antibiotic Resistance)[a]

Mutants	Spontaneous rho⁻	Induced rho⁻	Spontaneous Mitochondrial Mutations (Ant^S → Ant^R)	Remarks and References
pet	inc	–	–	At least 40 loci (Sherman and Slonimski, 1964)
cdc8	inc	–	N	Temperature sensitive, defective in the synthesis of nuclear and mtDNA (Newlon et al., 1979)
cdc21	inc	–	N	Temperature sensitive, defective in the synthesis of nuclear and mtDNA (Newlon et al., 1979)
tsp20,25,30	inc	–	–	Temperature sensitive, reduction in transmission of markers (Backaus et al., 1978)
mmc	inc	–	inc	Defective in "replicative complex" (?) (Marmiroli et al., 1980b)
pet-ts1,2, 10,52,53	inc	–	–	Temperature sensitive (Marmiroli et al., 1980b)
C92-2C	inc	–	–	Defect in nuclease activity (?) (Lusena and James, 1976)
A4,13	inc or dec	–	inc or N	14 mutants belong to two complementation groups (Johnston, 1979)
uvs 5,6,34	inc	inc (UV)	N	Sensitivity to UV induction of rho⁻ (Moustacchi et al., 1975) Foury and Goffeau (1979)
gam1,2,4	inc	N (γ-ray)	inc (for specific markers)	Cassier (personal communication); Foury and Goffeau (1979)
gam3	N	N (γ-ray)		
gam5	inc	N (γ-ray)	inc (for specific markers)	Cassier (personal communication); Foury and Goffeau (1979)
mut5,6,8,10	inc	–	–	Mutators for nuclear genes (Hastings et al., 1976; Heude, unpublished)
ebi2,8,20, 101	N or inc	dec (EB)	–	Dujardin and Dujon (1979)
rec4,5	N	dec (UV or EB)	–	Deficient in induced intragenic recombination (arg4) (Moustacchi, 1973; Rodarte-Ramon and Mortimer, 1972)
rad52-type	N	dec (psoralen + 365 nm)	–	Sensitive to the lethal effect of γ-ray and psoralen photoaddition (Henriques and Moustacchi, 1980b; Moustacchi, unpublished)
rad6-1, rev2	N	dec (UV)	–	Mutationally defective (Lemontt, 1970; Moustacchi, 1972)
rad18, rev3, umr1, pso1	N	inc (UV or psoralen + 365 nm)	–	Mutationally defective (Henriques and Chanet, unpublished; Lemontt, 1980; Moustacchi, 1972; Tuite and Cox, 1980)

[a] inc, increase in frequency compared to wild type; (–), not tested; N, same response as wild type; dec, decrease in frequency compared to wild type.

this idea. This observation was confirmed by the finding that photoreactivation of UV-exposed cells led to a reduction of T4 UV endonuclease-specific sites (dimers) in yeast mtDNA (Prakash, 1975).

Recently, the photorepair of rho⁻ induction in wild-type PHR⁺ S. cerevisiae was compared to that of a photoreactivation-deficient nuclear mutant phrl isolated by Resnick (1969). In order to distinguish between photoreactivation of lethal damage occurring in the nucleus and mitochondrial damage, diploid cells, which are more UV-resistant to cell killing than haploid cells, were used. Under conditions favoring relatively high survival, it is clear that the UV induction of rho⁻ mutants in phrl⁻/phrl⁻ strains is not photo-reversible, whereas it is in wild-type cells (Green and MacQuillan, 1980). This finding implies that the same photoreactivation enzyme, encoded by a single nuclear gene, is responsible for photorepair of UV-induced damage to DNA in the nucleus and in the mitochondria.

Excision Repair

One of the important dark-repair mechanisms known to remove UV-induced pyrimidine dimers (Py-Py), as well as a large spectrum of chemically induced DNA base damage, is excision repair. In this process incision of the lesion is followed by excision, the gap remaining in DNA being filled by resynthesis using the opposite undamaged strand as template. The capacity of yeast to remove UV-induced Py-Py in nuclear DNA by this mechanism has been demonstrated (Prakash, 1975; Reynolds and Friedberg, 1980; Unrau et al., 1971; Waters and Moustacchi, 1974); nine genetic loci governing Py-Py excision (rad3 group) are known (for review see Prakash and Prakash, 1980).

In contrast to the nuclear DNA of repair-proficient strains in which virtually all of the dimers are removed within the few hours of dark incubation following UV treatment, the dimers induced in mtDNA are retained under the same conditions (Prakash, 1975; Waters and Moustacchi, 1974). This is true at low, as well as at high, UV doses. Similarly excision repair of Py-Py is absent in mtDNA of mouse L cells, human KB and HeLa cells (Clayton et al., 1974).

Recently we have shown that in wild-type yeast cells treated in exponential growth phase, DNA interstrand cross-links induced by 8-methoxypsoralen photoaddition are all removed from nuclear DNA within the 2 h of dark incubation following treatment. This repair of cross-links is dependent upon the RAD3 gene product; rad3 mutants exhibit hypersensitivity to inactivation by psoralen photoaddition (Averback and Moustacchi, 1975), as well as by UV. In contrast to nuclear DNA, more than 80% of the cross-links induced in mtDNA from wild-type cells persist even after 6 h of incubation (Magana-Schwencke et al., 1981). The dose of 8-methoxypsoralen plus 365 nm light used led to 20% survival and induced about 30% rho⁻ mutants.

Thus at the nuclear level excision repair seems to cope with DNA lesions that are structurally different. On the other hand, there is apparently no excision repair in mtDNA irrespective of the molecular structure of the DNA lesions (dimers or cross-links). Since the frequency of rho⁻ production is similar in both repair-proficient RAD⁺ and excision-defective rad3 strains, rho⁻ induction is independent of excision repair.

Other Repair Processes

Since excision response of damaged mtDNA does not occur, it may be asked whether other dark-repair mechanisms can act on lesions in mtDNA. A number of observations support the notion that once lesions are produced on mtDNA the outcome can be modulated or reversed. Much attention has been given to recovery following UV and EB treatments. These observations are summarized below.

(a) We have seen above that several genetic determinants that are known to govern error-prone or recombination repair pathways acting on nuclear DNA lesions appear to also control rho⁻ induction. Although it can be argued that these repair pathways may indirectly alter the number of copies of the mitochondrial genome or their expression, it is also possible that by analogy with the nuclear DNA situation, the fate of lesions in mtDNA (repaired or unrepaired) will directly lead to variations in mitochondrial genetic alterations.

(b) If UV-irradiated cells are incubated in the dark in a non-nutrient medium before plating (liquid holding or LH), lethality is reduced (Patrick et al., 1964) and the frequency of induced rho⁻ decreases in the case of exponential-phase cells or increases in the case of stationary-phase cells (Heude and Moustacchi, 1973; Moustacchi and Enteric, 1970). Thus, lethal damage is repaired in both cases, whereas rho⁻ premutational damage is reversed only if growing cells are UV treated.

Nuclear- and mitochondrial-directed protein synthesis appears to be required for the restoration of both survival and of the rho⁺ phenotype in dark liquid-held exponential phase cells. The characteristic negative holding effect for rho⁻ induction in stationary-phase cells persists following treatments that inhibit nuclear or mitochondrial protein synthesis (Heude and Chanet, 1975; Heude et al., 1975).

When the loss of mitochondrial genetic markers distributed in different positions on the mitochondrial genome is examined under the same conditions, there is a dose-dependent loss of markers. Further, a rescue of markers is observed following dark holding of UV-treated exponential cells, whereas the loss is aggravated for stationary-phase cells (Heude and Moustacchi, 1979). Moreover, it has been shown that deletion of markers occurs in an ordered fashion around the mtDNA genome with a preferential loss of the oxi3 region (Heude and Moustacchi, 1979; Moustacchi et al., 1978), which is particularly AT-rich (Sanders et al., 1976).

The parallelism between the fate of markers and the response for rho$^-$ induction indicates that the recovery or the loss of the rho$^+$ phenotype is not due to modulations in segregation or expression of the total mitochondrial genome but more likely reflects specific modifications of mtDNA sequences.

(c) Other observations indicate that spontaneous and induced rho$^-$ mutagenesis involves a transient state during which a portion of cells may revert back to normal rho$^+$ phenotype. First, complementation has been observed in crosses between spontaneous rho$^-$ of recent origin yielding respiratory competent rho$^+$ diploids (Clark-Walker and Miklos, 1975). Second, various factors, including temperature, pH, visible light, addition of caffeine, antimycin A or nalidixic acid, and growth medium components during or after EB exposure, are effective to varying degrees in reversing steps that otherwise lead to the rho$^-$ mutation (Bastos and Mahler, 1976; Gardella and MacQuillan, 1977; Hall et al., 1976; Hixon et al., 1979; Mahler and Perlman, 1972; Whittaker et al., 1972; Wolf and Kaudewitz, 1976). It should be kept in mind, however, that some of these factors may decrease the level of drug binding to mtDNA and consequently modify the frequency of induced rho$^-$ (protective effect). Third, it was demonstrated that EB at high concentration (80 to 100 μg/ml) rapidly induces formation of rho$^-$ mutants. Yet upon continued exposure to the dye 60 to 70% of the cells regain the rho$^+$ phenotype and longer exposure times once more result in rho$^-$ formation (Wheelis et al., 1975). The general model for EB mutagenesis assumes that the dye becomes bound to mtDNA while scission and degradation of mtDNA occur. An EB-stimulated nuclease activity present in mitochondria (Paoletti et al., 1972) could account for the initial steps in the reaction sequence. The recovery phase described by Wheelis et al. (1975) would correspond to reassembly of the mtDNA fragments, while the final phase of rho$^-$ induction would result from further fragmentation of mtDNA.

Therefore, on the basis of these observations the existence of an accurate repair mechanism of damaged mtDNA is still possible. This is why we examined more closely the mtDNA of UV-treated exponential-phase cells during the LH period in nonnutrient medium (Hixon and Moustacchi, 1978) and at time intervals after a transfer to growth medium (Hixon et al., 1980).

Following different doses of UV, prelabeled DNA was examined by equilibrium cesium chloride gradients at time intervals during LH. The gradual decrease in rho$^-$ was accompanied by an ongoing degradation of mtDNA during the first 24 h followed by a stabilization. This is in contrast to the results with nuclear DNA, which is only slightly degraded (10% at most) under the same conditions.

The dose response for mtDNA degradation was biphasic. Since UV induction of rho⁻ also demonstrates a similar dual response, these phenomena are likely to be related to the differences in the rate of degradation at high and low doses. Such biphasic curves suggest a saturation of the enzymatic degradative reaction(s) in exponential-phase cells. The size of mtDNA as determined from both electron microscopy and the summation of restriction enzyme fragments is about 5×10^7 daltons. Since all isolation procedures yield broken molecules, it is difficult to determine accurately the average size changes of mtDNA after irradiation. The size of mtDNA from nonirradiated cells did not change during the 40 h of LH period (mol. weight 1.32×10^7 daltons), whereas it was reduced in irradiated cells (mol. weight 5.60×10^6 daltons for a dose leaving 20 to 30% survivors). The mtDNA remaining after degradation showed a slight shift to heavier buoyant density, indicating a possible degradation of AT-rich regions (Hixon et al., 1980). It should be noticed that in these experiments 1 µg/ml of cycloheximide was added to the growing culture prior to irradiation for preferential radioactive labeling of mtDNA. This did not interfere with the chase of the radioactive label and normal recovery after UV of both survival and rho⁺ cells (Hixon and Moustacchi, 1978). We recently reinvestigated whether any DNA synthesis occurred during the LH period by measuring the uptake of radioactive DNA precursors and isolating the DNA by equilibrium cesium chloride gradients. Cycloheximide was not added in these experiments. Under such conditions a preferential synthesis of mtDNA takes place during the LH in nongrowth medium in control and irradiated cells. Whether this synthesis reflects elongation or turnover of DNA is actually under study.

In order to mimic the second step of the LH recovery process, ^{3}H-prelabeled cells were resuspended after the LH period in growth medium containing a ^{14}C precursor of DNA. Double-labeled mtDNA (old ^{3}H and new ^{14}C) was isolated from samples removed during new growth. A recovery in the amount and size of mtDNA was observed in irradiated cells during growth resumption. A similar reassembly of mtDNA fragments occurred during the period of recovery of respiratory competence following EB treatment (Criddle et al., 1976). Treatment with S1 nuclease was applied in order to assay for nicks or gaps in mtDNA in control and irradiated cells. In purified mtDNA 0.15 cuts per molecule were found in nonirradiated old and new mtDNA, whereas 0.92 and 0.43 cuts per molecule were found in old and new mtDNA, respectively, from irradiated cells. Thus, the most sensitive DNA targets for cleavage were the irradiated ^{3}H-mtDNA parental molecules retained after degradation. It is suggested that the presence of dimers, especially in AT-rich regions, might cause sufficient distortion to melt out single-stranded S1 nuclease-sensitive regions.

In summary, the decrease in size and overall reduction in the
amount of mtDNA per cell followed by a slow restoration of the size
and amount of mtDNA agree with genetic experiments in exponential-
phase cells, i.e., the recovery of rho$^+$ cells and rescue of mitochondrial markers in remaining rho$^-$ cells. The restitution of molecular
weight of mtDNA molecules is concomitant with the resumption of DNA
synthesis. Whether it constitutes a postreplicative type of repair
that implies recombination between damaged and undamaged molecules
as in the bacterial model (Rupp and Howard-Flanders, 1968) or a
preferential replication of selected molecules that retain more
information than the majority of those molecules that are reduced
in size remains to be determined.

The fate of mtDNA from cells irradiated in stationary phase was
not followed in detail. Indeed, the major event that takes place
following LH in nonnutrient medium (direct shift of irradiated cells
in nutrient medium or LH followed by 12 h in nutrient medium) was
a considerable loss of mtDNA (Waters and Moustacchi, 1974). The
amount of recoverable radioactive label in the mtDNA was diminished
about tenfold following such posttreatments indicating that 80 to
90% of the mtDNA initially present at the time of irradiation was
degraded. Since the biological data show that the frequencies of
rho$^-$ increase and the loss of genetic markers is accentuated in such
conditions, it is tempting to conclude that mtDNA from irradiated
stationary-phase cells is not subject to repair. The overall reshaping of mtDNA genetic information can be envisaged as resulting
from an equilibrium between degradation and resynthesis, plus perhaps recombination of mtDNA sequences. The degradation of mtDNA
in UV-treated stationary-phase cells, which also occur in exponential-phase cells, could be viewed as an unsuccessful attempt at
repair, the equilibrium being disrupted in the former case in favor
of degradation. The degradative enzymatic activity is independent
of de novo protein synthesis and would be constitutive (Heude et al.,
1975), whereas the subsequent steps are the ones that are likely to
require novel protein synthesis. Table 2 summarizes the main features of exponential and stationary-phase cells with respect to
mtDNA and mitochondrial genetic response following UV treatment.
Why nuclease activity(ies), specifically acting on mtDNA, may be
favored in stationary-phase cells remains an open question.

CONCLUSIONS

A number of models have been proposed for the mechanism of rho$^-$
induction (Clark-Walker and Miklos, 1974; Gaillard et al., 1980;
Slonimski and Lazowkza, 1977). Illegitimate recombination involving
pseudo-homology between various AT-rich or GC-rich segments has been
evoked as the possible origin of the repeat sequence found in rho$^-$
mutants. How the primary repeat amplifies to yield the mitochondrial

Table 2. Influence of Growth Conditions at the Time of UV Treatment on Mitochondrial Genetic Events and on the Fate of mtDNA[a]

Culture	Survival	Frequency of rho⁻ Mutants	Loss of Mitochondrial Markers	Fate of mtDNA
Exponential phase	inc	dec	dec	Degradation Reduction in size in nonnutrient medium Restitution of size in nutrient medium
Stationary phase	inc	inc	inc	Extensive degradation

[a] Immediate plating after irradiation (doses up to 200 Jm^{-2}, i.e., around 1% survival) is compared to delayed plating on nutrient medium with a period of dark holding in nonnutrient medium (up to 48 h). Biochemical experiments mimic these conditions. For details see Heude and Moustacchi (1973, 1979), Hixon et al. (1980), Moustacchi and Enteric (1970), Waters and Moustacchi (1974); inc, increase in survival, induction of rho⁻ or loss of markers on delayed plating compared to immediate plating; dec, decrease, as above.

genome unit carried in a rho⁻ clone is still a matter of speculation. How the total amount of mtDNA per cell is kept constant, as well as how the size of the mtDNA is determined, is not yet clear. Understanding the steps of reshaping mtDNA molecules containing lesions that are closely dependent upon physiological conditions also relates to these questions. The rapid developments of techniques for DNA sequence determination and the fact that yeast mtDNA is a relatively simple molecule should provide answers to these questions in the near future. The battery of mutants that interfere with the production of spontaneous and induced rho⁻ mutants and the definition of the conditions that modulate the expression of mitochondrial genetic information might prove to be useful.

ACKNOWLEDGMENT

We thank the Commission des Communautés Européennes (Bruxelles), contract n° BIO-E-397F(S1) and the Centre National de la Recherche Scientifique (A.T.P. Microbiologie) for support.

REFERENCES

Averbeck, D., and Moustacchi, E., 1975, 8-Methoxypsoralen plus 365 nm light effects and repair in yeast, Biochim. Biophys. Acta, 395: 393.

Averbeck, D., and Moustacchi, E., 1979, Genetic effects of 3-carboxypsoralen, angelicin, psoralen and 8-methoxypsoralen plus 365 nm irradiation in Saccharomyces cerevisiae. Induction of reversions, mitotic crossing-over, gene conversion and cytoplasmic "petite" mutation, Mutat. Res., 68:133.

Averbeck, D., and Moustacchi, E., 1980, Decreased photo-induced mutagenicity of monofunctional as opposed to bifunctional furocoumarins in yeast, Photochem. Photobiol., 31:475.

Averbeck, D., Moustacchi, E., and Bisagni, E., 1978, Biological effects and repair of damage photo-induced by a derivative of psoralen substituted at the 3,4 reaction site. Photoreactivity of this compound and lethal effects in yeast, Biochim. Biophys. Acta, 518:464.

Backaus, B., Schweyen, R. J., Kaudewitz, F., and Dujon, B., 1978, On the formation of ρ⁻ petites in yeast. III. Effects of temperature on transmission and recombination of mitochondrial markers and on ρ⁻ cell formation in temperature sensitive mutants of Saccharomyces cerevisiae, Mol. Gen. Genet., 161:153.

Baranowska, H., and Putrament, A., 1979, Mitochondrial mutagenesis in Saccharomyces cerevisiae. III. Nitrous acid, Mutat. Res., 63:291.

Barclay, B. J., and Little, J. G., 1978, Genetic damage during thymidylate starvation in Saccharomyces cerevisiae, Mol. Gen. Genet., 160:33.

Bastos, R. N., and Mahler, H. R., 1976, Modulation of petite induction by low concentrations of ethidium bromide, Biochem. Biophys. Res. Commun., 69:528.

Blanc, H., and Dujon, B., 1980, Replicator regions of yeast mitochonDNA responsible for suppressiveness, Proc. Natl. Acad. Sci. U.S.A., 77:3942.

Bolotin-Fukuhara, M., Faye, G., and Fukuhara, H., 1977, Temperature-sensitive respiratory-deficient mitochondrial mutations: Isolation and genetic mapping, Mol. Gen. Genet., 152:295.

Borst, P., and Grivell, L. A., 1978, The mitochondrial genome of yeast, Cell, 15:705.

Caron, F., Jacq, C., and Rouvière-Yaniv, J., 1979, Characterization of a histone-like protein extracted from yeast mitochondria, Proc. Natl. Acad. Sci. U.S.A., 76:4625.

Chen, S. Y., Ephrussi, B., and Hottinguer, H., 1950, Nature génétique des mutants à déficience respiratoire de la souche BII de la levure de boulangerie, Heredity, 4:337.

Clark-Walker, G. D., and Miklos, G. L. G., 1974, Mitochondrial genetics, circular DNA and the mechanism of the petite mutation in yeast, Genet. Res. Camb., 24:43.

Clark-Walker, G. D., and Miklos, G. L., 1975, Complementation in cytoplasmic petite mutants of yeast to form respiratory competent cells, Proc. Natl. Acad. Sci. U.S.A., 72:372.

Clayton, D. A., Doda, J. N., and Friedberg, E. C., 1974, The absence of a pyrimidine dimer repair mechanism in mammalian mitochondria, Proc. Natl. Acad. Sci. U.S.A., 71:2777.

Criddle, R. S., Wheelis, L., Trembath, M. K., and Linnane, A. W., 1976, Molecular and genetic events accompanying petite induction and recovery of respiratory competence induced by ethidium bromide, Mol. Gen. Genet., 144:263.

De Zamaroczy, M., Baldacci, G., and Bernardi, G., 1979, Putative origins of replication in the mitochondrial genome of yeast, FEBS Lett., 108:429.

Dujardin, G., and Dujon, B., 1979, Mutants in yeast affecting ethidium bromide-induced ρ^- formation and their effects on transmission and recombination of mitochondrial genes, Mol. Gen. Genet., 171:205.

Egilsson, V., Evans, I., and Wilkie, D., 1979, Toxic and mutagenic effects of carcinogens on the mitochondria of Saccharomyces cerevisiae, Mol. Gen. Genet., 174:39.

Ejchart, A., and Putrament, A., 1979, Mitochondrial mutagenesis in Saccharomyces cerevisiae. I. Ultraviolet radiation, Mutat. Res., 60:173.

Ephrussi, B., 1953, Cytoplasmic heredity in yeast, in: "Nucleocytoplasmic Relations in Microorganisms," Clarendon Press, Oxford.

Ephrussi, B., and Hottinguer, H., 1951, On an unstable cell state in yeast, Cold Spring Harbor Symp. Quant. Biol., 16:75.

Ephrussi, B., Hottinguer, H., and Chimenes, A. M., 1949, Action de l'acriflavine sur les levures. I. La mutation "petites colonies," Ann. Inst. Pasteur, 76:351.

Faugeron-Fonty, G., Culard, G., Baldacci, G., Coursot, R., Prunell, A., and Bernardi, G., 1979, The mitochondrial genome of wild-type yeast cells. VIII. The spontaneous cytoplasmic "petite" mutation, J. Mol. Biol., 134:493.

Faye, G., Fukuhara, H., Grandchamp, G., Lazowska, J., Michel, F., Casey, J., Getz, G., Locker, H., Rabinowitz, M., Bolotin-Fukuhara, M., Coen, D., Deutsch, J., Dujon, B., Netter, P., and Slonimski, P., 1973, Mitochondrial nucleic acids in the petite colony mutants: Deletions and repetitions of genes, Biochimie, 55:779.

Foury, F., and Goffeau, A., 1979, Genetic control of enhanced mutability of mitochondrial DNA and γ-ray sensitivity in Saccharomyces cerevisiae, Proc. Natl. Acad. Sci. U.S.A., 76:6529.

Gaillard, C., Strauss, F., and Bernardi, G., 1980, Excision sequences in the mitochondrial genome of yeast, Nature, 283:218.

Gardella, S., and MacQuillan, A. M., 1977, Ethidium bromide mutagenesis in Saccharomyces cerevisiae: Modulation by growth medium components, Mutat. Res., 46:269.

Goldring, D. R., Grossman, L. I., Krupnick, D., Cryer, D. R., and Marmur, J., 1970, The "petite" mutation in yeast: Loss of mitochondrial deoxyribonucleic acid during induction of "petites" with ethidium bromide, J. Mol. Biol., 52:323.

Green, G., and MacQuillan, A. M., 1980, Photorepair of ultraviolet induction petite mutational damage in Saccharomyces cerevisiae requires the product of the PHR1 gene, J. Bacteriol., 144:826.

Hall, R. M., Trembath, M. F., Linnane, A. N., Wheelis, L., and Criddle, R. S., 1976, Factors affecting petite induction and the recovery of respiratory competence in yeast cells exposed to ethidium bromide, Mol. Gen. Genet., 144:253.

Handwerker, A., Schweyen, R. J., Wolf, K., and Kaudewitz, F., 1973, Evidence for an extrakaryotic mutation affecting the maintenance of the rho factor in yeast, J. Bacteriol., 113:1307.

Hastings, P. J., Quah, S. K., and von Borstel, R. C., 1976, Spontaneous mutation by mutagenic repair of spontaneous lesions in DNA, Nature (London), 264:719.

Henriques, J. A. P., Chanet, R., Averbeck, D., and Moustacchi, E., 1977, Lethality and "petite" mutation induced by photoaddition of 8-methoxypsoralen in yeast. Influence of ploidy, growth phases and stages in the cell cycle, Mol. Gen. Genet., 158:63.

Henriques, J. A. P., and Moustacchi, E., 1980, Sensitivity to photoaddition of mono- and bifunctional furocoumarins of X-ray sensitive mutants of Saccharomyces cerevisiae, Photochem. Photobiol., 31:557.

Henriques, J. A. P., and Moustacchi, E., 1980, Isolation and characterization of pso mutants sensitive to photoaddition of psoralen derivatives in Saccharomyces cerevisiae, Genetics, 95:273.

Heude, M., and Chanet, R., 1975, Protein synthesis and recovery of both survival and cytoplasmic "petite" mutation in ultraviolet-treated yeast cells. II. Mitochondrial protein synthesis, Mutat. Res., 28:47.

Heude, M., Chanet, R., and Moustacchi, E., 1975, Protein synthesis and recovery of both survival and cytoplasmic "petite" mutation in ultraviolet-treated yeast cells. I. Nuclear-directed protein synthesis, Mutat. Res., 28:37.

Heude, M., Fukuhara, H., and Moustacchi, E., 1979, Spontaneous and induced rho mutants of Saccharomyces cerevisiae: Patterns of the loss of mitochondrial genetic markers, J. Bacteriol., 139:460.

Heude, M., and Moustacchi, E., 1973, Influence de la croissance sur la réparation des radiolésions responsables de la mutation cytoplasmique "petite colonie" chez la levure, C. R. Acad. Sci. Paris, 277:1561..

Heude, M., and Moustacchi, E., 1979, The fate of mitochondrial loci in rho minus mutants induced by ultraviolet irradiation of Saccharomyces cerevisiae: Effects of different postirradiation treatments, Genetics, 93:81.

Hixon, S., Burnham, A. D., and Irons, R. L., 1979, Reversal on protection by light of the ethidium bromide induced petite mutation in yeast, Mol. Gen. Genet., 169:63.

Hixon, S., Franks, H. L., and Moustacchi, E., 1980, Yeast mitochondrial DNA characterization after ultraviolet irradiation, Mutat. Res., 73:267.

Hixon, S., and Moustacchi, E., 1978, The fate of mitochondrial DNA after ultraviolet irradiation. I. Degradation during post-UV dark liquid holding in non-nutrient medium, Biochem. Biophys. Res. Commun., 81:288.

Hollenberg, C. P., and Borst, P., 1971, Conditions that prevent ρ^- induction by ethidium bromide, Biochem. Biophys. Res. Commun., 45:1250.

Johnston, L. H., 1979, Nuclear mutations in Saccharomyces cerevisiae which increase the spontaneous mutation frequency in mitochondrial DNA, Mol. Gen. Genet., 170:327.

Lawrence, C. W., 1981, Mutagenesis in Saccharomyces cerevisiae, in: "Advances in Genetics," E. W. Caspari, ed., Academic Press, New York, in press.

Leff, J., and Eccleshall, R. T., 1978, Replication of bromodeoxyuridylate-substituted mitochondrial DNA in yeast, J. Bacteriol., 135:436.

Lemontt, J. F., 1970, Genetic control of mutation induction in Saccharomyces cerevisiae, Ph.D. thesis, University of California, UCLRL-20115.

Lemontt, J. F., 1980, Genetic and physiological factors affecting repair and mutagenesis in yeast, in: "DNA Repair and Mutagenesis in Eukaryotes," W. M. Generoso, M. D. Shelby, and F. J. de Serres, eds., Plenum Press, New York.

Lusena, C. V., and James, A. P., 1976, Alterations in mitochondrial DNA of yeast which accompany genetically and environmentally controlled changes in ρ^- mutability, Mol. Gen. Genet., 144:119.

Magana-Schwencke, N., Averbeck, D., Henriques, J. A. P., and Moustacchi, E., 1980, Absence de pontages inter-chaines dans l'ADN traité par le 3-carbéthoxypsoralène et une irradiation à 365 nm, C. R. Acad. Sci. Paris, 291:207.

Magana-Schwencke, N., Henriques, J. A. P., Chanet, R., and Moustacchi, E., 1981, The fate of 8-methoxypsoralen photo-induced cross-links in nuclear and mitochondrial yeast DNA. Comparison of wild type and repair deficient strains, submitted.

Mahler, H. R., and Bastos, R. N., 1974, Coupling between mitochondrial mutation and energy transduction, Proc. Natl. Acad. Sci. U.S.A., 71:2241.

Mahler, H. R., and Perlman, P. S., 1972, Mutagenesis by ethidium bromide and mitochondrial membrane, J. Supramol. Struct., 1:105.

Mahler, H. R., and Perlman, P. S., 1973, Induction of respiration-deficient mutants in Saccharomyces cerevisiae by Berenil. I. Berenil, a novel, nonintercalating mutagen, Mol. Gen. Genet., 121:285.

Marmiroli, N., Donnini, C., Restivo, F. M., and Puglisi, P. O., 1980a, Analysis of rho mutability in Saccharomyces cerevisiae. II. Role of mitochondrial protein synthesis, Mol. Gen. Genet., 177:589.

Marmiroli, N., Restivo, F. M., Donnini, C., Bianeti, L., and Puglisi, P. P., 1980b, Analysis of rho mutability in Saccharomyces cerevisiae. I. Effects of mmc and pet-ts alleles, Mol. Gen. Genet., 117:581.

McKee, R. H., and Lawrence, C. W., 1979, Genetic analysis of γ-ray mutagenesis in yeast. I. Reversion in radiation-sensitive strains, Genetics, 93:361.

Moustacchi, E., 1964, Facteurs de la sensibilité des levures à l'action létale des radiations ionisantes. Etude d'un mutant radiorésistant, Thèse Doctorat d'Etat (Paris), n° d'ordre 5172.

Moustacchi, E., 1969, Cytoplasmic and nuclear genetic events induced by UV light in strains of Saccharomyces cerevisiae with different UV sensitivities, Mutat. Res., 7:171.

Moustacchi, E., 1972, Evidence for nucleus-independent steps in control of repair of mitochondrial damages. I. UV induction of the cytoplasmic "petite" mutation in UV-sensitive nuclear mutants of Saccharomyces cerevisiae, Mol. Gen. Genet., 114:50.

Moustacchi, E., 1973, Cytoplasmic "petite" induction in recombination-deficient mutants of Saccharomyces cerevisiae, J. Bacteriol., 115:805.

Moustacchi, E., and Enteric, S., 1970, Differential liquid holding recovery for the lethal effect and cytoplasmic "petite" induction by UV light, Mol. Gen. Genet., 109:69.

Moustacchi, E., Heude, M., and Hixon, S., 1978, The fate of yeast mitochondrial DNA and mitochondrial genetic markers after an ultraviolet light treatment, in: "Biochemistry and Genetics of Yeast," M. Bacila, B. L. Horecker, and A. O. M. Stoppani, eds., Academic Press, New York.

Moustacchi, E., Perlman, P. S., and Mahler, H. R., 1976, A novel class of Saccharomyces cerevisiae mutants specifically UV-sensitive to "petite" induction, Mol. Gen. Genet., 148:251.

Moustacchi, E., Waters, R., Heude, M., and Chanet, R., 1975, The present status of DNA repair mechanisms in UV-irradiated yeast taken as a model eukaryotic system, in: "Radiation Research: Biomedical, Chemical, and Physical Perspectives," O. F. Nygaard, H. I. Adler, and W. K. Sinclair, eds., Academic Press, New York.

Nagai, S., Yanagishima, N., and Nagai, H., 1961, Advances in the study of respiration-deficient (RD) mutation in yeast and other microorganisms, Bacteriol. Rev., 25:404.

Nagley, P., and Linnane, A. W., 1970, Mitochondrial DNA deficient petite mutants of yeast, Biochem. Biophys. Res. Commun., 39:989.

Newlon, C. S., and Fangman, W. L., 1975, Mitochondrial DNA synthesis and cell cycle mutants of Saccharomyces cerevisiae, Cell, 5:423.

Newlon, C. S., Ludescher, P. D., and Walter, S. K., 1979, Production of petites by cell cycle mutants of Saccharomyces cerevisiae defective in DNA synthesis, Mol. Gen. Genet., 169:189.

Oakley, K. M., 1979, Studies on the petite mutation in Saccharomyces cerevisiae, Ph.D. thesis, Australian National University.

Oakley, K. M., and Clark-Walker, G. D., 1978, Abnormal mitochondrial genomes in yeast restored to respiratory competence, Genetics, 90:517.

Oliver, S. G., and Williamson, D. H., 1976, The molecular events involved in the induction of petite yeast mutants by fluorinated pyrimidines, Mol. Gen. Genet., 146:253.

Paoletti, C., Couder, H., and Guerineau, M., 1972, A yeast mitochondrial deoxyribonuclease stimulated by ethidium bromide, Biochem. Biophys. Res. Commun., 48:950.

Patrick, M. H., Haynes, R. H., and Uretz, R. B., 1964, Dark recovery phenomena in yeast. I. Comparative effects with various inactivating agents, Radiat. Res., 21:144.

Pinto, M., Guerineau, M., and Paoletti, C., 1975, Ethidium bromide mutagenesis in yeast: Protection by anaerobiosis, Mutat. Res., 30:219.

Pittman, D., and Pedigo, P. R., 1959, Reactivation studies on yeasts. I. Ultraviolet inactivation and photoreactivation of respiration-sufficient and respiration-deficient yeasts, Exp. Cell Res., 17:359.

Polakowska, R., and Putrament, A., 1979, Mitochondrial mutagenesis in Saccharomyces cerevisiae. II. Methyl methanesulphonate and diepoxybutane, Mutat. Res., 61:207.

Prakash, L., 1975, Repair of pyrimidine dimers in nuclear and mitochondrial DNA of yeast irradiated with low doses of ultraviolet light, J. Mol. Biol., 98:781.

Prakash, L., and Prakash, S., 1980, Genetic analysis of error-prone repair systems in Saccharomyces cerevisiae, in: "DNA Repair and Mutagenesis in Eukaryotes," W. M. Generoso, M. D. Shelby, and F. J. de Serres, eds., Plenum Press, New York.

Prakash, S., Prakash, L., Burke, W., and Montelone, B. A., 1979, Effects of the rad52 gene on recombination in Saccharomyces cerevisiae, Genetics., 94:31.

Prunell, A., and Bernardi, G., 1974, The mitochondrial genome of wild-type yeast cells, J. Mol. Biol., 86:825.

Putrament, A., Baranowska, H., and Prazmo, W., 1973, Induction by manganese of mitochondrial antibiotic resistance mutations in yeast, Mol. Gen. Genet., 126:357.

Radman, M., Villani, G., Boiteux, S., Kinsella, A. R., Glickman, B. W., and Spadari, S., 1979, Replication fidelity: Mechanism of mutation avoidance and mutation fixation, Cold Spring Harbor Symp. Quant. Biol., 43:937.

Raut, C., and Simpson, W. L., 1955, The effect of X rays and ultraviolet light of different wavelengths on the production of cytochrome-deficient yeasts, Arch. Biochem. Biophys., 57:218.

Resnick, M. A., 1969, A photoreactivationless mutant of Saccharomyces cerevisiae, Photochem. Photobiol., 9:307.

Reynolds, R. J., and Friedberg, E. C., 1980, Molecular mechanism of pyrimidine dimer excision in Saccharomyces cerevisiae. I. Studies with intact cells and cell-free systems, in: "DNA Repair and Mutagenesis in Eukaryotes," W. M. Generoso, M. D. Shelby, and F. J. de Serres, eds., Plenum Press, New York.

Rodarte-Ramon, U. S., and Mortimer, R. K., 1972, Radiation-induced recombination in Saccharomyces: Isolation and genetic study of recombination-deficient mutants, Radiat. Res., 49:133.

Rupp, W. D., and Howard-Flanders, P., 1968, Discontinuities in the DNA synthesized in an excision-defective strain of Escherchia coli following ultraviolet irradiation, J. Mol. Biol., 31:291.

Saeki, T., Machida, I., and Nakai, S., 1980, Genetic control of diploid recovery after γ-irradiation in the yeast Saccharomyces cerevisiae., Mutat. Res., 73:251.

Sanders, J. P. N., Heyting, C., Difranco, A., Borst, P., and Slonimski, P., 1976, The organization of genes in yeast mitochondrial DNA, in: "The Genetic Function of Mitochondrial DNA," C. Saccone, and A. M. Kroon, eds., Elsevier/North-Holland, Amsterdam.

Sarachek, A., 1959, The induction by UV and the photoreactivation of heritable respiratory deficiency in polyploid Saccharomyces, Cytologia, 24:507.

Schenberg-Frascino, A., and Moustacchi, E., 1972, Lethal and mutagenic effects of elevated temperature on haploid yeast, Mol. Gen. Genet., 115:243.

Schweyen, R. J., Steyer, U., Kaudewitz, F., Dujon, B., and Slonimski, P., 1976, Mapping of mitochondrial genes in Saccharomyces cerevisiae. Population and pedigree analysis of retention or loss of four genetic markers in rho⁻ cells, Mol. Gen. Genet., 146:117.

Sena, E. P., Welch, J. W., Halvorson, H. O., and Fogel, S., 1975, Nuclear and mitochondrial deoxyribonucleic acid replication during mitosis in Saccharomyces cerevisiae, J. Bacteriol., 123: 497.

Sherman, F., 1958, A study of the effects of elevated temperatures on the growth and inheritance of Saccharomyces cerevisiae, Ph.D. thesis, University of California, UCRL-8573.

Sherman, F., Jackson, M., Liebman, S. W., Schweingruber, A. M., and Stewart, J. W., 1975, A deletion map of cyc1 mutants and its correspondence to mutationally altered iso-1-cytochrome c of yeast, Genetics, 81:51.

Sherman, F., and Slonimski, P., 1964, Respiration deficient mutants of yeast. II. Biochemistry, Biochim. Biophys. Acta, 90:1.

Slonimski, P., and Lazowska, J., 1977, Transposable segments of mitochondrial DNA: A unitary hypothesis for the mechanism of mutation, recombination, sequence reiteration and suppressiveness of yeast "petite colony" mutants, in: "Mitochondria 1977," W. Bandlow, R. J. Schweyen, K. Wolf, and F. Kaudewitz, eds., De Gruyter, Berlin.

Slonimski, P., Perrodin, G., and Croft, J. H., 1968, Ethidium bromide-induced mutation of yeast mitochondria: Complete transformation of cells into respiratory-deficient nonchromosomal "petites", Biochem. Biophys. Res. Commun., 30:232.

Slonimski, P., and Tzagoloff, A., 1976, Localization in yeast mitochondrial DNA of mutations expressed in a deficiency of cytochrome oxidase and for coenzyme QH_2-cytochrome-c-reductase, Eur. J. Biochem., 61:27.

Stevens, B. J., 1981, Mitochondrial structure, in: "The Molecular Biology of the Yeast Saccharomyces," J. N. Strathern, E. W. Jones, and J. R. Broach, eds., Cold Spring Harbor Laboratory, New York, in press.

Stevens, B. J., and Moustacchi, E., 1976, Ultrastructural characterization of mitochondria from a yeast mutant sensitive to petite induction (uvsρ72), in: "Genetics, Biogenesis and Bioenergetics of Mitochondria," W. Bandlow, R. J. Schweyen, D. Y. Thomas, K. Wolf, and F. Kaudewitz, eds., De Gruyter, Berlin.

Storm, E. M., and Marmur, J. A., 1975, A temperature-sensitive mitochondrial mutation of Saccharomyces cerevisiae, Biochem. Biophys. Res. Commun., 64:752.

Subik, J., Takacsova, G., and Kovač, L., 1978, Intramitochondrial ATP and cell functions. I. Growing yeast cells depleted of intramitochondrial ATP are losing mitochondrial genes, Mol. Gen. Genet., 166:103.

Swandbeck, G., 1976, Site genetic effects of psoralens and light, in: "Photochemical Therapy Basic Techniques and Side Effects," E. G. Jung, and S. K. Schattauer, eds., Proc. of the German-Swedish Symp. of Photomedicine in Oberursel, Stuttgart, New York.

Trembath, M. K., Monk, B. C., Kellerman, G. M., and Linnane, A. W., 1975, Biogenesis of mitochondria 36. The genetic and biochemical analysis of a mitochondrially determined cold-sensitive oligomycin-resistant mutant of Saccharomyces cerevisiae with affected mitochondrial ATPase assembly, Mol. Gen. Genet., 141:9.

Tuite, M. F., and Cox, B. X., 1980, Ultraviolet mutagenesis studies of (psi), a cytoplasmic determinant of Saccharomyces cerevisiae, Genetics, 95:611.

Unrau, P., Wheatcroft, R., and Cox, B., 1971, The excision of pyrimidine dimers from DNA of ultraviolet irradiated yeast, Mol. Gen. Genet., 113:359.

von Borstel, R. C., Cain, K. T., and Steinberg, C. M., 1971, Inheritance of spontaneous mutability in yeast, Genetics, 69:17.

Waters, R., and Moustacchi, E., 1974, The fate of ultraviolet-induced pyrimidine dimers in the mitochondrial DNA of Saccharomyces cerevisiae following various postirradiation cell treatments, Biochim. Biophys. Acta, 366:241.

Wheelis, L., Trembath, M. K., and Criddle, R. S., 1975, Petite induction and recovery in the presence of high levels of ethidium bromide, Biochem. Biophys. Res. Commun., 65:838.

Whittaker, P. A., Hammond, R. C., and Luha, A. A., 1972, Mechanism of mitochondrial mutation in yeast, Nature (London), New Biol., 238:266.

Williamson, D. H., 1970, The effect of environmental and genetic factors on the replication of mitochondrial DNA in yeast, in: "Control of Organelle Development," Symp. Soc. Exp. Biol. XXIV, P. Miller, ed., The University Press, Cambridge.

Williamson, D. H., and Fennell, D. J., 1974, Apparent dispersive replication of yeast mitochondrial DNA as revealed by density labeling experiments, Mol. Gen. Genet., 131:193.

Williamson, D. H., Maroudas, N. G., and Wilkie, C. D., 1971, Induction of the cytoplasmic petite mutation in Saccharomyces cerevisiae by the antibacterial antibiotics erythromycin and chloramphenicol, Mol. Gen. Genet., 111:209.

Wintersberger, U., and Hirsch, J., 1973, Induction of cytoplasmic respiratory-deficient mutants in yeast by the folic acid analogue methotrexate. I. Studies on the mechanism of petite induction, Mol. Gen. Genet., 126:61.

Wolf, K., and Kaudewitz, F., 1976, Effect of caffeine on the rho$^-$ induction with ethidium bromide in *Saccharomyces cerevisiae*, *Mol. Gen. Genet.*, 146:89.

CHAPTER 20

ALTERATIONS IN CHROMATIN STRUCTURE DURING DNA EXCISION REPAIR

Michael W. Lieberman

Department of Pathology, Washington University School of Medicine, St. Louis, Missouri 63110

SUMMARY

Work from a number of laboratories recently has demonstrated that alterations in chromatin structure occur during excision repair in mammalian cells. It is now clear that when cells are damaged with a wide variety of chemical agents or ultraviolet radiation, almost all of the repair synthesis is initially sensitive to staphylococcal nuclease. With time, there is a redistribution of the counts incorporated during excision repair synthesis so that many of them become nuclease resistant and associated with nucleosome core length DNA. In our laboratory, we have demonstrated this phenomenon in human cells damaged with N-acetoxy-2-acetylaminofluorene, 7-bromomethylbenz[a]anthracene, and ultraviolet radiation. It is clear from the work of others that the phenomenon is not unique to human cells since African green monkey cells damaged with either ultraviolet radiation or angelicin also show an initial nuclease sensitivity of repair-incorporated nucleotides follow by rearrangement. Two models to explain these observations have been proposed; one suggests that there is an unfolding of nucleosomes during excision repair followed by a refolding, while the other suggests that sliding of core proteins with respect to DNA occurs during excision repair. These models, as well as recent data bearing on them, will be discussed.

INTRODUCTION

For the past few years, the analysis of how DNA excision repair occurs in chromatin has been a major preoccupation of my laboratory. This work originated as an outgrowth of earlier studies demonstrating that a great variety of different types of sequences in the mammalian genome were repairable (Lieberman and Poirier, 1974a,b; Meltz and Painter, 1973) and the realization that chromatin, and not DNA, is the real substrate for repair processes in mammalian cells. The almost explosive development of the chromatin field in the mid and late 1970s (Cold Spring Harbor Symposium on Quantitative Biology, 1977; Felsenfeld, 1978; Klug et al., 1981) provided not only a comprehensive understanding of the nucleosome, the fundamental unit of chromatin structure, but also a series of useful techniques. Our results and work from other laboratories also suggested that DNA excision repair in human cells is largely error-correcting and possibly error-free (Heflich et al., 1980; Lieberman, 1976; Lieberman and Poirier, 1974c; Maher et al., 1978, 1979). While these data indicated that all four deoxyribonucleotides were incorporated into repair patches in a template-directed fashion with the resulting restoration of the fidelity of DNA sequence, they left unanswered the important question of the restoration of nuclear function. Simultaneously, a literature was emerging on alterations in chromatin structure associated with changes in gene expression. A number of papers demonstrated changes in DNase I and micrococcal nuclease sensitivity of regions of the genome that were either expressed or potentially expressable, while others defined the fine structure of the nuclease sensitivity of individual genes (e.g., Bloom and Anderson, 1979; Garel and Axel, 1976; Samal et al., 1981; Weintraub and Groudine, 1976; Weisbrod and Weintraub, 1979; Wu et al., 1979a,b). Other explorations indicated that various proteins, and especially the high mobility group proteins, were associated with active regions of chromatin (e.g., Levy et al., 1979; Weisbrod and Weintraub, 1979; Weisbrod et al., 1980). All of these factors made it appropriate and even compelling to examine changes in chromatin structure during repair synthesis.

CHANGES IN CHROMATIN STRUCTURE DURING DNA EXCISION REPAIR SYNTHESIS

While early studies from a number of laboratories, including our own, analyzed DNA repair synthesis in heterochromatic and euchromatic regions of the nucleus (Berliner et al., 1975; Harris et al., 1974) and in "covered" and "uncovered" regions of chromatin (Wilkins and Hart, 1974), it was not until the application of nuclease digestion techniques and gel electrophoresis that detailed progress was made in understanding how repair synthesis occurs in chromatin. Within a year of one another, three groups (Bodell, 1977; Cleaver, 1977; Smerdon et al., 1978) demonstrated that following treatment of cells with ultraviolet radiation or methyl methanesulfonate most of the label incorporated during DNA repair synthesis was sensitive to micrococcal nuclease, i.e., it was released more rapidly by the nuclease from nuclei (chromatin) than was DNA as a whole. At the time this finding was felt to be due to the presence of repair-incorporated nucleotides in "linker" regions of chromatin. While it is true that linker regions of chromatin are more susceptible to attack by micrococcal nuclease than core regions, it does not follow that micrococcal nuclease sensitive DNA and linker DNA are synonymous. It is now clear that resistance to micrococcal nuclease is conferred by the presence of DNA in core regions of chromatin (or possibly the association of DNA with other proteins). Conversely, any DNA that is unprotected by core proteins including, but not limited to, linker DNA is staphylococcal nuclease sensitive.

A major finding to come out of our laboratory was the demonstration that, although most of the nucleotides incorporated during repair synthesis are initially nuclease sensitive, with time they become more uniformly distributed between sensitive and resistant regions of chromatin (Smerdon and Lieberman, 1978). This redistribution has been termed "rearrangement" and occurs largely, but not completely, during the first 3-4 h following repair synthesis. Gel electrophoretic analysis has demonstrated that the acquisition of nuclease resistance is associated with the appearance of newly repaired regions in nucleosome core length DNA. One of the intriguing aspects of these data is that while rearrangement and the acquisition of nuclease resistance is a well-defined phenomenon that has been observed under a wide variety of conditions, nevertheless, it is rarely complete. As late as 24 h following repair synthesis there is still a slight, but clearly demonstrable, micrococcal nuclease sensitivity of repair-incorporated nucleotides (see Section on Recent Findings from Our Laboratory for DNase I results).

The initial micrococcal nuclease sensitivity of repair-incorporated nucleotides and their subsequent "rearrangement" has been observed in human diploid fibroblasts damaged not only with ultraviolet radiation but also with two direct-acting chemical carcinogens, N-acetoxy-2-acetylaminofluorene and 7-bromomethylbenz[a]anthracene (Oleson et al., 1979; Tlsty and Lieberman, 1978). Thus, although there may be subtle differences among these agents, the phenomenon seems to be a general one. All of these agents, however, induce a "long patch" type of DNA repair in which somewhere between 30 and 140 nucleotides are incorporated into the gap (Edenberg and Hanawalt, 1972; Regan and Setlow, 1974; Smith and Hanawalt, 1978). Agents that produce so-called "short patch" repair, such as the DNA methylating and ethylating agents, have not been examined thoroughly in terms of rearrangement. Other data have demonstrated that xeroderma pigmentosum cells that retain partial repair competence (XP complementation groups C, D, and E) also show rearrangement during repair synthesis (Smerdon et al., 1979; Williams and Friedberg, 1979); this finding is of interest since it has been suggested that one of the defects in xeroderma pigmentosum cells is that they have some defect in the ability to repair DNA damage in chromatin (Mortelmans et al., 1976).

MODELS OF EXCISION REPAIR SYNTHESIS IN CHROMATIN

As discussed previously (Lieberman et al., 1979; Oleson et al., 1979), one may envision two general kinds of models that might explain these results. The first involves a "sliding" of protein cores along the DNA during repair synthesis. One might envision, for instance, that removal of damage and repair synthesis occur only, or primarily, in linker regions. This postulate would explain why repair-incorporated nucleotides are always initially nuclease sensitive. If either a constituitive or induced sliding mechanism of some type occurred, many of these nucleotides might eventually become nuclease resistant. Although a number of studies have shown sliding under artificial conditions of incubation (Beard, 1978; Spadafora et al., 1979; Steinmetz et al., 1978; Tatchell and Van Holde, 1978), there is no evidence that it is a physiologic process occurring in living cells. In fact, studies of nucleosome phasing of the tRNA genes (Louis et al., 1980; Wittig and Wittig, 1979) and of the structure of the histone genes and the heat shock loci in Drosophila (Samal et al., 1981; Wu et al., 1979a) indicate that under many conditions the structure of chromatin remains constant.

A model that is intrinsically more appealing to us suggests that during excision repair there is a change in the conformation of chromatin in the region undergoing repair (Lieberman et al., 1979; Oleson et al., 1979). The model suggests that the change, termed "unfolding" by us, results from core DNA in this region becoming partially "dissociated" from core histones. This change might be obligatory when repair synthesis is initiated in core regions of nucleosomes. It is not clear whether "unfolding" would be obligatory for repair synthesis that is initiated in linker regions. This unfolding would render any nucleotides newly incorporated during repair synthesis nuclease sensitive. If there were a subsequent reestablishment or partial reestablishment of chromatin structure (i.e., a "refolding"), many of the repair-incorporated nucleotides would become nuclease resistant as DNA-protein contacts were reestablished. While there is no direct evidence that distinguishes between the sliding and the unfolding-refolding rearrangement models, nevertheless, they provide a framework for thinking about alterations in chromatin structure during repair and a strong incentive for additional studies in this area. Although there has been much speculation about the role of histone modification [e.g., poly(ADP)ribosylation, phosphorylation, methylation, and acetylation] in relation to changes in chromatin structure during repair synthesis, to date there is no solid evidence implicating any modification of nuclear proteins in the changes in micrococcal nuclease sensitivity of repair-incorporated nucleotides.

DATA ON REARRANGEMENT FROM OTHER LABORATORIES

Although initially Bodell and Cleaver, working independently (Bodell, 1977; Bodell and Banerjee, 1979; Cleaver, 1977, 1979), did not observe "rearrangement," a subsequent study by them in African green monkey cells damaged with ultraviolet radiation has confirmed our findings (Bodell and Cleaver, 1981). Williams and Friedberg (1979) have also demonstrated rearrangement in human diploid fibroblasts damaged with ultraviolet radiation. Although the amount of rearrangement they observed was <u>apparently</u> less than the amount we observed, the difference is due largely to the longer pulse time employed by them (see Smerdon and Lieberman, 1980). Other studies by Zolan and Hanawalt (1981) have demonstrated rearrangement in African green monkey cells damaged with ultraviolet radiation or psoralen and long wavelength ultraviolet light. Thus, although little is known about the molecular mechanism by which rearrangement occurs, the phenomenon is now solidly established, having been observed independently in four different laboratories.

DNA DAMAGE AND REMOVAL IN RELATION TO REARRANGEMENT

Initially, it seemed as if a precise analysis of the distribution of damage within chromatin and its subsequent removal was crucial for an analysis of rearrangement: it was argued that if most of the damage occurred in "linker" regions of the genome then it would be logical that most of the repair synthesis should also occur there. While this supposition is reasonable, the emphasis is misplaced in that too little attention is given to the phenomenon of major interest (i.e., "rearrangement"). In fact, an initial study looked at the distribution of [^{3}H]methylbenzanthracene in DNA in nuclease-sensitive and -resistant regions of chromatin following treatment of human diploid fibroblasts with [^{3}H]7-bromomethylbenz-[a]anthracene (Oleson et al., 1979). It was found that on a per nucleotide basis, linker DNA is somewhat more likely to be damaged; however, because there are more total nucleotides in core regions, there was more total damage in core DNA than in linker DNA. Furthermore, the rate of removal from core and linker regions appeared to be approximately the same. Other studies have looked at the distribution of cyclobutylpyrimidine dimers (Snapka and Linn, 1981; Williams and Friedberg, 1979) and found that on a per-nucleotide basis dimers were at least as likely, and perhaps more likely, to be formed in core DNA as in linker. Again, because there are more total nucleotides in core DNA more total damage occurred in core DNA than in linker. In one of these studies (Williams and Friedberg, 1979), the rate of removal of pyrimidine dimers from core and total DNA was found to be the same. These studies are in contrast to work from Cerutti's lab that has shown on a per-nucleotide basis much more [^{3}H] was bound to linker DNA than to core DNA following the administration of [^{3}H]N-acetoxy-2-acetylaminofluorene to cultured human diploid fibroblasts. Furthermore, these investigators found that removal occurred more rapidly from linker DNA than from core DNA (Kaneko and Cerutti, 1980). In general, a great many studies have looked at the distribution of damage in linker and core DNA, and some types of agents such as the benzopyrene-diolepoxide and psoralens plus long wave length ultraviolet light apparently damage linker primarily (Cech and Pardue, 1977; Jahn and Littman, 1979; Kootstra et al., 1979; Wiesehahn et al., 1977; Yamasaki et al., 1978), as 7-bromomethylbenz[a]anthracene, ultraviolet radiation, and some of the methylating agents (Kaneko and Cerrutti, 1980; McGhee and Elsenfeld, 1979; Mirzabekov et al., 1977; Oleson et al., 1979; Williams and Friedberg, 1979) apparently make much less distinction between core and linker DNA.

Although the definitive study remains to be done, the implication of much of these data is that rearrangement occurs regardless of whether the damage is initially in linker or core DNA. Certainly the preliminary data of Zolan and Hanawalt (1981) on psoralens support this concept as do the aggregate data from our lab on the distribution of repair synthesis following treatment of cells with N-acetoxy-2-acetylaminofluorene (Tlsty and Lieberman, 1978) and those from Cerutti's laboratory on removal (Kaneko and Cerrutti, 1980). A totally satisfactory analysis of the repair synthesis results and the removal results has been hampered not only by a shortage of studies of removal and rearrangement from the same laboratory under the same conditions, but also by a lack of good data on patch size. Clearly if patch sizes are in the range of 30, as suggested by data from Hanawalt's laboratory (Edenberg and Hanawalt, 1972; Smith and Hanawalt, 1978), the implications are very different for rearrangement following linker specific agents than they are if the average patch size is between 100 and 140, as suggested by the data of Regan and Setlow (1974). These numbers should be compared with a maximal average "linker" length of about 47 for human fibroblasts (192 = the repeat length and 145 = the length of "core" DNA).

RECENT FINDINGS FROM OUR LABORATORY

Recent studies from our laboratory have confirmed and extended our previous findings (Smerdon and Lieberman, 1980). It has been known for over a decade that following damage repair synthesis occurs for many hours or even days (e.g., Kantor and Setlow, 1981; Smerdon et al., 1978). It was, therefore, of interest to determine whether or not the same initial nuclease sensitivity followed by rearrangement occurred in sequences being repaired at long times after damage. Our findings indicate that the rearrangement process in cells damaged at time 0 but not labeled with [^{3}H]dThd until 23½ h following damage was similar, though not identical, to that in cells pulsed immediately following damage. While at both times there was an initial nuclease sensitivity of the repair-incorporated nucleotides, the rate of rearrangement appeared more rapid in the 23½-h protocol. The meaning of this finding is unclear at present, and it serves to underscore how little we really know about the detailed molecular events taking place during repair in chromatin.

The reestablishment of core particles during repair was analyzed by looking at the restoration of DNase I digestion patterns during rearrangement. Interestingly, the data indicated that, while much of the characteristic ten base-pair repeat was reestablished during rearrangement, there appeared to be differences between bulk chromatin and the chromatin containing repair-incorporated nucleotides. Whether these differences are real and represent an "imperfect" or altered rearranged state or are the result of experimental error is unclear at this point. Nevertheless, it is intriguing to think that following rearrangement the restoration of chromatin structure might not be perfect; this is especially so when it is remembered that rearrangement, while initially rapid, might be incomplete even 24 h following repair synthesis.

What is the current status of our understanding of the role of chromatin structure in repair synthesis? On the one hand, several aspects of the process are well established. First, it is clear that initially the nucleotides incorporated during excision repair synthesis are micrococcal nuclease and DNase I sensitive; second, it is now well established that these nucleotides that are initially nuclease sensitive acquire within a matter of hours a degree of nuclease resistance characteristic of the DNA in chromatin as a whole. Third, the acquisition of nuclease resistance is associated with the appearance of many repair-incorporated nucleotides in nucleosome core length DNA, and thus, one may infer that resistance is acquired by the packaging of repair-incorporated nucleotides in core particles. Fourth, rearrangement appears to be incomplete since some residual nuclease sensitivity may persist even 24 h after repair synthesis; furthermore, the rate of rearrangement may vary depending on when after damage the repair occurs.

In contrast, we are completely ignorant of the molecular processes involved. It is not clear whether the unfolding model or the sliding model or, for that matter, some other possibility, explains rearrangement. It is also uncertain whether only the immediate small area undergoing repair (i.e., 50 to 200 nucleotides) becomes nuclease sensitive or whether a much larger region is involved. We do not know whether the process is the same or different in different types of chromatin (e.g., heterochromatin vs euchromatin; genes being actively transcribed vs quiescent genes). The fact that the time course of rearrangement appears different following repair at early times after damage and late times is tantalizing but unexplained. Likewise, the possible differences in DNase I sensitivity following "rearrangement" raise the issue of whether or not core nucleosomes are accurately and faithfully reconstituted during rearrangement.

One of the general problems hindering further analysis is the fact that damage and repair, if not random, are certainly widespread throughout the genome and involve repair of a great many different types of DNA sequences that undoubtedly are packaged in different ways within chromatin. Thus, one is faced with the problem of the simultaneous analysis of repair in many different types of chromatin. It is not immediately obvious what strategies may be used to sort out all of our questions about how repair occurs in chromatin; however, with many of the new tools available from molecular biology and genetics, undoubtedly some of these questions may be framed in answerable terms.

REFERENCES

Beard, P., 1974, Mobility of histones on the chromosome of simian virus 40, Cell, 15:955.
Berliner, J., Himes, S. W., Aoki, C. T., and Norman, A., 1975, The sites of unscheduled DNA synthesis within irradiated human lymphocytes, Radiat. Res., 63:544.
Bloom, K. S., and Anderson, J. N., 1979, Conformation of ovalbumin and globin genes in chromatin during differential gene expression, J. Biol. Chem., 254:10532.
Bodell, W. J., 1977, Nonuniform distribution of DNA repair in chromatin after treatment with methyl methanesulfonate, Nucleic Acids Res., 4:2619.
Bodell, W. J., and Banerjee, M. R., 1979, The influence of chromatin structure on the distribution of DNA repair synthesis studied by nuclease digestion, Nucleic Acids Res., 6:359.
Bodell, W. J., and Cleaver, J. E., 1981, Transient conformation changes in chromatin during excision repair of ultraviolet damage to DNA, Nucleic Acids Res., 9:203.
Cech, T., and Pardue, M. L., 1977, Cross-linking of DNA with trimethylpsoralen is a probe for chromatin structure, Cell, 11:631.
Cleaver, J. E., 1977, Nucleosome structure controls rates of excision repair in DNA of human cells, Nature (London), 270:451.
Cleaver, J. E., 1979, Similar distributions of repair sites in chromatin of normal and xeroderma pigmentosum variant cells damaged by ultraviolet light, Biochim. Biophys. Acta, 565:387.
Cold Spring Harbor Symp. Quant. Biol., 1977, 42.
Edenberg, H., and Hanawalt, P. C., 1972, Size of repair patches in the DNA of ultraviolet-irradiated HeLa cells, Biochim. Biophys. Acta, 272:361.
Felsenfeld, G., 1978, Chromatin, Nature, 271:115.

Garel, A., and Axel, R., 1976, Selective digestion of transcriptionally active ovalbumin genes from oviduct nuclei, Proc. Natl. Acad. Sci. U.S.A., 73:3966.

Harris, C. C., Connor, P. J., Jackson, F. E., and Lieberman, M. W., 1974, Intranuclear distribution of DNA repair synthesis induced by chemical carcinogens or ultraviolet light in human diploid fibroblasts, Cancer Res., 34:3461.

Heflich, R. H., Hazard, R. M., Lommel, L., Scribner, J. D., Maher, V. M., and McCormick, J. J., 1980, A comparison of the DNA binding, cytotoxicity, and repair synthesis induced in human fibroblasts by reactive derivatives of aromatic amide carcinogens, Chem.-Biol. Interact., 29:43.

Jahn, C. L., and Littman, G. W., 1979, Accessibility of deoxyribonucleic acid in chromatin to the covalent binding of the chemical carcinogen benzo[a]pyrene, Biochemistry, 18:1442.

Kaneko, M., and Cerutti, P. A., 1980, Excision of N-acetoxy-2-acetylaminofluorene-induced DNA adducts from chromatin fractions of human fibroblasts, Cancer Res., 40:4313.

Kantor, G. J., and Setlow, R. B., 1981, Rate and extent of DNA repair in nondividing human diploid fibroblasts, Cancer Res., 41:819.

Klug, A., Rhodes, D., Smith, J., Finch, J. T., and Thomas, J. O., 1981, A low resolution structure for the histone core of the nucleosome, Nature (London), 287:509.

Kootstra, A., Slaga, T. J., and Olins, D. E., 1979, Interaction of benzo[a]pyrene diol-epoxide with nuclei and isolated chromatin, Chem.-Biol. Interact., 28:225.

Levy, W. B., Connor, W., and Dixon, G. H., 1979, A subset of trout testis nucleosomes enriched in transcribed DNA sequences contains high mobility group proteins as major structural components, J. Biol. Chem., 254:609.

Lieberman, M. W., 1976, Approaches to the analysis of fidelity of DNA repair in mammalian cells, Int. Rev. Cytol., 45:1.

Lieberman, M. W., and Poirier, M. C., 1974a, Intragenomal distribution of DNA repair synthesis: Repair in satellite and mainband DNA in cultured mouse cells, Proc. Natl. Acad. Sci. U.S.A., 71:2461.

Lieberman, M. W., and Poirier, M. C., 1974b, Distribution of deoxyribonucleic acid repair synthesis among repetitive and unique sequences in the human diploid genome, Biochemistry, 13:3018.

Lieberman, M. W., and Poirier, M. C., 1974c, Base pairing and template specificity during deoxyribonucleic acid repair synthesis in human and mouse cells, Biochemistry, 13:5384.

Lieberman, M. W., Smerdon, M. J., Tlsty, T. D., and Oleson, F. B., 1979, The role of chromatin structure in DNA repair in human cells damaged with chemical carcinogens and ultraviolet radiation, in: "Environmental Carcinogenesis," P. Emmelot, and E. Kriek, eds., Elsevier/North-Holland Biomedical Press, Amsterdam.

Louis, C., Schedl, P., Samal, B., and Worcel, A., 1980, Chromatin structure of the 5S RNA genes of D. melanogaster, Cell, 22:387.

Maher, V. M., Curren, R. D., Ouellette, L. M., and McCormick, J. J., 1979, Role of DNA repair in the cytotoxic and mutagenic action of physical and chemical carcinogens, in: "In Vitro Metabolic Activation in Mutagen Testing," F. J. deSerres, J. R. Fouts, J. R. Bend, and R. M. Philpot, eds., Elsevier/North-Holland Biomedical Press, Amsterdam.

Maher, V. M., Dorney, D. J., Heflich, R. H., Levinson, J. W., Mendrala, A. L., and McCormick, J. J., 1978, Biological and biochemical evidence that DNA repair processes in normal human cells act to reduce the lethal and mutagenic effects of exposure to carcinogens, in: "DNA Repair Mechanisms, " P. C. Hanawalt, E. C. Friedberg, and C. F. Fox, eds., Academic Press, New York.

McGhee, J. D., and Felsenfeld, G., 1979, Reaction of nucleosome DNA with dimethyl sulfate, Proc. Natl. Acad. Sci. U.S.A., 76:2133.

Meltz, M. L., and Painter, R. B., 1973, Distribution of repair replication in the HeLa cell genome, Int. J. Radiat. Biol., 23:637.

Mirzabekov, A. D., Shick, V. V., Belzavsky, A. V., Karpov, V. L., and Bavykin, S. G., 1977, The structure of nucleosomes: The arrangement of histones in the DNA grooves and along the DNA chain, Cold Spring Harbor Symp. Quant. Biol., 42:149.

Mortelmans, K., Friedberg, E. C., Slor, H., Thomas, G., and Cleaver, J. E., 1976, Defective thymidine dimer excision by cell-free extracts of xeroderma pigmentosum cells, Proc. Natl. Acad. Sci. U.S.A., 73:2757.

Oleson, F. B., Mitchell, B. L., Dipple, A., and Lieberman, M. W., 1979, Distribution of DNA damage in chromatin and its relation to repair in human cells treated with 7-bromomethylbenz[a]-anthracene, Nucleic Acids Res., 7:1343.

Regan, J. D., and Setlow, R. B., 1974, Two forms of repair in the DNA of human cells damaged by chemical carcinogens and mutagens, Cancer Res., 34:3318.

Samal, B., Worcel, A., Louis, C., and Schedl, P., 1981, Chromatin structure of the histone genes of D. melanogaster, Cell, 23:401.

Smerdon, M. J., Kastan, M. B., and Lieberman, M. W., 1979, Distribution of repair-incorporated nucleotides and nucleosome rearrangement in the chromatin of normal and xeroderma pigmentosum human fibroblasts, Biochemistry, 18:3732.

Smerdon, M. J., and Lieberman, M. W., 1978, Nucleosome rearrangement in human chromatin during UV-induced DNA repair synthesis, Proc. Natl. Acad. Sci. U.S.A., 75:4238.

Smerdon, M. J., and Lieberman, M. W., 1980, Distribution within chromatin of deoxyribonucleic acid repair synthesis occurring at different times after ultraviolet radiation, Biochemistry, 19:2992.

Smerdon, M. J., Tlsty, T. D., and Lieberman, M. W., 1978, Distribution of ultraviolet-induced DNA repair synthesis in nuclease sensitive and resistant regions of human chromatin, Biochemistry, 17:2377.

Smith, C. A., and Hanawalt, P. C., 1978, Phage T4 endonuclease V stimulates DNA repair replication in isolated nuclei from ultraviolet-irradiated human cells, including xeroderma pigmentosum fibroblasts, Proc. Natl. Acad. Sci. U.S.A., 75:2598.

Snapka, R. M., and Linn, S., 1981, Efficiency of formation of pyrimidine dimers in SV40 chromatin in vitro, Biochemistry, 20:68.

Spadafora, C., Oudet, P., and Chambon, P., 1979, Rearrangement of chromatin structure induced by increasing ionic strength and temperature, Eur. J. Biochem., 100:225.

Steinmetz, M., Streeck, R. E., and Zachau, H. G., 1978, Closely spaced nucleosome cores in reconstituted histone DNA complexes and histone-H1-depleted chromatin, Eur. J. Biochem., 83:615.

Tatchell, K., and Van Holde, K. E., 1978, Compact oligomers and nucleosome phasing, Proc. Natl. Acad. Sci. U.S.A., 75:3583.

Tlsty, T. D., and Lieberman, M. W., 1978, The distribution of DNA repair synthesis in chromatin and its rearrangement following damage with N-acetoxy-2-acetylaminofluorene, Nucleic Acids Res., 5:3261.

Weintraub, H., and Groudine, M., 1976, Chromosomal subunits in active genes have an altered conformation, Science, 193:848.

Weisbrod, S., Groudine, M., and Weintraub, H., 1980, Interaction of HMG 14 and 17 with actively transcribed genes, Cell, 19:289.

Weisbrod, S., and Weintraub, H., 1979, Isolation of a subclass of nuclear proteins responsible for conferring a DNase I-sensitive structure on globin chromatin, Proc. Natl. Acad. Sci. U.S.A., 76:630.

Wiesehahn, G., Hyde, J. E., and Hearst, J. E., 1977, The photoaddition of trimethylpsoralen to Drosophila melanogaster nuclei: A probe for chromatin substructure, Biochemistry, 16:925.

Wilkins, R. J., and Hart, R. W., 1974, Preferential DNA repair in human cells, Nature (London), 247:35.

Williams, J. I., and Friedberg, E. C., 1979, Deoxyribonucleic acid excision repair in chromatin after ultraviolet irradiation of human fibroblasts in culture, Biochemistry, 18:3965.

Wittig, B., and Wittig, S., 1979, A phase relationship associates tRNA structural gene sequences with nucleosome cores, Cell, 18:1173.

Wu, C., Bingham, P. M., Livak, K. J., Holmgren, R., and Elgin, S. C. R., 1979a, The chromatin structure of specific genes: I. Evidence for higher order domains of defined DNA sequence, Cell, 16:797.

Wu, C., Wong, Y-C., and Elgin, S. C. R., 1979b, The chromatin structure of specific genes: II. Disruption of chromatin structure during gene activity, Cell, 16:807.

Yamasaki, H., Rousch, T. W., and Weinstein, I. B., 1978, Benzo[a]-7,8-dihydrodiol-9,10-oxide modification of DNA: Relation to chromatin structure and reconstitution, Chem.-Biol. Inter., 23:201.

Zolan, M. E., and Hanawalt, P. C., 1981, Repair of DNA-containing furocoumarin adducts in mammalian chromatin, *J. Supromol. Struct. Suppl.*, 5:185 (abstract).

A similar summary of this work is in press in a volume entitled Proceedings of International Meeting on DNA-Repair, Chromosome Alterations and Chromatin Structure, edited by A. T. Natarajan, H. Altmann, and G. Obe, to be published by Elsevier/North-Holland, Amsterdam.

CHAPTER 21

NEW APPROACHES TO DNA DAMAGE AND REPAIR: THE ULTRAVIOLET LIGHT EXAMPLE

>
> William A. Haseltine, Lynn K. Gordon, Christina L[...]
> Judith Lippke, Douglas Brash, Kwok Ming Lo, and
> Brigitte Royer-Pokora
>
> Sidney Farber Cancer Institute, 44 Binney Street
> Boston, Massachusetts 02115

SUMMARY

DNA fragments of defined sequence are used as probes to [...] DNA damage and repair. The case of ultraviolet light is pre[...] and includes the following:

(a) Description of the distribution of cyclobutane pyri[...] dimers within defined DNA sequences. Considerations of the [...] of neighboring base composition, dose rate, and double- or s[...] stranded property of the DNA are discussed.

(b) Dissection of the anatomy of the incision event and s[...] quent repair steps. A three-step incision model for repair of [...] cyclobutane dimers by the <u>Micrococcus luteus</u> repair enzymes wi[...] presented. The steps are (1) recognition of the lesion and N-glycosylase scission, (2) cleavage of the phosphodiester bond [...] the newly created apyrimidinic site, and (3) scission of the a[...] dinic sugar on the 5' side.

(c) Use of human alphoid sequences as indicators of D[...] in intact human cells.

(d) Biological significance of a novel ultraviolet [...] photoproduct. This photoproduct occurs at pyrimidine-cy[...] quences and may have a significant biological role.

INTRODUCTION

Ultraviolet light (UV) induces cytotoxic, mutagenic, and carcinogenic changes in cells. Biological changes result from damage to DNA. The primary forms of DNA damage result from chemical reactions that occur when the bases reach excited states after absorption of the light. Reactions between neighboring bases and between the excited bases and surrounding molecules occur. An understanding of the biological effects of UV requires knowledge of the chemistry of the changes that occur in DNA, of the enzymology of recognition and repair of specific DNA lesions, and of the genetics that governs expression of the repair enzymes.

For the past few years, my laboratory has been involved in new approaches to the study of DNA damage and repair. These studies have taken advantage of the remarkable progress that has been made in nucleic acid biochemistry. It is now possible to obtain fragments of DNA of known sequence. We have used such fragments as substrates to study DNA damage and its repair by a variety of chemical and physical agents. The methods developed prove to be general and can be illustrated by reference to our work on UV-induced DNA damage and the repair of such damage.

THE DISTRIBUTION OF ULTRAVIOLET LIGHT-INDUCED PHOTODAMAGE IN DEFINED DNA SEQUENCES

The initial work on UV damage was motivated by an attempt to understand the distribution of UV-induced mutations within a gene of known sequence. Miller and his co-workers (Coulondre et al., 1978; Miller and Coulondre, 1977) measured the frequency of UV-induced mutations in the lacI gene of Escherichia coli in a repair-proficient genetic background. They found that about 60% (about 600 of 1,000) of all UV-induced amber, ochre, and UGA mutations could be grouped into five "hot spots," that is, mutations of the same bases of amber, ochre, or UGA sequences. Thirty percent of the mutations were accounted for by "warm spots," whereas the remaining 10% were scattered over a number of different sites. We reasoned that the mutation frequency at a particular base within a gene must reflect the frequency of damage to the base multiplied by the probability of error induction at that altered base. We sought to define the first parameter, frequency of damage at a particular base, so as to arrive at a reasonable estimate of the magnitude of the second parameter.

The initial experiments were designed to reveal the distribution of cyclobutane pyrimidine dimers within a defined DNA sequence. Cyclobutane pyrimidine dimers have long been suspected to be the primary mutagenic and cytotoxic lesion induced by UV. It has been known for some time that UV induces many photochemical changes in

DNA. However, attention has been focused on cyclobutane dimers. The reasons for this focus, as gleaned from a review of the recent literature, seem to be the chemical stability of the lesion, the frequency of the lesion, and, most importantly, photoreversal of the lesion by visible light. Many organisms contain enzymes that upon exposure to visible light revert the cyclobutane dimers to the original bases. The major biological effects of UV exposure are also photoreversible by visible light. Hence, the association of biological effect of UV and the formation of cyclobutane dimers was made.

The method we devised (Gordon and Haseltine, 1981a,b; Haseltine et al., 1980) to determine the sites of cyclobutane dimers utilizes fragments of DNA of known sequence labeled with ^{32}P at one terminus. Treatment of such DNA fragments with enzymes that cleave the phosphodiester bonds at sites of cyclobutane pyrimidine dimers (purified from either Micrococcus luteus or phage T4-infected E. coli) produces a series of DNA fragments. The lengths of the ^{32}P-labeled fragments reflect the distance in nucleotides of the cyclobutane dimers from the labeled terminus. The exact location of the scission event was determined by comparison of the length of the scission product with the lengths of DNA products produced by the Maxam-Gilbert DNA sequencing method (Maxam and Gilbert, 1977). The results of such an experiment are shown in Fig. 1.

Treatment of UV-irradiated DNA resulted in a discrete series of scission products that occurred only in the positions of possible pyrimidine dimers. The percentage of incision at each potential dimer site was calculated by measurement of the fraction of total input radioactivity in each scission product. A correction factor for end-labeling effects was included in the final calculation.

Distribution of cyclobutane dimer damage to DNA was determined for many independent dimer sites (Gordon and Haseltine, 1981b; Haseltine et al., 1980); Lippke et al., 1981). The dose response of dimer damage was also determined. (Figure 2 is a representative sample.)

From these studies we concluded:

1. The number of cyclobutane dimers formed upon irradiation with 254 nm light is dose dependent. The damage increases as a function of dose and reaches a plateau value well below 100%. The plateau value reflects the equilibrium Py Py $\overset{uv}{\rightleftharpoons}$ Py<>Py.

2. The plateau value of dimer formation at a potential dimer site reflects the composition of the dimer (T<>T > C<>T or T<>C > C<>C). The plateau values expressed as % dimer formation at a potential site vary from 5 to 10% for T<>T, 2 to 4% for C<>T and T<>C, 0.1 to 0.8% for C<>C.

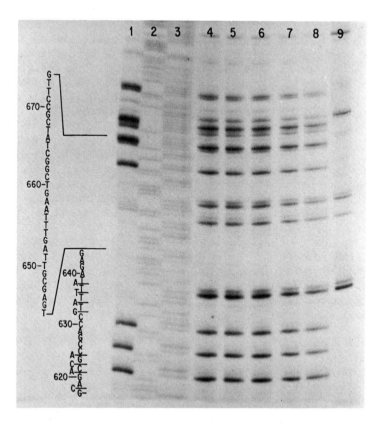

Fig. 1. Disposition of UV-induced damage in a bacterial DNA fragment. A segment of the lacI gene extending from base pair 561 to a sequence flanking lacI was purified from the pMC1 plasmid by digestion of the plasmid with restriction endonuclease BstEII, labeled at the 3' termini as described previously (Gordon and Haseltine, 1980), and redigested with restriction endonuclease HincII. The fragment extending from base pair 561 to 885 was irradiated with 5,000 J/m^2 and layered on a urea containing 8% polyacrylamide gel after the treatments indicated below. Lane 1 - Treatment with 1 M piperidine at 90°C for 30 min. Lanes 2,3 - Unirradiated DNA treated with the G+A and C+T DNA sequencing reactions, respectively. Lanes 4 to 8 - Treatment with 1, 2, 5, 10, 25 µl of M. luteus pyrimidine dimer endonuclease, respectively, followed by treatment with 1 M piperidine at 90°C for 30 min. Lane 9 - Treatment with 25 µl, M. luteus pyrimidine dimer endonuclease only.

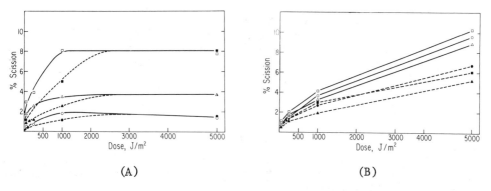

Fig. 2. Dose response for UV-induced damage to human alphoid DNA irradiated as naked or cellular DNA. (A) Dose response of pyrimidine dimer damage. The dose response for pyrimidine dimer damage was determined by treatment of the UV-irradiated DNA with the M. luteus pyrimidine dimer endonuclease followed by resolution of the scission products on polyacrylamide gels. The amount of radioactivity of individual scission products was determined by Cerenkov counting of gel fragments. The percentage of incision of a sequence was computed as described previously (Haseltine et al., 1980). This calculation includes a correction factor to account for multiple cleavage events within a single molecule. DNA prepared from unirradiated and irradiated CEM cells was used. For naked DNA (open symbols) the 3' end-labeled 92 base-pair alphoid DNA fragment was treated with the indicated dose of UV, then treated with enzyme and layered on the gel. For cellular DNA (solid symbols) intact CEM cells were irradiated, the DNA extracted and cleaved with the restriction endonuclease EcoRI.* The 341 base-pair alphoid DNA fragment was labeled at the 3' termini and cleaved with the restriction endonuclease EcoRI.* The 3' end-labeled 92 base-pair alphoid DNA fragment was isolated and treated with the M. luteus pyrimidine dimer endonuclease. The symbols (open-naked DNA, closed-cellular DNA) represent scission at the (O,●) G T*T G (37'-34'); (△,▲) G T*T*C A (51'-47'); (□,■) G T*T*T*C A (79'-74'). The positions of the sequences within the alphoid DNA are indicated by the numbers in parentheses. The percentage of scission is the fraction of the input molecules broken at the sequences indicated. The asterisks indicate the actual sites of the DNA breaks within the sequences. (B) Dose response for alkali-labile lesions. Parallel samples of the DNA preparations used in the experiment described in (A) were used. However, DNA was treated with 1 M piperidine at 90°C for 20 min prior to layering on the gel. The percentage of scission at a given sequence was computed as above. The symbols (open-naked

DNA, closed-cellular DNA) represent scission at the sequences indicated: (○,●) ACT*CTG (46'-41'); (△,▲) GTT*CA (51'-47'); (□,■) GTTT*CA (79'-74'). The asterisks in (A) and (B) indicate the actual site of the alkali labile break within the sequence. Similar results were obtained using alphoid DNA extracted from HeLa cells for both types of DNA damage.

3. There are neighboring base effects that modulate the plateau level, i.e., ATTA greater than GTTC.

4. There are some long range (several bases) effects of neighboring sequences in the plateau level.

The measurements of percentage of scission at an individual site probably reflect the actual level of cyclobutane dimer formation because:

1. The same distribution of damage is determined using both the T4 and M. luteus enzymes.

2. A similar distribution is calculated using an independent method, i.e., exonuclease III inhibition by cyclobutane dimers (Royer-Pokora et al., in press).

Do these rules for dimer formation explain the hot spots for UV mutation in the lacI gene? The answer is clearly no. Using these methods we have measured the formation of cyclobutane dimers in the lacI DNA sequence at low fluences, ~20 J/m^2, a dose similar to that used to induce the mutations. The hot spots and warm spots do not correspond to TT sequences not to the most frequent sites of cyclobutane dimer formation at codons that do occasionally mutate to amber, ochre, or UGA sequences.

Two interpretations of these data are possible:

1. The cyclobutane dimers are the mutagenic lesions, but the rate of mutation is very different at different cyclobutane dimer sites.

2. There are other UV-induced changes in DNA that are highly mutagenic. In the course of our experiments to determine the distribution of cyclobutane dimers in DNA, we also noticed that UV-induced lesions that resulted in strand scission at specific sites upon treatment of the DNA at high pH and high temperature (Gordon and Haseltine, 1981b; Haseltine et al., 1980; Lippke et al., 1981).

Further investigation revealed the following features of such damage:

1. Alkaline-labile lesions are chemically stable. Once formed, they are stable to incubation of the DNA at 90°C at neutral pH for several hours and are stable for weeks at 25°C at neutral pH.

2. They are formed at approximately equal frequency in double- and single-stranded DNA.

3. The DNA does not break at these sites if incubated with apurinic/apyrimidinic (AP) endonucleases purified from M. luteus or human placenta.

4. Breakage of the phosphodiester bonds requires elevated temperatures at alkaline pH. Incubation of the DNA at 90°C at neutral pH does not result in strand scission, nor does incubation of DNA for 30 min at 25°C in 0.1 N or 0.3 N NaOH.

5. The alkaline-labile lesions do not reach a plateau as a function of dose but increase over the dose range of 10 to 10,000 J/m^2 (Figure 2B).

6. The 5' terminus of the nick is phosphorylated.

This set of properties rules out both simple 5,6 pyrimidine photohydrates or AP sites as the origin of the lesion.

To investigate the sequence dependence of the lesion, the mobility of the DNA alkaline-induced scission products was compared to that of the products produced by the Maxam-Gilbert DNA sequencing reactions and the products produced by M. luteus enzyme treatment of the same DNA fragment (Fig. 1). Experiments of this sort demonstrate that alkaline-labile lesions occur at cytosines located 3' to pyrimidines. They do not occur at single cytosines flanked by purine bases. At sequences of adjacent cytosines, the break occurs only at the 3' cytosine. The frequency of breaks at CT sequences is at least 50-fold less than at TC or CC sequenced. We call the lesion that gives rise to the alkaline lability the PyC lesion.

It is evident from Fig. 1 that the M. luteus enzyme does not cleave the DNA at sites of the PyC lesion. The T4 UV-specific endonuclease also does not act at this lesion.

What is the chemistry of the PyC lesion? We suspect that the PyC lesion may be that which also yields Thy(6-4)Pyo and Cyd(6-4)Pyo upon acid hydrolysis of UV-irradiated DNA (Varghese, 1971; Varghese and Wang, 1967; Wang, 1976; Wang and Varghese, 1967). These photoproducts were identified by Wang and his co-workers some years ago. They are thought to arise from excited states of pyrimidines that result in formation of bonds between the 5 and 6 positions of thymine or cytosine and the exocyclic amino group, and the 4 position of cytosine, respectively. However, the original photoproducts that give rise to the Thy(6-4)Pyo and Cyd(6-4)Pyo products were never identified. These products are of interest as their formation may be asymmetric with respect to the polarity of the DNA.

We have carried out a series of studies on a model compound, TpC. UV irradiation of this compound results in the formation of at least 15 photoproducts that can be resolved by high pressure, reversed-phase liquid chromatography. Many of these products are not stable to heating and are probably photohydrates. However, one family of photoproducts has the following interesting properties.

1. The absorbance maximum is shifted to about 320 nm.

2. The products fluoresce with a maximum of about 440 nm.

3. The products are stable to treatment at 90°C at neutral pH.

4. The products become much more polar upon alkaline hydrolysis (90°C at 0.1 M NaOH for 30 min).

5. The polarity of the products is decreased by treatment with bacterial alkaline phosphatase after alkaline hydrolysis. Phosphatase treatment does not alter the products prior to alkaline hydrolysis.

6. Treatment of the products with 313 nm light results in loss of absorbance at 315 nm, loss of fluorescence, and a shift in elution properties. The elution properties of the photoreversion product are different from the parent compound TpC.

These results are consistent with the formation of a dimer between the 5,6 positions of the T and the exocyclic amino group and 4 positions of the C to give rise to a compound that possesses an alkaline-labile glycosylic bond. The red shift indicates that some dimer has been formed, and the fluorescence indicates a Pyo structure is present. Sensitivity of the phosphate to phosphatase after alkaline hydrolysis suggests rupture of a phosphodiester bond.

Loss of 315 nm absorbance and fluorescence upon 313 nm irradiation suggests that the dimer linkage can be broken by direct photoreversion. These observations suggest that the fluorescent products absorbing at 315 nm might be the same as those that give rise to alkaline-labile lesions in DNA. Moreover, the properties of these compounds are similar to those that are suspected to yield Thy(6-4)Pyo upon acid hydrolysis. However, further work on the chemistry of these PyC lesions must be done to elucidate the structure of the lesion.

What is the potential biological significance of the PyC lesion? We note that all the UV-induced mutagenic hot spots and warm spots occur at sites of potential PyC lesions in the lacI gene. We also note that the relative rate of formation of the PyC lesion is sequence dependent and varies by a factor of 10 to 20, depending on the local sequence. We are currently measuring the rate of PyC lesions within the lacI gene to determine if the mutagenic hot spots correspond to hot spots of PyC damage. We find that the PyC lesions that form most rapidly as a function of dose are located at the exact site of the hot spots for UV. These results suggest that the distribution of the PyC lesions may be the primary determinant of UV mutagenesis.

The quantum yield of PyC damage is also consistent with a possible mutagenic role. At low fluences of UV, 15 to 30 J/m^2, the extent of formation of some of the PyC lesions is at least as great as the most reactive TT sequences. The photoreversion and photoreactivation properties of these lesions are a current topic of investigation.

ENZYMATIC REPAIR OF ULTRAVIOLET LIGHT-INDUCED CYCLOBUTANE DIMERS

The mechanism for excision repair of cyclobutane dimers by bacterial enzymes has long been codified in textbooks on the subject. The model derives, for the most part, from studies of UV-specific endonucleases purified from M. luteus or phage T4-infected E. coli (the denV gene product). The excision repair model held that the first step of the process was recognition of the cyclobutane dimer, incision of a phosphodiester bond 5' to the dimer site to yield a 5' phosphoryl group and a 3' hydroxyl group, followed by initiation of DNA synthesis by a polymerase I-like activity, and displacement of the dimer-containing DNA fragment by action of the polymerase. Excision of the dimer was thought to be accomplished by the 5' to 3' exonucleolytic activity of the Pol I enzyme. It was evident from our first experiments, using 5' end-labeled DNA fragments as substrates, that this model was not correct. The electrophoretic mobility of the 5' end-labeled fragment was inconsistent with cleavage 5' to the dimer site. Indeed, the electrophoretic mobility of the scission product was paradoxical. The product migrated on the gels as if the DNA was cleaved between the dimerized pyrimidines, a surprising result because the M. luteus and T4 endonucleases do not cleave the cyclobutane bond. The resolution of this apparent paradox was that the electrophoretic mobility of the scission products was the result of an unusual 3' terminus. Treatment of the scission products with 0.1 N NaOH resulted in a shift of the electrophoretic mobility of scission products. From these results it was deduced that the M. luteus and T4 UV-specific endonucleases must contain two enzymatic activities, an N-glycosylase that cleaves the N-glycosyl bond between the 5' pyrimidine of the cyclobutane dimer and the corresponding sugar and an AP endonuclease that cleaves the newly created AP site on the 3' side (Fig. 3, steps 1 and 2).

This model was subsequently confirmed (Haseltine et al., 1980) by demonstration that an activity could be purified from M. luteus that carried out only step 1. The DNA treated with this enzyme fraction was not broken. However, treatment of such DNA with either 0.1 N NaOH or an AP endonuclease resulted in strand scission. A further proof of the N-glycosylase action was the release of free thymine upon either photoreversal or photoreactivation of the DNA treated with the M. luteus enzyme. Free thymine could only be released from the DNA in such an experiment if the N-glycosyl bond had been cleaved. The presence of the 3' AP deoxyribose posed questions regarding subsequent steps of DNA repair. Although it was reported that the lesion generated by the M. luteus and T4 endonucleases was a substrate for DNA ligase, we were unable to confirm this observation (Gordon and Haseltine, 1981a). This was not surprising as it is not anticipated that an AP sugar can serve as a substrate for DNA ligase. In another series of experiments, we demonstrated that the 5' terminus of the nick generated by the M. luteus enzyme was phosphorylated. Hence the M. luteus enzyme preparation used contained both an N-glycosylase and an AP endonuclease that cleaved the phosphodiester bond between the AP sugar and the 3' phosphate (Fig. 3).

To determine whether or not the nick generated by the M luteus enzyme could serve as a substrate for the activity of E. coli DNA polymerase I, nicked DNA was used as a substrate for the enzyme (both the holoenzyme and the Klenow fragment) in reactions that contained either no triphosphates, TTP, or all four triphosphates. Double-stranded DNA fragments were used as the substrates for these reactions. These experiments showed that the cleavage products of the M. luteus reaction were not substrates for DNA polymerase I. The experiment pictured in Fig. 4 shows that the electrophoretic mobility of the scission products was unaffected by incubation with the polymerases in the presence or in the absence of triphosphate precursors. The observation that the mobility of the scission products was unaffected by the polymerase in the absence of triphosphate also shows that the 3' terminus of the nick is not a substrate for the 3' to 5' exonuclease of polymerase I.

It seemed likely that removal of the AP deoxyribose would convert the nicked DNA to a substrate for DNA polymerase. To test this hypothesis, double-stranded DNA that had been nicked with the M. luteus enzyme was then treated with an AP endonuclease (human AP endonuclease) that cleaves the phosphodiester bond 5' to the AP site leaving a 3' hydroxyl group. Such DNA is a substrate for DNA

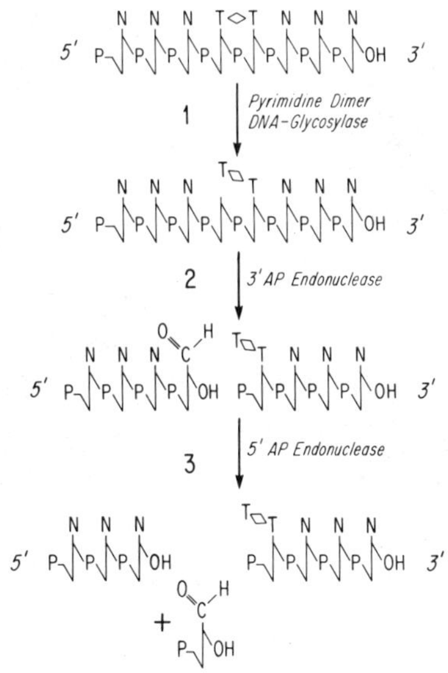

Fig. 3. Three-step model for incision of DNA at pyrimidine dimers. The first step in the incision of DNA at sites of pyrimidine dimers is cleavage of the N-glycosyl bond between the 5' pyrimidine of the dimer and its corresponding sugar. The second step is cleavage of the phosphodiester bond on the 3' side of the apyrimidinic deoxyribose leaving a 3' hydroxyl group. The third step is postulated to be cleavage of the phosphodiester bond on the 5' side of the apyrimidinic deoxyribose leaving a 3' terminus which is a substrate for DNA polymerase.

polymerase I (Fig. 4B). Both the holoenzyme and Klenow fragment of Pol I were active, suggesting that the 3' fragment was displaced from the template in these reactions. Based on these data, we suggest that the early steps of excision repair of cyclobutane pyrimidine dimers by the M. luteus and T4 UV-specific endonuclease involves three steps (Fig. 3).

1. Scission of the N-glycosyl bond of the 5' pyrimidine of the dimer by an activity that is specific for these lesions.

2. Scission of the phosphodiester bond 3' to the newly created AP site by an AP endonuclease that need not be specific for cyclobutane dimers.

3. Scission of the phosphodiester bond between the base 5' to the cyclobutane dimer and the AP site by a second AP endonuclease

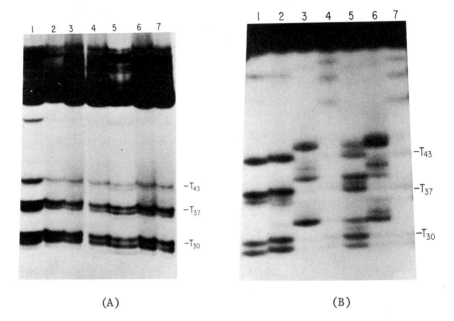

Fig. 4. (A) Polymerase action on the products of cleavage with the M. luteus UV-specific endonuclease. 5' end-labeled lac p-o 117 DNA fragments were UV-irradiated with 5000 J/m^2 and treated with the M. luteus UV-specific endonuclease prior to incubation with either Pol I or Klenow enzymes. The enzyme used and deoxynucleotide triphosphate precursors included in the reaction mixture are described below. After treatment, the DNA fragments were analyzed after electrophoresis on a 12% urea-polyacrylamide gel. Lane 1 - No further treatment. Lane 2 - Pol I. Lane 3 - Pol I, TTP. Lane 4 - Pol I, TTP, dATP, dCTP, dGTP. Lane 5 - Klenow. Lane 6 - Klenow, TTP. Lane 7 - Klenow, TTP, dATP, dCTP, dGTP. (B) Polymerase action on the products of cleavage with the M. luteus UV-specific endonuclease and the human placental AP endonuclease. 5' end-labeled lac p-o 117 DNA fragments were UV-irradiated with 5000 J/m^2 and treated with both the M. luteus UV specific endonuclease and human placental AP endonuclease. The DNA was then incubated with either Pol I or Klenow in the presence or absence of deoxynucleotide triphosphate precursors as listed below. After treatment, the DNA fragments were analyzed by electrophoretic mobility on a 12% urea-polyacrylamide gel. Lane 1 - No further treatment. Lane 2 - Pol I. Lane 3 - Pol I, TTP. Lane 4 - Pol I, TTP, dATP, dCTP, dGTP. Lane 5 - Klenow. Lane 6 - Klenow, TTP. Lane 7 - Klenow, TTP, dATP, dCTP, dGTP.

to generate a 3' OH terminus suitable for initiation of repair synthesis by DNA polymerase. The second AP endonuclease also need not be specific for the cyclobutane pyrimidine dimer. The net result of this proposed pathway would be complete removal of the 5' pyrimidine nucleotide from the DNA. Identification of subsequent enzymatic steps, i.e., removal of the cyclobutane dimer itself, awaits further experimentation.

USE OF THE ALPHOID SEQUENCE OF HUMAN DNA AS AN INDICATOR OF DNA DAMAGE IN INTACT CELLS

We wished to develop a means of determining the distribution of DNA damage in intact human cells at the level of individual nucleotides. It seemed likely that the human alphoid sequence (Maio et al., 1978; Manuelidis, 1967; Manuelidis and Wu, 1978; Wu and Manuelidis, 1980) could serve as a convenient indicator of such damage. The human alphoid sequence is a 341-nucleotide sequence that is highly reiterated in human DNA. It is present in tandem arrays within the chromosome and comprises approximately 1% of the total cellular DNA. The alphoid sequence can be obtained in homogeneous form by treatment of total cellular DNA extract with restriction enzyme EcoRI. The bulk of EcoRI-digested cellular DNA is 3000 to 7000 nucleotides long and is easily separated from the 341 nucleotide alphoid DNA. For use as an indicator, the alphoid DNA fragment is isolated, labeled at the termini with ^{32}P, and cleaved into the asymmetric 92- and 249-base fragments. Each fragment contains a single end label. The sequence of the 92-base fragment prepared in this way is homogeneous as judged by the Maxam-Gilbert sequencing method (Fig. 5). The alphoid sequence should provide a convenient tool for comparison of the effects of DNA damage to naked DNA relative to the effects of such agents on DNA within intact cells. Accordingly, the purified end-labeled 92-base DNA fragment was exposed to UV and then treated with the M. luteus enzyme, 1 M piperidine at 90°C for 20 min, or a combination of both treatments. Similar treatment was performed on a preparation of ^{32}P end-labeled 92-base alphoid DNA that had been extracted from intact HeLa cells exposed to the same dose of UV. Figure 6 demonstrates that damage to the naked alphoid DNA can be compared to damage to alphoid DNA irradiated within intact cells. The relative extent of cyclobutane dimer formation and the PyC lesion was measured as a function of dose in these two experimental situations. Figure 2 demonstrates that, except for the apparent dose, the distribution of the PyC and cyclobutane dimer lesions is similar in both experimental circumstances. The cellular environment apparently screens out about half the UV damage.

It is our hope that the human indicator sequence can be used to elucidate repair pathways in intact cells. For this purpose we

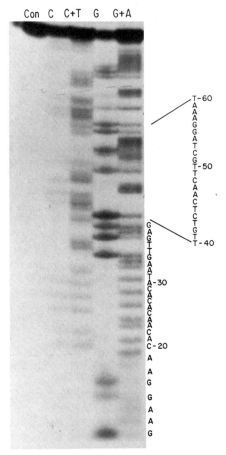

Fig. 5. Sequence of human alphoid DNA. A total HeLa cellular DNA extract was treated with restriction endonuclease EcoRI, and the material corresponding to the 341 base-pair alphoid DNA collected from a sucrose gradient. The DNA of this fraction was labeled at the 3' termini in a reaction that contained (α-^{32}P)dATP and (α-^{32}P)TTP and the Klenow fragment of DNA polymerase I. The labeled DNA was treated with restriction endonuclease HaeIII and the 92 base-long fragment labeled at a single terminus was recovered from a nondenaturing 10% polyacrylamide gel. This DNA was subject to the Maxam-Gilbert DNA sequencing protocol. The lanes are from left to right Con, no treatment; C, the cytosine specific reaction; C+T, the cytosine-thymidine reaction; G, the guanosine reaction; G+A, the guanosine-adenosine reaction.

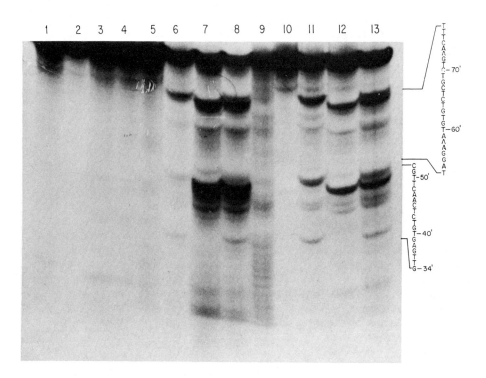

Fig. 6. Comparison of UV-induced damage to the human alphoid sequence irradiated as naked DNA or as cellular DNA. The 342 base-pair alphoid DNA fragment of human DNA was prepared from HeLa cells before (Lanes 1 to 9) or after (Lanes 10 to 13) UV irradiation with 5,000 J/m^2. The DNA was labeled at the 3' termini in reactions that included (α-^{32}P)dATP and (α-^{32}P)TTP and the Klenow fragment of E. coli DNA polymerase I. The DNA was digested with the restriction endonuclease EcoRI* and a 92 base-pair fragment separated from other labeled DNA fragments by electrophoresis on a nondenaturing polyacrylamide gel using methods described previously (Gordon and Haseltine, 1980). The DNA was treated as described below prior to layering on a urea-containing 8% polyacrylamide gel. The sequence of the un-irradiated DNA fragment determined by the chemical DNA sequencing reactions is indicated. Lanes 1 to 4: DNA prepared from unirradiated cells untreated (Lane 1), treated with M. luteus pyrimidine dimer endonuclease (Lane 2), treated with 1 M piperidine at 90°C for 20 min (Lane 3), treated with M. luteus pyrimidine dimer endonuclease followed by treatment with 1 M piperidine at 90°C for 20 min (Lane 4). Lanes 5 to 8: DNA extracted from unirradiated cells, exposed to 5000 J/m^2, and then subjected to the same four treatments in the same order prior to layering as

described for Lanes 1 to 4. Lane 9: DNA purified from unirradiated cells treated with neocarzinostatin. Lanes 10 to 13: DNA purified from cells exposed to 5000 J/m^2 treated as described for DNA of Lanes 1 to 4 prior to layering.

intend to analyze intermediates in the DNA repair pathway by exposure of cells to doses of agents which they can be expected to repair. The sensitivity of the method, one break per 2×10^3 bases, should be sufficient for this purpose.

As a preliminary test of this approach, we measured the stability of the PyC lesion in the cellular environment. HeLa cells were exposed to a relatively high dose of UV, 200 J/m^2 or 7000 J/m^2. DNA was extracted either immediately after irradiation or after the irradiated cells had been incubated for 24 h in complete medium. The amount and distribution of the PyC and cyclobutane dimer damage in the alphoid DNA were determined as before. These experiments demonstrated that both types of lesions were stable in the cellular environment since the extent of damage observed was the same in the two samples. In this experiment we suspect that the doses of UV produced much more DNA damage, by a considerable margin, than could be accommodated by the repair capacity of the cells.

CONCLUSIONS

Application of new methodologies to the problem of DNA damage and repair offers the opportunity to advance our knowledge of these problems. The application of some of these methods to the problem of UV-induced damage is explored here. The results of these studies have yielded some surprising answers that alter, in a fundamental way, the current notions regarding the mechanisms of UV-induced biological damage and the enzymatic repair of such damage.

The most surprising observation is that the mutagenic distribution of hot spots in the lacI gene does not correlate with the hot spots for formation of cyclobutane pyrimidine dimers. Rather, it correlates with the distribution of a novel lesion, which we call the PyC lesion. The chemistry of this lesion has not been fully elucidated, but it is clearly different from that of the cyclobutane dimer, pyrimidine photohydrates, and AP sites. The enzymology of repair of these lesions is unknown but does not seem to involve either the M. luteus or the T4 UV-specific endonucleases.

A second unanticipated result of these studies has been the realization that the initial steps of excision repair of cyclobutane dimers are considerably more complex than originally proposed. It seems likely that the early steps of excision repair involve the combined action of a pyrimidine dimer N-glycosylase and two AP endonucleases that cleave the phosphodiester bonds 5' to each of the phosphates that flank the AP sugar.

Further studies of DNA damage and repair in human cells should be facilitated by use of the human alphoid sequence as an indicator. Experiments discussed here demonstrate that this sequence can be used to obtain useful information of DNA lesions within intact human cells.

REFERENCES

Coulondre, C., Miller, J. H., Farabaugh, P. J., and Gilbert, W., 1978, Molecular basis of base substitution hotspots in Escherichia coli, Nature, 274:775.

Gordon, L. K., and Haseltine, W. A., 1980, Comparison of the cleavage of pyrimidine dimers by the bacteriophage T4 and Micrococcus luteus UV-specific endonucleases, J. Biol. Chem., 255:12047.

Gordon, L. K., and Haseltine, W. A., 1981a, Early steps of excision repair of cyclobutane pyrimidine dimers by the Micrococcus luteus endonuclease: A three-step incision model, J. Biol. Chem., 256:6608.

Gordon, L. K., and Haseltine, W. A., 1981b, Quantitation of cyclobutane pyrimidine dimer formation in double- and single-stranded DNA fragments of defined sequences, Radiat. Res., in press.

Haseltine, W. A., Gordon, L. K., Lindan, C. P., Grafstrom, R. H., Shaper, N. L., and Grossman, L., 1980, Cleavage of pyrimidine dimers in specific DNA sequences by a pyrimidine dimer DNA-glycosylase of M. luteus, Nature, 285:634.

Lippke, J. A., Gordon, L. K., Brash, D. E., and Haseltine, W. A., 1981, The distribution of ultraviolet light-induced damage in a defined sequence of human DNA: Detection of alkaline sensitive lesions of pyrimidine-cytosine sequences, Proc. Natl. Acad. Sci. U.S.A., 78:3388.

Maio, J. J., Brown, F. L., and Musich, P. R., 1977, Subunit structure of chromatin and the organization of eukaryotic highly repetitive DNA: Recurrent periodicities and models for the evolutionary origins of repetitive DNA, J. Mol. Biol., 117:637.

Manuelidis, L., 1967, Repeating restriction fragments of human DNA, Nucleic Acids Res., 3:3063.

Manuelidis, L., and Wu, J. C., 1978, Chemical homology between human and simian repeated DNA, Nature, 276:92.

Maxam, A. M., and Gilbert, W., 1977, A new method for sequencing DNA, Proc. Natl. Acad. Sci. U.S.A., 74:560.

Miller, J. H., and Coulondre, C., 1977, Genetic studies of the lac repressor, J. Mol. Biol., 117:577.

Royer-Pokora, B., Gordon, L. K., and Haseltine, W. A., Use of exonuclease III to determine the site of stable lesions in defined sequences of DNA: The cyclobutane pyrimidine dimer and cis- and trans-dichlorodiammine platinum II examples, Nucleic Acids Res., in press.

Varghese, A. J., 1971, Photochemical reactions of cytosine nucleosides in frozen aqueous solution and in deoxyribonucleic acid, Biochemistry, 10:2194.

Varghese, A. J., and Wang, S. Y., 1967, Ultraviolet irradiation of DNA in vitro and in vivo produces a third thymine-derived product, Science, 156:955.

Wang, S. Y., 1976, Pyrimidine biomolecular photoproducts, in: "Photochemistry and Photobiology of Nucleic Acids," S. Y. Wang, ed., Academic Press, New York.

Wang, S. Y., and Varghese, A. J., 1967, Cytosine-thymine addition product from DNA irradiated with ultraviolet light, Biochem. Biophys. Res. Commun., 29:543.

Wu, J. C., and Manuelidis, L., 1980, Sequence definition and organization of a human repeated DNA, J. Mol. Biol., 142:363.

CHAPTER 22

CHROMOSOMAL AND NONCHROMOSOMAL DNA:

A SUMMARY AND PERSPECTIVE

>   Ada L. Olins
>
>   University of Tennessee-Oak Ridge Graduate School of
>   Biomedical Sciences and the Biology Division
>   Oak Ridge National Laboratory, Oak Ridge, Tennessee
>   37830

Mutagenesis and DNA repair are probably modulated by the intrinsic characteristics of the DNA such as its sequence, its secondary structure (i.e., B-DNA, Z-DNA, hairpin loops, etc.), and its ploidy in the cell. Of equal importance is the DNA environment: histones forming nucleosomes, nonhistone proteins altering the nucleosomal configuration, and proteins which determine the higher-order packing of nucleosomes.

Mitochondrial mutagenesis and repair are certainly not indifferent to the structural constraints of DNA, and in addition they are highly sensitive to the presence of numerous nuclear gene products. Yeast mitochondrial DNA is a 25-μ circle, present in 50 copies per haploid cell, and is not nucleosomal. The complexities of mitochondrial mutation and repair, as well as its dependence on nuclear functions such as replication and recombination, were presented by Dr. Moustacchi.

The role of nucleosomal organization in eukaryotic DNA repair was explored by Dr. Lieberman. Nucleosomal conformation is thought to alter when chromatin is activated for transcription. The suggestion presented here that nucleosomal conformation is also altered during DNA repair is therefore extremely important, as it falls in line with an earlier proposal that nucleosomes act much like large globular proteins which undergo specific allosteric changes to allow specific biological functions of the DNA.

The third subject presented in this Session was directed at exploring the biochemistry of UV-induced mutagenesis and DNA repair.

Using modern DNA technology, Dr. Haseltine was able to make a very important correlation between known mutational hot spots in the E. coli lacI gene and a new type of UV photoproduct called a PyC lesion, thus casting doubt on the previously proposed role of UV-induced cyclobutane pyrimidine dimers. Further interest in this paper was generated by the careful step-by-step analysis of cyclobutane dimer repair and the enzymes involved in the complicated process.

There is no dearth of unanswered questions in the various topics discussed in this Symposium. The most emphatically established point was the need for a basic understanding of DNA structure and function as it is normally modulated by metal ions, specific proteins, nucleic acids, and membranes. As new basic concepts in the biology of DNA emerge, they demand a reconsideration of our present perspectives in mutagenesis.

CHAPTER 23

COMPARISON OF THE INDUCTION OF SPECIFIC LOCUS MUTATIONS

IN WILD-TYPE AND REPAIR-DEFICIENT STRAINS OF NEUROSPORA CRASSA

F. J. de Serres

Office of the Associate Director for Genetics
National Institute of Environmental Health Sciences
Research Triangle Park, North Carolina 27709

SUMMARY

A comparison of mutation induction between wild-type and excision repair-deficient strains has shown that, after treatment with four of the five mutagens tested, an enhanced recovery of induced mutants was found in the excision repair-deficient strains. In this sense we have confirmed for Neurospora Ames' (1977) observations with Salmonella. Furthermore, genetic analysis of the mutants induced in Neurospora in both wild-type and excision repair-deficient strains has shown that in some cases the enhanced recovery of mutants can be attributed to the recovery of a spectrum of genetic alterations in the excision repair-deficient strains that is qualitatively different from that found in the wild-type strain. This qualitatively different spectrum appears to arise as a result of a preferential increase in frameshift mutations. Thus, it appears that in excision repair-deficient strains of Neurospora genetic lesions are processed differently than in wild-type strains both to give enhanced yield as well as a selective increase in frameshift mutations. In this regard, the apparent difference between Salmonella and Neurospora data with regard to the characterization of the genetic effects of chemical carcinogens is most likely attributed to a difference in the genetic background of the strains. The present data with Neurospora suggest that this difference not only results in an enhanced recovery of revertants in Salmonella strains carrying the uvrB mutation but also that a different spectrum of genetic damage was detected from that which would have been observed in the standard wild-type strain G46.

Observations from the present experiments with Neurospora have important implications not only for comparative mutagenesis, where the effects of the same mutagen are studied in different laboratory organisms, but also for risk estimation since the spectrum of genetic damage produced by a given agent may not be uniform in our genetically heterogeneous human population.

INTRODUCTION

A recurrent theme during the first day of this Symposium, especially in the papers of Singer (1981) and Chambers (1981), relates to the mutagenicity of chemical carcinogens. The question that was raised was whether chemical carcinogens produce some lesion in the DNA that results in some characteristic change at the molecular level in the resulting mutants. This point has been the subject of a great deal of controversy during the past 10-15 years. Ames et al. (1971), for example, have claimed that chemical carcinogens are frameshift mutagens in Salmonella, whereas research in my own laboratory (Malling and de Serres, 1969) has indicated that in Neurospora carcinogens produce predominantly base-pair substitutions. Expressed in more general terms, there is often a fundamental disagreement when we compare the spectra of genetic alterations produced by the same chemical mutagen in different organisms, and the disagreement is not confined to Salmonella-Neurospora comparisons alone.

Other papers in this Symposium have shown quite clearly that mutation induction at specific loci is under genetic control, and we know now that striking differences can be obtained both quantitatively and qualitatively when, for example, we study reversion of particular mutants in the presence or absence of various mutations that affect DNA repair.

One possible explanation for Ames' data with carcinogens in Salmonella and our data with Neurospora could be quite simply that his strains were excision repair deficient whereas ours were excision repair proficient. The excision-repair deficiency in Salmonella has a mutator effect and gives enhanced recovery of mutants after treatment with many, but not all, mutagens. To investigate this possibility, we began experiments to determine to what extent mutation induction at specific loci in Neurospora is under genetic control and can be modified quantitatively and qualitatively. Rather than studying reversion of particular alleles, we used a forward-mutational assay where effects on all types of genetic alterations could be measured. In addition to the standard wild-type strain, we have used two strains, <u>upr-1</u> and <u>uvs-2</u>, which are excision repair-deficient, (Worthy and Epler, 1973). These two mutants are not alleles (Schroeder, 1975) and presumably they block different steps in the same excision-repair pathway.

The specific locus assay system that we have used is the induction of ad-3 mutants. These mutants accumulate a reddish-purple pigment in the mycelium and also have a requirement for adenine. ad-3 mutants result from mutations at two loci, ad-3A and ad-3B, which are closely linked in Neurospora (de Serres, 1956). Because these mutants both accumulate pigment, they can be recovered by a direct method based on a pigment accumulation rather than their requirement for adenine (de Serres and Kølmark, 1958; de Serres and Malling, 1971). This technique makes it possible to determine their frequencies in the total treated population precisely and to obtain dose-response curves for their induction. Control and treated populations are incubated in individual 12-liter Florence flasks in 10 liters of medium and are inoculated to give a total population of about 10^6 survivors. Each cell surviving treatment forms a small colony, about 2 mm in diameter, after about 7 days incubation, with aeration, in the dark. At the end of this period, flasks are harvested by placing aliquots in white photographic developing trays. During this harvesting procedure, the total volume of medium in control and treated samples in each flask is measured and aliquots of background colonies are counted to determine the total number of colonies in each flask. These numbers are used to determine total viability and survival. The reddish-purple mutants are isolated and counted to determine the frequencies of ad-3 mutants in each flask and to calculate dose-response curves for the overall induction of ad-3 mutants.

By using this simple and direct assay, we can make precise comparisons between strains with regard to changes in sensitivity to inactivation. Quantitative changes in the frequency of ad-3 forward mutations induced in different strains can be detected readily by comparing the dose-response curves for mutation induction.

ad-3B mutants show allelic complementation and specify a linear complementation map of 17 complementation subgroups or complons (de Serres et al., 1971). On this map, mutants have three types of patterns: complementing with either nonpolarized (NP) or polarized (P) patterns or noncomplementing (NC). Mutants can be leaky or nonleaky, but the leaky mutants always have NP patterns. These leaky mutants are believed to result from missense mutations that specify complete polypeptide chains with only a single erroneous amino acid. P and NC mutants result from nonsense or frameshift mutations that specify polypeptide fragments or, in the case of NC mutants, no polypeptide chain at all (Malling and de Serres, 1967, 1968). Thus, the ad-3 mutants that are picked up on the basis of pigment accumulation can be characterized by simple heterokaryon tests as ad-3A mutants or ad-3B mutants with three types of complementation patterns. A small portion of the mutants are so leaky that they cannot be characterized at all.

As expected, the frequencies of ad-3B mutants in the three complementation classes are mutagen dependent (de Serres et al., 1971). N-methyl-N'-nitro-N-nitrosoguanidine (MNNG), which makes predominantly base-pair transitions in Neurospora (from $A \cdot T \rightarrow G \cdot C$), produces the highest frequency of complementing mutants, whereas 2-methoxy-6-chloro-9-(3-[ethyl-2-chloroethyl]aminopropylamino) acridine dihydrochloride (ICR-170), which produces predominantly frameshift mutations, has the lowest frequency of complementing mutants (Malling and de Serres, 1970).

In experiments in which we have examined the specific revertibility of mutants with NP, P, or NC patterns after treatment with specific chemical mutagens, we have shown a striking correlation between complementation pattern and genetic alteration at the molecular level (Malling and de Serres, 1967, 1968). Thus, by performing an analysis for allelic complementation among different samples of ad-3B mutants, we can determine whether there has been a change in the spectrum of genetic alterations at the molecular level.

To determine what type of differences might be found in the induction of ad-3 mutants between our wild-type and two excision repair-deficient strains, we have selected five mutagens that produced very different spectra of genetic alterations, ultraviolet light (UV), gamma rays, MNNG, ICR-170, and 4-nitroquinoline 1-oxide (4NQO). The main questions in this study then were (1) would there be any quantitative differences in the induction of ad-3 mutants, and (2) would any quantitative differences in recovery be due to qualitative differences in the spectrum of genetic alterations at the molecular level?

QUANTITATIVE COMPARISON OF THE INDUCTION OF ad-3 MUTANTS IN WILD-TYPE AND EXCISION REPAIR-DEFICIENT STRAINS

A comparison of the spontaneous mutability of the ad-3A and ad-3B loci in wild-type and the two excision repair-deficient strains (de Serres et al., 1980) showed no significant differences. However, higher frequencies of ad-3 mutants were found in the excision repair-deficient strains than in the wild-type strain after treatment with UV (de Serres, 1980), gamma rays, (Schüpbach and de Serres, 1981), MNNG, and 4NQO (Inoue et al., 1981). However, after treatment with ICR-170, the same frequency of ad-3 mutants was found with upr-1 as with wild type, whereas a lower frequency was found with uvs-2 as shown in Table 1. The data from these same experiments show that the excision repair-deficient strains were more sensitive to inactivation than wild type after treatment with all five mutagens but that the effect on the induction of ad-3 mutants was both mutagen and strain dependent. The most important and noteworthy observation is that in four out of five experiments the introduction of an

Table 1. Comparison of Sensitivity of Excision Repair-Deficient Strains with Wild Type with Regard to Inactivation and Induction of ad-3 Mutants[a]

Effect	UV	Gamma rays	MNNG	ICR-170	4NQO
Inactivation					
upr-1	++	+	+	+	++
uvs-2	+++	+	++	+	+++
Mutation induction					
upr-1	++	++	+	0	++
uvs-2	++	++	++	R	+++

[a] 0, same sensitivity as wild type; +→+++, increasing degrees of sensitivity relative to wild type; R, less sensitive than wild type. Adapted from Inoue et al., 1981.

excision-repair mutant has resulted in enhanced recovery of forward mutations at the ad-3A and ad-3B loci. In this respect we have made the same observation in Neurospora that led Ames (1971) to introduce uvrB into strain G46, namely that higher mutant yields are obtained with excision repair-deficient strains than with normal wild-type strains.

QUALITATIVE COMPARISONS OF THE SPECTRA OF ad-3 MUTATIONS IN WILD-TYPE AND EXCISION REPAIR-DEFICIENT STRAINS

The data from comparisons of the spectra of complementation patterns among ad-3B mutants induced in wild-type and the two excision repair-deficient strains after treatment with different mutagens are presented in Table 2. If we consider first the data on the wild-type strain, we can see that the percentages of ad-3B mutants in each class are mutagen dependent. After UV, 4NQO, and MNNG treatment, the most notable effect is a decrease in the percentages of mutants with NP complementation patterns and a concomitant increase in the percentages of NC mutants. In most cases no significant change occurs among the mutants with P patterns. The magnitude of the change among mutants with NP patterns or in the NC group is mutagen dependent with the most striking effect among 4NQO-induced ad-3B mutants. Unfortunately, the percentages of mutants with NP patterns are so low in wild type after gamma-ray treatment or after ICR-170 treatment that much larger samples of ad-3B mutants would probably be required to detect a small change such as that found

Table 2. Comparison of the Spectra of ad-3B Mutations Between Wild-Type and Excision Repair-Deficient Strains uvs-2 and upr-1

Mutagenic Agent	Strain	Total Mutants Tested	Percentage and Number of Each Type of Complementation Pattern		
			Nonpolarized	Polarized	Noncomplementing
UV	WT	202	39.1 (79)	12.9 (26)	48.0 (97)
	uvs-2	141	14.9 (21)[a]	9.9 (14)	75.2 (106)[a]
	upr-1	144	9.0 (13)[a]	9.0 (13)	81.9 (118)[a]
MNNG	WT	231	61.9 (143)	3.9 (9)	34.2 (79)
	upr-1	158	45.6 (72)[a]	3.8 (6)	50.6 (80)[a]
4NQO	WT	125	44.8 (56)	22.4 (28)	32.8 (41)
	uvs-2	39	7.7 (3)[a]	5.1 (2)[a]	87.2 (34)[a]
	upr-1	88	7.9 (7)[a]	11.5 (10)	80.7 (71)[a]
Gamma rays	WT	174	10.3 (18)	13.8 (24)	75.9 (132)
	uvs-2	147	4.1 (6)	16.3 (24)	79.6 (117)
	upr-1	120	5.0 (6)	15.8 (19)	79.2 (95)
ICR-170	WT	129	6.3 (8)	16.5 (21)	77.2 (98)
	uvs-2	78	10.3 (8)	11.5 (9)	78.2 (51)
	upr-1	204	2.0 (4)	13.2 (27)	84.8 (173)

[a] $p < 0.01$.

with MNNG. The percentages of complementation patterns found among the ad-3B mutants in the two excision repair-deficient strains do not differ significantly from wild type in the samples induced by gamma rays or ICR-170.

The data from this portion of the genetic analysis show that the spectrum of genetic changes among ad-3B mutants can be markedly altered in excision repair-deficient strains as compared with the wild type. In this sense the spectrum of genetic alterations induced in the ad-3B locus can be modified markedly by mutations at other loci in the genome. It is not clear whether these results can be generalized to other organisms. When the biochemistry of the excision-repair pathway in Neurospora has been elucidated, then comparisons can be made with other organisms. The data in the present experiments show that there are quantitative differences in the overall induction of ad-3 mutants in wild-type, upr-1, and uvs-2 strains, and that there are qualitative differences in the spectra of genetic alterations at the molecular level. Brychy and von Borstel (1977), however, found a marked influence on spontaneous mutability in excision repair-defective strains of Saccharomyces that is not found with excision repair-defective strains of Neurospora (de Serres et al., 1980).

COMPARISON OF THE EFFECTS OF CHEMICAL MUTAGENS IN WILD-TYPE AND EXCISION REPAIR-DEFICIENT STRAINS

The data from the present experiments with Neurospora show that in those cases where enhanced recovery of ad-3B mutants was found in excision repair-deficient strains (UV, MNNG, 4NQO) a qualitatively different spectrum of ad-3B mutants was recovered. Thus, the enhanced recovery of ad-3 mutants found after certain mutagenic treatments can be attributed to a preferential increase in a particular type of genetic alteration rather than all types of genetic alterations produced by that mutagen. The most likely explanation is a preferential increase in frameshift mutations which would increase the frequencies of NC mutants and decrease the frequencies of mutants with NP patterns (resulting from missense mutations).

In Salmonella the uvrB gene was introduced into the standard wild-type strain G46 to enhance the recovery of histidine revertants in strains TA1530, TA1535, and TA100. If this enhanced recovery of mutants in Salmonella also results from a preferential increase in frameshift mutations, this would provide a simple explanation for the differences in the genetic effects of chemical carcinogens in Salmonella and Neurospora. In fact, now that we know that the process of mutation induction is under genetic control, extreme caution must be exercised in making comparisons of the genetic effects of a given agent in different laboratory organisms. Such attempts at comparative mutagenesis (de Serres and Shelby, 1981) are extremely important not only for evaluating the genetic effects of exposure to a given agent but also for estimation of the risk of exposure to a given agent for the human population.

New data from experiments on two-component heterokaryons (de Serres, 1980) of Neurospora have shown that an excision-repair deficiency affects not only the production of point mutations at the ad-3A and ad-3B loci but also the frequency of multilocus deletions covering one or both loci. This is a more serious problem for risk estimation since the same exposure to an environmental chemical can give a quantitatively higher yield in excision repair-deficient strains, and as a result there is a high probability of an enhanced recovery of multilocus deletions. The work in mice with specific locus mutations of the latter type show quite clearly that they can have immediate genetic effects on the F_1 progeny and are not completely recessive as are specific locus mutations believed to result from point mutation (Russell, 1971).

REFERENCES

Ames, B. N., 1971, The detection of chemical mutagens with enteric bacteria, in: "Chemical Mutagens: Principles and Methods for Their Detection," A. Hollaender, ed., Vol. 1, Plenum Press, New York.

Ames, B. N., Gurney, E. G., Miller, J. A., and Bartch, H., 1972, Carcinogens as frameshift mutagens: Metabolites and derivatives of 2-acetylaminofluorene and other aromatic amine carcinogens, Proc. Natl. Acad. Sci. U.S.A., 69:3128.

Brychy, T., and von Borstel, R. C., 1977, Spontaneous mutability in UV-sensitive excision-defective strains of Saccharomyces, Mutat. Res., 45:185.

Chambers, R. W., 1981, Site-directed mutagenesis: A new approach to molecular mechanisms of mutations produced by covalent adducts of DNA, this volume.

de Serres, F. J., 1956, Studies with purple adenine mutants in Neurospora crassa. I. Structural and functional complexity in the ad-3 region, Genetics, 41:668.

de Serres, F. J., 1980, Mutation-induction in repair deficient strains of Neurospora, in: "DNA Repair and Mutagenesis in Eukaryotes," W. Generoso, M. D. Shelby, and F. J. de Serres, eds., Plenum Press, New York.

de Serres, F. J., Brockman, H. E., Barnett, W. E., and Kølmark, H. G., 1971, Mutagen specificity in Neurospora crassa, Mutat. Res., 12:120.

de Serres, F. J., and Kølmark, H. G., 1958, A direct method for determination of forward-mutation rates in Neurospora crassa, Nature, 182:1249.

de Serres, F. J., and Malling, H. V., 1971, Measurement of recessive lethal damage over the entire genome and at two specific loci in the ad-3 region of a two-component heterokaryon of Neurospora crassa, in: "Chemical Mutagens: Principles and Methods for Their Detection," A. Hollaender, ed., Vol. 2, Plenum Press, New York.

de Serres, F. J., and Shelby, M. D., 1981, "Comparative Chemical Mutagenesis," Plenum Press, New York.

de Serres, F. J., Inoue, H., and Schüpbach, M. E., 1980, Mutagenesis at the ad-3A and ad-3B loci in haploid UV-sensitive strains of Neurospora crassa. I. Development of isogenic strains and spontaneous mutability, Mutat. Res., 71:53.

Inoue, H., Ong, T. M., and de Serres, F. T., 1981, Mutagenesis at the ad-3A and ad-3B loci in haploid UV-sensitive strains of Neurospora crassa. IV. Comparison of dose-response curves for MNNG, 4NQO, and ICR-170 induced inactivation and mutation induction, Mutat. Res., 80:27.

Malling, H. V., and de Serres, F. J., 1967, Relation between complementation patterns and genetic alterations in nitrous acid-induced ad-3B mutants of Neurospora crassa, Mutat. Res., 4:425.

Malling, H. V., and de Serres, F. J., 1968, Identification of genetic alterations induced by ethyl methanesulfonate in Neurospora crassa, Mutat. Res., 6:181.

Malling, H. V., and de Serres, F. J., 1969, Mutagenicity of alkylating carcinogens, Ann. N. Y. Acad. Sci., 163:788.

Malling, H. V., and de Serres, F. J., 1970, Genetic effects of N-methyl-N'-nitro-N-nitrosoquanidine in Neurospora crassa, Mol. Gen. Genet., 106:195.

Russell, L. B., 1971, Definition of functional units in a small chromosomal segment of the mouse and its use in interpreting the nature of radiation-induced mutation, Mutat. Res., 11:107.

Schroeder, A. L., 1975, Genetic control of radiation sensitivity and DNA repair in Neurospora, in: "Molecular Mechanisms for Repair of DNA, Part B," P. C. Hanawalt, and R. B. Setlow, eds., Plenum Press, New York.

Schupbach, M. E., and de Serres, F. J., 1981, Mutagenesis at the ad-3A and ad-3B loci in haploid UV-sensitive strains of Neurospora crassa. III. Comparison of dose-response curves for inactivation and mutation induced by gamma rays, Mutat. Res., 81:49.

Singer, B., 1981, Mutagenesis from a chemical perspective: Nucleic acid reactions, repair, translocation, and transcription, this volume.

Worthy, T. E., and Epler, J. L., 1973, Biochemical basis of radiation-sensitivity in mutants of Neurospora crassa, Mutat. Res., 19:167.

CHAPTER 24

MAMMALIAN MUTAGENESIS: FUTURE DIRECTIONS

Liane B. Russell and E. G. Bernstine

Biology Division, Oak Ridge National Laboratory
Oak Ridge, Tennessee 37830

SUMMARY

Future research in mammalian germ-line mutagenesis will benefit from the reciprocal relationship that can exist between the production of mutations and their analysis. Thus, mutagenesis experiments supply altered genes, and thereby tools for basic studies; while, in turn, information on gene structure and function helps in understanding mechanisms of mutagenesis.

The nature of genetic alterations recovered in specific-locus tests is illustrated by an account of the analysis of 112 radiation-induced mutations involving the c locus on chromosome 7. Using phenotypic characterizations of various kinds, deficiency mapping with nearby markers, and complementation studies, the mutants could be classed into 13 groups, and 8 functional units could be identified in the c-locus region. Twelve different deficiencies overlapping at c range in length from <<2 to 6-11 cM. Using the deficiencies, as well as a tandem duplication that involves the c region, and several T(X;7)'s in which c is inactivated in a mosaic fashion, it is possible to generate gene doses from 0 to 3, in steps of 0.5, not only for c but also for other genes included in various deficiencies. $\underline{Cis}$ and $\underline{trans}$ configurations can also be compared. This array of genetic materials is now being used for the isolation of DNA sequences from genetically-defined regions.

Results from analyses of mutations can contribute answers to some of the pragmatic questions in germ-line mutagenesis and risk assessment. Areas in which contributions may be expected are: the relative roles of intracellular conditions (e.g., nature of chromatin, presence of repair enzymes) and secondary circumstances (e.g., selection) in determining the quantity and quality of transmitted mutations; the validity of quantitative extrapolations, such as projections to low doses and calculations of doubling dose; and the relation between measures of mutation rate and projections of phenotypic damage.

INTRODUCTION

There is a symbiotic relationship between investigations into mutagenesis and basic studies of gene structure and function. Mutagenesis experiments supply altered genes, and thus tools for the basic studies. In turn, information on gene structure helps in understanding mechanisms of mutagenesis. In the case of whole-mammal germ-line mutagenesis, there are added reasons for exploiting this reciprocal relationship. The structure and expression of mammalian genes is, of course, of special interest; and mutations transmitted in the germ line are particularly convenient tools for basic studies in these areas, because they permit easy multiplication and maintenance of the genetic material for future use and, beyond that, allow genetic manipulation. In the reciprocal direction, information about the nature of mutations is potentially useful not only for understanding mechanisms of mutagenesis; it also allows analytical approaches to several parts of the very complex pathway that intervenes between mutagen application and expression of mutated genes in the next generation.

This paper briefly summarizes investigations into the nature of the genetic alterations recovered in certain mammalian germ-line mutagenesis studies. Specific-locus tests, which are usually conducted primarily to provide quantitative information on mutation rates, yield genetic material of potential value to the future directions of research (which are the subject of this Session). Methods will be proposed by which the mutants can be used as tools for studies of genomic organization and gene expression in mammals. Finally, the paper will also briefly examine some ways in which analysis of mutations can be made to relate to pragmatic questions of mutagenesis — questions which will continue to be asked particularly in relation to risk.

CHARACTERIZATION OF SPECIFIC-LOCUS MUTATIONS TO PROVIDE
TOOLS FOR STUDIES ON GENOME ORGANIZATION AND FUNCTION

The Mouse Specific-Locus Test

The mouse specific-locus test (SLT) with seven visible markers (W. L. Russell, 1951), which has for many years been used for genetic risk assessment in mutagenesis studies, yields a valuable by-product, namely, mutations that can be propagated in stocks and analyzed to the extent of present, or future, capabilities. In SLTs, in general, parent strains are chosen that differ with respect to alleles present at a number of selected loci, such that a genetic change involving one of these alleles leads to a first-generation phenotype distinguishable from that of the standard heterozygote.

Regardless of what type of marker is used (whether visible, biochemical, or immunological), the method can detect changes occurring within the marked locus, as well as some outside the boundaries of that locus (referred to as "extragenic" in Fig. 1). It picks up any kind of intragenic alteration that leads to a detectable change in the gene product, either its modification [P_x] or its absence [P_0]. It further picks up deficiencies that cut out the marker, resulting in absence of the product [P_0]. It may also detect alterations near (rather than at) the marked locus (e.g., rearrangements) which influence production of the gene product by position effects. All three types of genetic change have been found to be represented among specific-locus mutants and to provide potential tools for various types of studies. Obviously, the mere knowledge of whether the product is null or altered, a knowledge that is readily obtainable for some of the visible markers (Russell, 1971; Russell et al., 1979) and most of the biochemical ones (Johnson, 1981), is not sufficient for drawing conclusions about the genetic nature of a mutation.

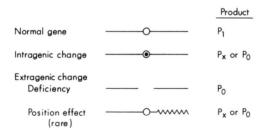

Fig. 1. Types of genetic changes detected by the specific-locus method. Intragenic changes of many kinds (frameshifts, base-pair substitutions, etc.) may be detected if they produce an alteration of the normal product (P_x in lieu of P_1) or absence of the product (P_0). Deletion of the locus (sometimes along with neighboring loci) causes absence of the product. Position effects affecting expression of the marked locus may be brought about by nearby rearrangements. It is obvious that the distinction between P_0 and P_x by itself provides only little information concerning the nature of the genetic change.

Analysis of Induced Mutations

Genetic methods available for analyzing the mutations include first, and most obviously, an allelism test to insure that one is, in fact, dealing with an alteration that involves the marked locus, rather than with a dominant mimic at another locus. The position effect shown in Fig. 1 (bottom line) formally qualifies as a subclass of the latter situation. Here, a linked rearrangement acts dominantly to affect the expression of an unaltered (wild-type) gene at the marked locus. With one possible exception, all cases of this type that have been discovered have involved X-chromosome material that had been moved into the vicinity of the marked locus as a result of rearrangement.

The inactivity of one of the X chromosomes of female mammals extends, to some degree, to cases of X-autosome translocations. Studies on several such translocations have indicated that in female heterozygotes one segment of the rearranged X and the intact X are subject to the alternate states that normally pertain to two intact X's (Russell and Cacheiro, 1978). Further, in the rearranged chromosome, portions of nearby autosomal material, including the marked locus, go along with the X's behavior, i.e., they are inactive when the contiguous X material is inactive, and active when the contiguous X material is active (Russell, 1963). Since, in a standard SLT, the rearranged chromosome bears the wild-type allele at the marked locus, and the homologous (intact) autosome bears the recessive, such animals are mosaics for the hemizygous expression of the marker. Examples of X-autosome translocations that affect the c (albino) locus and nearby regions are shown in Fig. 2.

Such position effects have been exceedingly rare. Most often, the allelism test on specific-locus mutants has indicated that the marked locus itself is genetically altered. Further procedures that have then been used for analysis, some routinely and others for special groups of mutants, are phenotypic characterization (e.g., tests of homozygous and heterozygous viability and fertility), deficiency mapping with nearby markers, and complementation studies. The most extensively tested groups are radiation-induced mutations, recovered at Oak Ridge, which involve 3 of the 7 loci of the standard SLT, namely, 122 d-locus mutants, 43 se-locus mutants, 37 d se's[*] and 112 c-locus mutants (Russell, 1971, 1979; Russell et al., 1979, 1982; Russell and Raymer, 1979; Russell and Russell, 1960). Some conclusions of the c-locus-mutant analysis will be summarized below, since they have a bearing on future work.

---

[*]The d (dilute) and se (short-ear) markers are separated by only 0.15 cM.

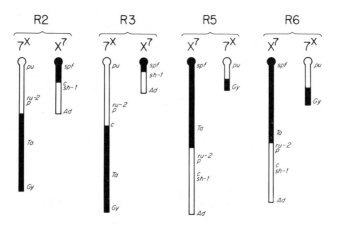

Fig. 2. Reciprocal X-autosome translocations involving chromosome 7, on which the c (albino) locus is located. X-chromosome material is shown in black. In R3, R5, and R6, the autosomal portion contiguous with translocated X-chromosomal material, and containing the c-locus, is inactive when the intact X (not shown) is active. In R2, the c locus is contiguous with a short portion of X material which is not subject to alternative states, and remains active at all times. (By permission from Russell and Cacheiro, 1978).

Mutants at the c-Locus

The Oak Ridge c-clous mutants were classified in a two-dimensional matrix according to (a) whether homozygotes are fully viable or lethal at a specific developmental stage (before implantation of the embryo in the uterus, at implantation, neonatally, or during the juvenile period); and (b) whether or not they involve nearby markers* (thus, some involved no other markers; some overlapped Mod-2; some overlapped both Mod-2 and sh-1, but none extended beyond sh-1 to Hbb, and none involved the proximal marker tp). By such classification and limited complementation results, a subset of 34 independent nonviable c-locus mutants fell into 12 groups (Russell et al., 1982), and the properties reported for three Harwell c-locus mutants (Gluecksohn-Waelsch, 1979) indicate that these, too, fit into the same grouping.

---

*The c (albino) locus lies on chromosome 7 in the vicinity of convenient markers, the following of which were used in the studies: tp (taupe), Mod-2 (mitochondrial malic enzyme), sh-1 (shaker-1), and Hbb (hemoglobin β chain). Map relations between these loci, from proximal to distal, are tp (3) c (2) Mod-2 (4) sh-1 (2) Hbb; the numbers in parentheses refer to genetic distance between loci expressed in cM (Roderick and Davisson, 1981).

A full-scale complementation grid of all 435 possible combinations of 30 nonviables was developed (Russell et al., 1982). Abnormal phenotypes utilized in this study included deaths at various stages of development, weaning weights, the reduction or absence of various enzymes, and the uncovering of recessive markers. The 12 mutant groupings that had been derived on the basis of the above matrix held up in the full-scale complementation tests. Little has yet been done with the 13th group, the homozygous viables, which actually contains almost two-thirds of all $\underline{c}$-locus mutants. Some of these mutations (designated $\underline{c}^{av}$) resemble the standard albino, while others (designated $\underline{c}^{xv}$) produce some pigment but less than wild type.

It is possible to postulate an alignment of functional units by which all analyzed $\underline{c}$-locus mutations fit a linear pattern (Fig. 3). [By contrast, analysis of the $\underline{d}$-$\underline{se}$ region had revealed a few mutations that could not be fitted into a linear complementation map (Russell, 1971)]. The results are consistent with the hypothesis that all of the nonviable albino mutations are deficiencies overlapping at $\underline{c}$, and ranging in size from <2 cM to 6-11 cM. Since combinations of lethal deficiencies that extend beyond both ends of the $\underline{c}$ locus produce viable albino animals that resemble the standard $\underline{c/c}$ type, it may be concluded that the $\underline{c}$ locus contains no sites essential for survival. For similar reasons, the easiest explanation for viable intermediate alleles, $\underline{c}^{xv}$'s (see above), is that they are the result of mutations within the $\underline{c}$ cistron.

It is of interest that starting with one marker, $\underline{c}$, we were able to identify 13 distinct groups of mutations and eight functional units in the $\underline{c}$-locus region. Some functions could not be separated by complementation. Thus, glucose-6-phosphatase (G6Pase) and neonatal survival went together (both absent or both present) in 157 separate combinations of independent mutations (DeHamer, 1975; Russell et al., 1982); and, in a more limited group of six combinations, two other liver-specific enzymes, serum protein, and the structure of subcellular membrane organelles all paralleled the results for G6Pase (Gluecksohn-Waelsch et al., 1974). On the other hand, complementation tests were successful in separating the MOD-2 function from the juvenile-survival function, and, taken together with earlier findings (Bernstine et al., 1978), there is now no evidence that mitochondrial malic enzyme influences survival at any age.

The map of functional units could undoubtedly be greatly enriched (a) by future inclusion of new mutants in the complementation grid, and (b) by studies of more detailed functions in the existing grid. For example, it was possible by means of embryological studies (Russell et al., 1982) to separate what had originally been described merely as "prenatal death" into preimplantation and implantation

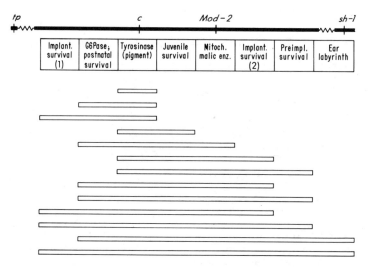

Fig. 3. Complementation map of c-locus mutants. With the postulated alignment of functional units shown below the genetic map, 13 mutant groups established on the basis of homozygous viability and involvement of nearby markers (see text) fit a linear complementation map.

death and to define three separate units controlling these functions (Fig. 3). Similarly, more detailed studies of biochemical, morphological, and developmental characters in the combination types would undoubtedly produce finer subdivisions. Refinements of this nature might eventually result in a series of functional units that represent coding sequences for a series of identifiable polypeptides.

Additional Genetic Tools for Studying Genes in the c-Locus Region

Even the relatively limited type of characterization that has been generated by the existing complementation map helps to make the mutants into potential tools for studies that involve not only the c locus but other nearby loci as well, including already known ones such as Mod-2. Among possible uses is the generation of a series of gene dosages. A tandem duplication of a region that includes the c-Hbb segment of chromosome 7 adds to the usefulness of the other types in that respect (Russell et al., 1976). By combining the various chromosomal conditions (duplication, deficiencies, rearrangements, and normal chromosomes) one can produce a great array of dosages (0, 0.5, 1, 1.5, 2 trans, 2 cis, 2.5, and 3) for the c locus, the Mod-2 locus, or any other locus that might become identified within given deficiencies (Fig. 4). Trans and cis actions can also be compared. The dosage series has already been useful in the analysis

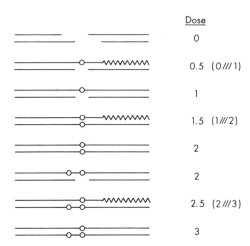

Fig. 4. Combinations of chromosome 7 that can yield various doses of the c locus (shown as a circle) and nearby loci (e.g., Mod-2). Overlapping c-locus deficiencies that complement for viability can produce a zero dose. A tandem duplication involving c (shown as two circles in a row) can produce three doses, if opposite a normal chromosome, or two doses in cis, if opposite a deficiency. Doses of 0.5, 1.5, and 2.5 can be achieved by using X-autosome translocations (which inactivate the c locus in approximately half the cells of the body) opposite, respectively, a deficiency, a normal chromosome 7, and the tandem duplication.

of Mod-2 gene action and the study of the associated regulatory gene Mdr-1 (Bernstine, 1979; Bernstine et al., 1978, 1979). The characterized deficiencies and the translocations, moreover, should provide potentially valuable tools for molecular studies.

ISOLATION OF DNA SEQUENCES FROM GENETICALLY DEFINED REGIONS

In addition to their intrinsic interest as the products of mutagenesis experiments, chromosome aberrations in the mouse provide material for studies designed to identify and isolate DNA sequences from genetically defined regions of the mouse genome. Such studies will provide the basis for estimates of the molecular extent of mutagen-induced deletions, allow the extension of deletion mapping of well-studied regions (such as the c locus), and eventually permit the identification of gene products associated with small chromosomal regions. In this section, we will outline one approach currently in progress (L. Albritton and E. G. Bernstine, unpublished data) to identify DNA restriction fragments which are missing in a large c-locus deletion (at least 4 cM in length).

Our strategy makes use of two chromosome aberrations: a large deletion including the c locus on chromosome 7 and a reciprocal translocation which involves a large part of both the X chromosome and chromosome 7 (from about 20 cM proximal to c to the telomere). In addition, we have generated by genetic recombination (in vivo) a translocation chromosome whose chromosome-7 moiety carries the large deletion. The principle is to detect restriction fragments that are present in the "normal" translocation chromosome, i.e. T(X; but missing from the translocation chromosome which carries the c-locus deletion, i.e. T[X;7(Dfc)]. In order to accomplish this goal, it is necessary to enrich for DNA from the T(X;7) and T[X; 7(Dfc)] chromosomes and to devise a screening procedure that will display a sufficient number of restriction fragments to enable the comparison to be made. {It is expected that 3 to 6% of the DNA in the T(X;7) chromosome will be absent from the T[X;7(Dfc)] chromosome.}

Since both translocation chromosomes carry wild-type HPRT genes on their X-chromosome portions, they may be selected in interspecific somatic cell hybrids with HPRT-deficient cell lines. Following cell fusion and chromosome loss in HAT-selective medium, we have isolated one line which, karyologically, appears to have retained a translocation chromosome as its sole mouse chromosome. Given the screening method described below, such lines constitute "purification" of the chromosomes in question. In the ideal case then, we will have available for comparison two chromosomes that differ only in that one is lacking some sequences present on the other.

In order to identify DNA restriction fragments missing in the deletion-bearing chromosome, we will digest DNA from appropriate cell lines with a restriction enzyme (any relatively infrequent cutter) and run the digested sample into an agarose gel. Several divisions will be made in the lane from which the various size classes of DNA will be electroeluted and digested with a second restriction endonuclease. Each digest will then be run in a separate lane on a second gel, which will then be transferred to nitrocellulose and hybridized to a ^{32}P-labeled cloned mouse (intermediately) repetitive probe. The autoradiogram of this hybridization should show many bands representing fragments of mouse DNA that contain the (interspersed) repeated element. As an example, let us assume that we have a cloned sequence that is repeated 1000 times in the mouse genome, and that we have 5% of the mouse genome in our hybrid cell lines. Thus, we expect our DNA to contain about 50 copies of the repeat. If we have taken ten size cuts, then our two-dimensional gels would have, on average, five bands per lane. Actual results, of course, will depend on the frequency of the repetitive element in the mouse genome and on the fraction of the genome contained in the translocation chromosomes. When DNA from a T(X;7)-containing cell line is compared to DNA from a line retaining the T[X;7(Dfc)] chromosome, the missing bands in the latter should correspond to the deleted region.

So far, we have obtained one of the required cell lines. Moreover, we have prepared a library of mouse DNA (from a complete digestion with Sau3A) in the plasmid pBR322 in which we have identified several clones whose hybridization characteristics appear to be those of repetitive sequences, i.e., they hybridize to total mouse DNA which has been denatured and reassociated sufficiently to renature very highly repetitive (satellite) sequences. Further characterization of these clones is in progress.

The methods proposed to attack this problem promise to provide powerful tools for detailed biochemical analysis of chromosome aberrations in the mouse. The labor involved in obtaining the proper set of cell hybrids is counterbalanced by the gain of permanent cell lines which carry the desired chromosomes. Since analysis is possible without physically purifying the chromosomes, due to the species specificity of repeated elements, considerable effort will eventually be saved. The variety of aberrations available for study provides exciting prospects for expanding these and other approaches to explore other regions of the mouse genome.

CONTRIBUTIONS TO THE ASSESSMENT OF GENETIC RISKS FROM INFORMATION ON GENE STRUCTURE AND EXPRESSION

A highly complex pathway intervenes between mutagen exposure of a mammalian organism and expression of a mutated gene in the next generation. An understanding of some of the steps in this pathway requires investigations in the fields of pharmacokinetics, metabolism, gonadal microanatomy, and reproductive cell kinetics. The interpretation of other steps, however, will be aided by basic findings concerning the nature of genetic events and gene expression. Examples of problem areas in which such contributions may be expected are: the role of biological variables, such as germ-cell type, in determining what kind of mutations are induced; the relation between initial lesions and secondary influences, such as repair or selection at various stages; the evaluation of various quantitative extrapolations, such as projections to low doses and calculations of "doubling dose"; and the relating of quantitative measures of mutation rate to projections of phenotypic damages.

Initial Lesions and Secondary Influences

The effect of such biological variables as sex, germ-cell stage, and cell-cycle characteristics on transmitted mutation rates has been demonstrated extensively for radiations (W. L. Russell, 1965), and, to a more limited extent, for chemicals (Russell et al., 1981;

W. L. Russell, 1981). Underlying these quantitative differences, there are qualitative ones. For example, with ionizing radiation as the mutagen, very different spectra of mutations are found from exposure of different germ-cell types (Fig. 5). Mature oocytes and postspermatogonial stages clearly yield relatively more multilocus deficiencies than do spermatogonial stem cells; and there is a suggestion that spermatogonial populations whose cycling characteristics may have been altered by certain dose distributions yield slightly different mutational spectra than do previously undisturbed stem cells (Russell, 1971). Is the difference in recovered mutations the result of cellular conditions that exist at the time the mutagen attacks, e.g., different states of the chromatin in the different cell types, or the existence of different repair systems? Or is it due to secondary circumstances, e.g., gonadal selection (before, during, or after meiosis) against the multilocus deficiencies? The extent to which these possibilities are involved may be a little troublesome to determine, but analysis of the nature of mutations produced by a few model mutagens with different basic actions should help shed some light on this.

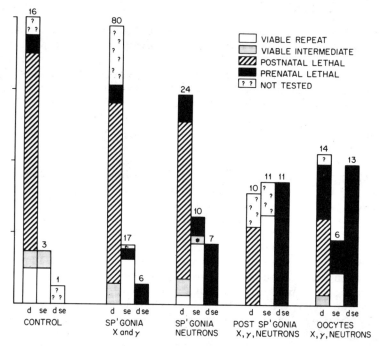

Fig. 5. Relative numbers of d, se, and d se mutations obtained following irradiation of different germ-cell stages. Complementation studies established that lethal mutations, shown in black, as well as d se's, are multilocus deletions. (By permission from Russell, 1971).

If mutational lesions of certain kinds but not of others can find expression during the brief haploid phase of the mammalian life cycle (in spermatids and spermatozoa), relative transmitted mutation rates will be affected. From the point of view of human grief caused, selection acting by such a mechanism is preferrable to selection acting on the whole organism. From the point of view of basic research opportunities, the demonstration of this type of selection can provide a handle for studying expression of genes in single dose. Figure 6 illustrates a proposed method for using mutants that are being maintained as stocks to detect the occurrence of selection during the haploid phase. In crosses of males that are heterozygous for the mutation to wild-type females, the finding of reduced transmission ratios <u>without</u> reduced litter size will provide evidence for haploid selection, whereas a reduction in both parameters indicates selection against heterozygous embryos. Although only a few genetically characterized specific-locus mutations have so far been studied by these criteria, indications of haploid selection were obtained in the case of certain multilocus deletions (Russell and Russell, 1960).

Quantitative Extrapolations

Mutation-rate experiments are generally carried out at doses much higher than those to which human beings are likely to be exposed, and extrapolation is therefore an unavoidable component of risk projections. Assuming the dose of agent that reaches the genetic target is known, mutation-rate extrapolation from high to low dose levels can be made more meaningful if it is known whether one is dealing with a single-hit or multi-hit curve. Knowledge of the nature of the mutations can provide pertinent information in this regard.

Another example of how basic information can contribute to estimates of human risk concerns spontaneous mutations. Risk estimates have commonly used a "doubling-dose" approach, i.e., determination of the dose of mutagen which would induce a mutation frequency equal to the spontaneous rate per generation. This quantitative approach would lose validity if there were qualitative differences between spontaneous and induced mutations and thus probably differences between the two sets in the degree and type of phenotypic damage produced. Analysis of mutations at three of the seven loci marked in the standard SLT indicates that very few of the spontaneous mutations show evidence of being multilocus deletions, while many of the mutations induced by radiation in oocytes and postspermatogonial stages are of that type (Russell, 1971, 1979; Russell et al., 1979). Doubling-dose calculations would seem to be of doubtful validity for radiation exposures of these germ-cell stages but are probably justifiable when spermatogonia are exposed.

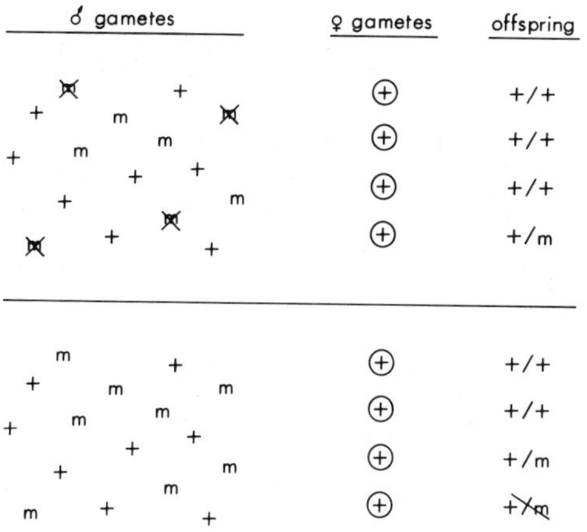

MAMMALIAN MUTAGENESIS

♂ gametes	♀ gametes	offspring
⊠ +	⊕	+/+
+ m m ⊠	⊕	+/+
+ m + +	⊕	+/+
+ m		
⊠ + ⊠ +	⊕	+/m
m +	⊕	+/+
+ m m	⊕	+/+
+ m + +	⊕	+/m
m m		
m + +	⊕	+/⊠m

Fig. 6. Consequences of selection occurring during haploid phases, i.e., in spermatids, spermatozoa (top portion), or in early diploid embryos (bottom portion), when crosses of +/+ ♀ × +/m ♂ are used for study (where m is a mutant allele of the wild-type, +). If, in a population of 50% m and 50% + sperm, some m sperm are eliminated (crossed out in the diagram), the total number of sperm left should still be more than sufficient to fertilize all available eggs. This should result in the normal number of offspring, which, however, contain a < 50% proportion of +/m. By contrast, if selection acts on +/m embryos (shown diagrammatically by the crossing out of one +/m offspring), the total number of offspring, as well as the proportion of +/m will be reduced. Note that since there is no haploid phase in the female life cycle, the opposite cross cannot be used.

Mutation Rates and Phenotypic Damage

One of the most important tasks in estimating human risk from potential mutagens is to relate quantitative parameters, i.e., mutation rates, to qualitative ones, i.e., types of phenotypic damage. Some promising empirical approaches are under development that start from the phenotypic end of this relationship, namely, through detection of certain classes of morphological damages (skeletal defects, cataracts) in first-generation offspring, and the subsequent demonstration of heritability (Kratochvilova and Ehling, 1979; Selby, 1982).

The opposite approach is to go from genotype to phenotype, i.e., to analyze the genetic nature of a mutational lesion and then to determine the mode of its expression. It would be useful to know, e.g., whether mutations that give no evidence of being small multilocus deletions and could be intragenic can, in the heterozygous condition, affect vital characters. It is already known that heterozygous small deficiencies can have viability effects, and there is some evidence that the specific content of a deleted segment is more important than its length. For example, a deficiency involving the se locus, which is at most 2 cM in length (since flanking markers d and sv are not involved), clearly depresses viability in the heterozygous state (Russell, 1972); but several c Mod-2 deficiencies that must be over 2 cM long have no obvious effects on vital characters (Russell et al., 1979). In the vicinity of the c locus, but outside the c-Mod-2 segment, several functional units controlling survival at various stages of development have been identified by complementation studies. Total absence of one of these (e.g., in overlapping deficiencies or in homozygotes) leads to juvenile death, absence of another causes neonatal death, and absences of yet others cause implantation or preimplantation death (Fig. 3). One or more of these functional units can be made hemizygous by combining a normal chromosome with one having an appropriate deficiency. It will be of interest to determine whether the identity of the hemizygous unit(s) has any influence on whether, and if so to what degree, viability is affected. Because heterozygous effects on vital characters are likely to be extremely subtle, their study requires elimination of heterogeneity of the genetic background. Coisogenic stocks for some of the genetically characterized mutants have been constructed to permit the future study of subtle phenotypic differences, and answers to some of the questions raised above may be forthcoming.

ACKNOWLEDGMENT

This research was sponsored by the Office of Health and Environmental Research, U.S. Department of Energy, under contract W-7405-eng-26 with the Union Carbide Corporation.

REFERENCES

Bernstine, E. G., 1979, Genetic control of mitochondrial malic enzyme in mouse brain, J. Biol. Chem., 254:83.

Bernstine, E. G., Koh, C., and Lovelace, C. C., 1979, Regulation of mitochondrial malic enzyme expression in mouse brain, Proc. Natl. Acad. Sci. U.S.A., 76:6539.

Bernstine, E. G., Russell, L. B., and Cain, C. S., 1978, Effect of gene dosage on the expression of mitochondrial malic enzyme activity in the mouse, Nature, 271:748.

DeHamer, D. L., 1975, A biochemical study of lethal mutations at the c locus in the mouse, Ph.D. Dissertation, The University of Tennessee, Knoxville.

Gluecksohn-Waelsch, S., 1979, Genetic control of morphogenetic and biochemical differentiation: Lethal albino deletions in the mouse, Cell, 16:225.

Gluecksohn-Waelsch, S., Schiffman, M. B., Thorndike, J., and Cori, C. F., 1974, Complementation studies of lethal alleles in the mouse causing deficiencies of glucose-6-phosphatase, tyrosine aminotransferase, and serine dehydratase, Proc. Natl. Acad. Sci. U.S.A., 71:825.

Johnson, F. M., 1981, Mutation-rate determinations based on electrophoretic analysis of laboratory mice, Mutat. Res., 82:125.

Kratochvilova, J., and Ehling, U. H., 1979, Dominant cataract mutations induced by γ-irradiation of male mice, Mutat. Res., 63:221.

Roderick, T. H., and Davisson, M. T., 1981, Linkage map of the mouse, Mouse News Letter, 64:5.

Russell, L. B., 1963, Mammalian X-chromosome action: Inactivation limited in spread and in region of origin, Science, 140:976.

Russell, L. B., 1971, Definition of functional units in a small chromosomal segment of the mouse and its use in interpreting the nature of radiation-induced mutations, Mutat. Res., 11:107.

Russell, L. B., 1972, Heterozygous effects on viability of d or se mutations that involve a known minimum number of lethal functional units, in: "Oak Ridge National Laboratory Biology Div. Annu. Progr. Rep.," ORNL-4817, Oak Ridge.

Russell, L. B., 1979, Analysis of the albino-locus region of the mouse. II. Fractional mutants, Genetics, 91:141.

Russell, L. B., and Cacheiro, N. L. A., 1978, The use of mouse X-autosome translocations in the study of X-inactivation pathways and nonrandomness, in: "Genetic Mosaics and Chimeras in Mammals," L. B. Russell, ed., Plenum Publishing Co., New York and London.

Russell, L. B., Montgomery, C. S., and Raymer, G. D., 1982, Analysis of the albino-locus region of the mouse. IV. Characterization of 34 deficiencies, Genetics, in press.

Russell, L. B., and Raymer, G. D., 1979, Analysis of the albino-locus region of the mouse. III. Time of death of prenatal lethals, Genetics, 92:205.

Russell, L. B., and Russell, W. L., 1960, Genetic analysis of induced deletions and of spontaneous nondisjunction involving chromosome 2 of the mouse, J. Cell. Comp. Physiol., Suppl. 1, 56:169.

Russell, L. B., Russell, W. L., and Kelly, E. M., 1979, Analysis of the albino-locus region of the mouse. I. Origin and viability, Genetics, 91:127.

Russell, L. B., Russell, W. L., Popp, R. A., Vaughan, C., and Jacobson, K. B., 1976, Radiation-induced mutations at the mouse hemoglobin loci, Proc. Natl. Acad. Sci. U.S.A., 73:2843.

Russell, L. B., Selby, P. B., von Halle, E., Sheridan, W., and Valcovic, L., 1981, The mouse specific-locus test with agents other than radiation: Interpretation of data and recommendations for future work, Mutat. Res., in press.

CHAPTER 25

PERSPECTIVES IN MOLECULAR MUTAGENESIS

John W. Drake

Laboratory of Molecular Genetics, National Institute of
Environmental Health Sciences, Research Triangle Park
North Carolina 27709

SUMMARY

The models and paradigms that underlie a vigorously developing science may tend to stifle progress or may serve to sharpen the knife edge of paradox. Working out mutagenic mechanisms is a conceptually and technologically demanding task, and we are accumulating an increasingly uncomfortable number of experimental and theoretical inconsistencies. First, there continue to be widespread difficulties in specifying the chemical nature of mutagenic DNA alterations, both because of the multitude of DNA reaction products induced by many mutagens and because of the intrinsic rarity of most mutational responses. For instance, alkylation of the O^6 position of guanine to generate adducts of modest dimensions is widely believed to form the basis for the mutagenic and carcinogenic actions of numerous chemicals. However, while this scheme is supported by in vitro evidence, it has failed to explain why bacteriophages can be thus alkylated in vitro by N-methyl-N'-nitro-N-nitrosoguanidine without the production of mutations, or why microbial eukaryotes alkylated by ethyl methanesulfonate or N-methyl-N'-nitro-N-nitrosoguanidine display no mutagenic response when their "error-prone repair systems" are mutationally inactivated. Second, a base pair is typically mutated at vastly different rates, and with different directional specificities, when it resides at different positions within a gene; whereas very little of this variability is explained by current theories that aim to describe the determinants of fidelity in DNA replication. (Some sizable portion of this variation now appears to depend not only upon neighboring base-pair influences but also upon much more subtle and distant effects.) Third, experimental modifications of enzymatic fidelity by means of amino acid substitutions, and perhaps also cation replacements, lead to such a diversity of modified mutation rates as to seriously challenge the ability of any simple theory to organize the experimental observations into a coherent and predictive network.

INTRODUCTION

A topic such as this provides virtual carte blanche. I shall respond with comments directed to what seem to be the central areas of discomfort in our grasp of the mutagenic process and the more obviously promising directions for progress in the near future.

THE PROBLEM OF INHERENT COMPLEXITY

Thriving at the center of the bramble in which most of us have chosen to scratch out a living is a giant thornbush: inherent complexity. Most inquiries into genetical and chemical processes in biological systems are reductionist in nature and seek to determine how the organism or the cell carries out some process. We generally anticipate, with reasonable confidence based on historical results, that only one or at most a few mechanisms will turn out to be operating: the number of right ways to do something well is limited. The science of mutagenesis, on the other hand, encompasses the sum of all possible ways in which a highly evolved and extremely intricate process can go wrong. We can, therefore, anticipate that all possible routes to error actually occur with finite probabilities; conversely, we can also anticipate that our imaginations alone are unlikely to suggest each of these routes and that we will constantly be presented with delightful surprises.

The problem, then, is not simply one of how mutagenesis occurs but, more realistically, is that of imagining all the ways in which it might occur, and then designing experimental attacks on these possibilities powerful enough not only to discredit the inappropriate answers but also to ferret out the as yet unimagined possibilities. Furthermore, it is unwise to accept too blithely "the answer" in a given situation as definitive, since there are probably other good answers, alternative mechanisms, that may actually operate at lesser frequencies under the given conditions but may even predominate at other sites, in other genes, or in different organisms. It will often be difficult, for instance, to interchange conclusions among a mouse, a fruit fly, a bacterium, a complex DNA bacteriophage, and a simple RNA virus, to say nothing of a pinch of enzyme together with a touch of template and a dollop of precursors in a test tube. The situation with mutagenesis, then, may be analogous to that of academic medicine, which aims to understand all the ways in which the human body can go wrong. Both disciplines, of course, reflect a profound underlying need to understand the normal, unperturbed process.

PERSPECTIVES IN MOLECULAR MUTAGENESIS

THE PROBLEM OF SENSITIVITY

Genetical assays demonstrate a marked tendency to run well ahead of chemical assays in sensitivity and specificity. Sufficiently powerful selective techniques are often available to detect events occurring at frequencies of 10^{-8} to 10^{-12} per gene in microbial systems, for instance, although such numbers certainly do raise problems for mouse geneticists. Furthermore, such rare events should also be viewed as being distributed among target macromolecules that are themselves in extremely dilute solutions; in a fully grown culture of bacteria, for instance, a typical gene is only some 10^{-14} molar and it is within this dilute solution that we measure rare mutations. Even further sensitivity is readily available to the microbial geneticist but rarely to the chemist.

The significance of this exquisite sensitivity becomes uncomfortably clear when one recalls that even the most apparently simple classical mutagens, such as ultraviolet light and monofunctional

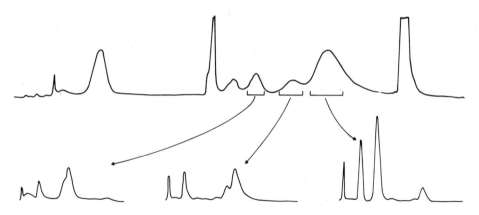

Fig. 1. The multiplicity of products from heated dGMP (J. A. Grunau, unpublished data). A sample of 5'-dGMP was heated for 1.5 h in 1 N acetic acid. The soluble fraction was adsorbed to and eluted from Dowex 1-X8 with a 0.009 → 0 17 N HCl gradient. Material from the three indicated peaks was then treated with alkaline phosphatase and adsorbed to and eluted from Dowex 50-X8 with 0.1 M ammonium formate, pH 4.0. Somewhat more than a score of products can be discerned in the top elution profile, the rightmost large peak of which is dGMP itself. Each of the dephosphorylated samples contained about seven to ten components. It is, therefore, likely that the more than two-score of products revealed here are a substantial underestimate of the total number.

methylating and ethylating agents, introduce a score or more of products apiece into DNA (the actual number swelling yearly, and monthly in California). As a result, we rarely know just which products are the true premutagenic lesions. This confusion is enhanced by the multiplicity and overlap of repair systems operating upon such lesions and by the number of enzymes within each system of repair or replication that may be involved in the processing of the damage.

As an example, this kind of problem currently confounds our attempts to understand heat mutagenesis, a process that has been suggested, on the basis of modest extrapolations from phage T4 to humans, to constitute a major challenge to large, warm genomes (Baltz et al., 1976). It is clear from the T4 work that cytidine (in the guise of 5-hydroxymethylcytidine) and guanosine are targets for heat mutagenesis, the mutational pathways being C → T and G → Py, the latter perhaps specifically G·C → C·G (Bingham et al., 1976). A number of heat-induced modifications of guanosine can readily be imagined that might cause it to mispair as a pyrimidine. The question, then, is which of these are actually produced. The result of a preliminary survey was appalling; when dGMP was heated in solution and then subjected to fractionation, a remarkably large number of species appeared (Fig. 1). It is reasonable to guess from the approximately 22 peaks in the initial fractionation, plus the peak splitting that occurred upon dephosphorylation and refractionating, that several score of products are produced by the heating of dGMP. Which one or more of these is the mutagenic culprit? Or will it turn out to be an as yet still undetected species?[*]

In principle, the general solution to the multi-lesion problem (and a number of related questions as well) has for some time been technically feasible, although by no means easy (see, for example, Bhanot et al., 1979). The method is to synthesize templates containing a high frequency of a single type of perturbed base, to subject these templates to in vitro and in vivo replication, and to characterize the products both chemically and genetically. Given the obvious power and practicality of this approach, it would seem useful to sequester further speculation about a host of currently popular problems and get on with the art of solving the soluble.

---

[*] In a related study, the claim has been laid that the depurination of adenine is mutagenic (Schaaper and Loeb, 1981). Presumably adenine, like guanine, is modified by heat in many ways. In addition, the ratio of depurinations to mutations at the responding site appeared to be roughly 100. Sorting out the mutagenic species is, therefore, likely to be difficult.

INDIRECT MUTAGENESIS

The actions of mutagens are commonly studied and interpreted in terms of their interactions with DNA (except for the trivial exception of RNA viruses). The logical possibility remains, however, that some mutagens work indirectly or by both direct and indirect mechanisms. Indeed, with the exception of systems in which viruses or nucleic acids are treated extracellularly, it is quite possible that a substantial fraction of mutagenesis is indirect.*

Two potential targets need to be examined much more closely than in the past: enzymes involved in maintaining fidelity, and direct precursors of DNA such as the deoxyribonucleoside triphosphates (dNTPs). Although definitive experiments are lacking, there exist provocative hints that both alkylating agents (e.g., Kondo and Ichikawa, 1973) and mutagenic metals (e.g., Kunkel and Loeb, 1979) sometimes act upon DNA polymerases so as to reduce their effectiveness in error avoidance. There is less evidence, but even more reason to expect, that dNTPs may often be the targets of chemical mutagens. The base analogue N^4-hydroxycytidine, for example, is a potent mutagen in both Escherichia coli and phage T4 systems (e.g., L. S. Ripley, unpublished data; Salganik et al., 1973). The corresponding dNTP should also be formed in vivo when cellular systems are treated with hydroxylamine, providing a much wider mutagenic specificity than when extracellular DNA is the target, and exactly such a broadening of specificity was observed (Tessman et al., 1965). Furthermore, recent evidence suggests that dNTPs may be superior targets for alkylation compared with cellular DNA (Baker and Topal, 1981).

POLYMERASE FIDELITY

This and the following two Sections will focus on three crucial mechanisms that have now been well documented by genetic studies to be major determinants of the mutagenic response but that remain either inadequately or, in two cases, virtually completely

---

*"Indirect mutagenesis" has acquired two distinct meanings. The first refers to the mechanism of action of agents that interact with primary targets other than DNA itself, resulting ultimately in mispairing due to the presence of modified precursors of DNA replication or to reduced fidelity on the part of enzymes of DNA replication or repair. The second refers to the mechanism of action of agents that damage DNA by generating lesions that block DNA replication (e.g., bulky adducts, cyclobutane pyrimidine dimers) and thereby induce ultraviolet-like mutagenesis, as discussed in a subsequent Section.

uncharacterized at the enzymological level. They consist of the
discriminatory powers of DNA polymerases, the putative process of
mismatch repair of mutational heteroduplexes, and the process that
will be called simply ultraviolet mutagenesis.

The very existence of variant DNA polymerases exhibiting mutator
or antimutator properties, mostly in phage T4 but to some extent in
bacteria, and the considerable differences among wild-type DNA polymerases from different organisms when compared for in vitro fidelity
define a central role for these enzymes in maintaining the accuracy
of DNA synthesis. Biochemical studies from the Bessman and Nossal
laboratories (Hershfield, 1973; Muzyczka et al., 1972) first demonstrated two discriminatory phases, namely one occurring during or
immediately before the covalent addition of a new residue to the end
of the primer strand and, at least in the case of enzymes possessing
an associated 3'-exonuclease, one acting to remove terminal mismatches after covalent binding. The genetic evidence, however,
implies a complexity of discrimination that has not yet even begun
to be attacked at the enzymological level. It has been known for
some time now, for instance, that mutant T4 DNA polymerases first
characterized as "antimutators" because of (often quite large) reductions in spontaneous and chemically induced transition rates may
simultaneously be mutators when assayed for effects on transversion
rates (Ripley, 1975). An equally complex pattern has recently
emerged from assays of the effects of variant T4 DNA polymerases
upon frameshift mutation rates (Ripley and Shoemaker, 1981). However, while reasonably sensitive assays are available for the further
in vitro analysis of the role of polymerases in base-substitution
avoidance, no corresponding assay is available to probe their effects
upon base addition and deletion frequencies.

It is not easy to devise even general models, based simply on
tightness-of-fit between the coding face of a progeny dNTP and that
of the parental template residue, that are capable of explaining the
evident complexity of polymerase fidelity. On the other hand, biochemical evidence does not require that we limit ourselves to models
based solely on the thermodynamics of base pairing: it has long been
obvious that a large number of discriminatory contacts are potentially available between a purine or a pyrimidine and the polymerase
molecule and that the strategy of allostery could, therefore, be used
in the process of base selection. In contrast to the appealing simplicity of direct-base-pairing models (e.g., Clayton et al., 1979),
the advantage of polymerase-intervention models is their ability to
explain great complexity in the fidelity process.

Although this Section has stressed the role of DNA polymerase in
determining mutation rates, the polymerase does not operate as an
independent entity. A considerable number of "associated proteins"
accompany it, in both the functional and the physical senses. In

phage T4, where the spectrum of such proteins is best characterized, virtually each one is also involved in the fidelity process since each gives rise to mutator alleles (see Drake, 1973). However, it remains unclear whether these effects are mediated primarily by physical contacts between polymerase and associated proteins, or more indirectly, for instance, by effects upon the topography and concentration of substrates. In addition, there clearly exists a set of "less-associated" enzymatic activities whose functions are likely to affect fidelity; any of the determinants of relative and absolute dNTP pool sizes, for instance, would be expected to fulfill such a role.

MISMATCH REPAIR

Unlike the preceding example, the enzymes of the putative process of antimutagenic mismatch repair are not only not well characterized: they are hardly even discovered. Impressive genetic evidence suggests the operation of a mutation-reversal process, wherein the orientation of base mispairing is signaled by DNA methylation patterns in E. coli, so that repair can efficiently distinguish the incorrect and correct bases (Glickman, 1981). However, a number of attempts to detect the requisite E. coli enzymes have already failed.[*] Here, too, it is possible that unconventional mechanisms should be considered. Perhaps it is only a problem of unstable or multicomponent endonucleases, glycosylases, or insertases; or perhaps a more radical kind of plastic surgery is carried out, operating insufficiently frequently to have eluded activity-hunters thus far but nevertheless adequate to the task of curing rare mistakes. Might there, for instance, exist systems for the direct in situ interconversion of the requisite diagnostic portions of adenine and guanine, thymine and cytosine, and even purine and pyrimidine? (Note that interconversion would not necessarily have to be total, only sufficient to ensure subsequent accurate replication.)

It should also be noted that evidence implicating mismatch repair in the reversal of mutational lesions is not only primarily genetic but is also somewhat restricted to E. coli, with lesser but strongly suggestive evidence from the pneumococcal (Tiraby and Fox, 1973) and Haemophilus (Bagci and Stuy, 1980) systems. There is limited and controversial evidence for recombinational mismatch repair in phage T4 (Berger and Pardoll, 1976), but even if the process does occur there its relevance to mutagenesis remains unclear. In E. coli, mutations tend to appear in a homozygous state (e.g.,

---

[*]On the other hand, an enzymatic activity detected in Ustilago maydis may fulfill the same role (Ahmad et al., 1975; Pukkila, 1978).

Witkin and Sicurella, 1964), whereas in T4 they tend to appear as heterozygotes (e.g., Lindstrom and Drake, 1970), and if the former result were due to disoriented mismatch "repair" that mistakenly removed the parental rather than the progeny base, then the latter could be explained as the absence of mismatch repair.

"ULTRAVIOLET" MUTAGENESIS

There exists a set of mutagenic responses characterized by dependence both upon types of DNA damage that would appear more likely to render a base totally nontemplating than susceptible to mispairing and upon specifically encoded mutagenic responses on the part of the organism. Because mutational inactivation of such mutagenic responses is usually accompanied by reduced survival, terms such as "error-prone repair" (Witkin), "misrepair" (Drake), or "mutagenic repair" (von Borstel) have been applied to it. However, these mechanism-laden appellations may be premature since there is still no proof that the mutagenic response is constituted primarily for purposes of repair. Since ultraviolet irradiation appears to depend wholly upon this kind of mechanism among organisms that employ DNA to encode their genomes, becoming nonmutagenic in the appropriate mutants of phage T4, several bacteria, and fungi, and being altogether absent from, for instance, wild-type pneumococcus and Haemophilus, I will refer to the process simply as ultraviolet mutagenesis for the present.*

As in the case of mismatch repair, information about the biochemical basis of ultraviolet mutagenesis is virtually nonexistent despite some imaginative attempts to ferret out appropriate activities. Indeed, even the nature of the inducing lesion is obscure. Cyclobutane pyrimidine dimers are popular candidates, since both photoreactivation and excision repair conjointly decrease dimer frequencies in DNA and decrease mutation frequencies in progeny. Furthermore, there is at least in vitro evidence that dimers block DNA replication and might, therefore, act as noncoding lesions requiring some type of mutagenic passover synthesis. However, the evidence that dimers are the exclusive domain of photoreactivating enzymes is very thin indeed, to say nothing about the allied problem of the contamination of enzymatic photoreactivation by the non-enzymatic phenomenon of photoreversal, whose specificity may be even less than that of photoreactivation. In addition, the E. coli dimer excision system is already known to excise a rather broad range of

---

*The inducibility of this response in some organisms has further enriched terminology at the expense of explanation, so that it is sometimes unclear whether the cell is attempting to save original sequences or sacrifice outdated strategies.

base damages, and to have sometimes curious effects upon mutation rates: it excises $\underline{O}^6$-ethylguanine, for instance, but $\underline{uvrA}$, $\underline{B}$ and $\underline{C}$ mutants greatly enhance ethyl methanesulfonate (EMS) mutagenesis (Todd et al., 1981). Finally, in the case of phage T4 there is evidence from mutational specificity to suggest that the chemical target need not correspond very well to the mutational target (Meistrich and Drake, 1972).

Given such confusion about the premutagenic lesion, it is hardly surprising that the processing of such lesions is even less well understood. The suggestion that lesions that block DNA replication can be circumvented by a passover mechanism of intrinsically low fidelity (Radman, 1974) has stimulated some brave attempts to detect passover in vivo but not, thus far, with any clear success. In any case, the power of passover models has been somewhat undercut by the apparent brevity of delay of chain elongation induced by ultraviolet irradiation in mammalian cells and by the enigma raised by an apparently large ratio between dimers and mutations. For instance, data from the $\underline{E}$. $\underline{coli}$ $\underline{lacI}$ system suggest ratios ranging from ten dimers per mutation at a hot spot to about 300 at sites of average mutability (B W. Glickman, personal communication).

In contrast to ultraviolet mutagenesis, certain mutagens appear to induce direct mispairing in some organisms but to act through the ultraviolet-like mutability system in others. EMS is a good example. Its mutagenicity is independent of the $\underline{WXY}$ system in phage T4 (J. W. Drake and R. R. Green, unpublished data), is partially dependent upon the $\underline{recA}$ system in $\underline{E}$. $\underline{coli}$ (Kondo et al., 1970; Todd et al., 1981), but is strongly dependent upon the $\underline{rad6}$ system in $\underline{Saccharo}$-$\underline{myces}$ $\underline{cerevisiae}$ (Lemontt, 1971; Prakash, 1974). However, it is clear both from structural considerations and from the effects of antimutagenic DNA polymerases upon EMS mutagenesis (Drake and Greening, 1970) that putative intermediates in EMS mutagenesis such as O^6-ethylguanine may exhibit dual-pairing properties, that is, can pair both correctly and incorrectly (in contrast, for instance, to thymine in DNA produced by the deamination of 5-methylcytosine). It is possible that the DNA polymerases of bacteria have evolved increased discrimination against ambivalent pairings, compared to a relatively faithless T4 enzyme, and that eukaryotic polymerases (perhaps abetted by as yet unidentified accessory proteins) have developed even further in this direction. In that case, if ultraviolet mutagenesis were to depend upon relaxed polymerases fidelity in order to bypass a nontemplating DNA lesion (Radman, 1974), then it might also depend upon relaxed fidelity for synthesis past ambiguously or weakly templating lesions. These possibilities should soon be susceptible to experimental attack.

With the exception of phage T4, the ultraviolet-mutagenesis response has generally exhibited inducibility when that trait has been carefully sought.* More specifically, in the well-studied example of the "SOS" response in E. coli (Radman, 1974), mutability is but one aspect of an induced pleiotropic response to DNA damage. The adaptive value of inducibility might be as simple as conservation of cellular resources or might instead reflect a need to avoid unpleasant side effects. One component of the SOS response, for instance, consists of half-chromatid recombinational repair, which might well interfere with DNA replication were it constitutive. Another component, prophage induction, is unequivocally lethal to the cell, although making good sense from the standpoint of the virus itself. The adaptive value of the mutagenic component, however, is less obvious.†

For some time it was fashionable to suppose that the organism, faced with potentially lethal lesions unrepaired by excision or recombination, would choose highly untemplated (and hence mutagenic) synthesis rather than termination of DNA synthesis (the "better read than dead" model). It turns out, however, that the increment of survival associated with the mutagenic response is actually quite small, as can be seen from the properties of appropriate mutants in both E. coli (Kato and Shinoura, 1977; Volkert et al., 1976) and phage T4 (M. A. Conkling, unpublished data). Indeed, a number of bacterial species lack this mutagenic response altogether. Perhaps then the adaptive significance of the optimal mutagenic response is not primarily survival but mutation itself; increasing genetic diversity would certainly be an appropriate response to a harsh environment. Thus the mutagenic response may have evolved via selection for more fit mutant progeny rather than simply by differential resistance to killing itself.

---

*Constitutivity in T4 is inferred from the linearity of the dose-response curve (see, for instance, Conkling et al., 1976) and from the general lack of examples of environmentally inducible responses in lytic viruses.

†B. W. Glickman (personal communication) has pointed out that the multiplicity of SOS responses might not be due solely to their control by a common regulatory element but might result in part from the multiplicity of lesions produced by ultraviolet irradiation. Recombinational repair and prophage induction, for instance, might be triggered by blocked DNA replication, while the mutagenic response might be triggered by a different lesion altogether. In that case, the ultraviolet immutability of some wild-type bacteria might simply be due to an efficient excision repair system operating upon such lesions.

MACROLESIONS

Another enzymological mystery, perhaps not fully distinct from some aspects of the preceding several, is that of macrolesion formation. Cytogeneticists continue to attest with confidence that most chromosomal rearrangements (e.g., translocations, transpositions, and inversions) are the result of break-reunion events following upon two (or more) damage-induced chromatid or chromosome breaks. The origins of these breaks remain unclear, however, since rather few clastogens (chromosome-breaking agents) directly break single-stranded DNA, and a cytologically visible break probably involves the introduction of discontinuities into either two or four DNA strands. Equally unclear is the rejoining process, which may or may not be the same as that involved in the normal repair of double-stranded breaks: is blunt-end ligating involved or are limited overlaps required? Are repeated sequences essential, and if so, need they be only a few bases long or hundreds of bases in extent?

Even more mysterious is the generation of deletions (and perhaps also duplications, although here the evidence is thin) with linear dose-response kinetics, as in the example of phage T4 (Conkling et al., 1976) and Neurospora (Webber and de Serres, 1965) with ionizing radiations. Clearly some "one-hit" events can subsequently lead to deletions, although it is not necessary to postulate that the energy from a single event becomes distributed between two distant sites. It is instead conceivable that certain types of damage induce the DNA polymerase to separate from the template but not from the primer, later to reassociate with a template strand farther along in the 3' direction. (An analogous mechanism can easily be imagined for duplications.) At least in the case of T4, such a mechanism would have to be triggered by damage to DNA rather than to the polymerase itself since free particles were irradiated.

GENETIC INSTABILITIES

Next I wish to turn to a set of phenomena clearly diverse in mechanism but all generating mutations at rates well out on the high end of the distribution of mutation rates, that is, localized hypermutability.*

---

*For the purpose of this discussion, high mutation rates observed throughout the genome will not be considered. (Their determinants are usually called mutator mutations.) Some overlap between the two phenomena is inevitable, however, particularly in the case of mobile elements.

Classically, genetic instabilities have long been recognized to occur at several levels. The resurrection of the concept of mutagenesis, indeed the very word itself, grew out of the highly unstable inheritance of color patterns in the evening primrose Oenothera (de Vries, 1901); in this case, however, the explanation was conceptually simple, namely frequent rearrangements and aberrant segregation within a genetic system addicted to curious chromosome mechanics. With the discovery of chemical mutagenesis in the 1940s, there appeared the concept of "replicating instabilities" in Drosophila and yeasts (see Auerbach, 1976, for examples), wherein either the wild-type or a mutant allele acquired a high probability of mutation to the alternative state that persisted for many cell generations, perhaps indefinitely; but the molecular basis of this behavior has never been solved. By about 1960 two new classes of instability had become well documented: intragenic sites at which forward mutations arose at frequencies up to thousands of times greater than at other sites ("mutational hot spots") (e.g., Benzer, 1957), and a kind of mobile instability first characterized in maize in which specific regions of the chromosome suddenly acquired hypermutability (e.g., McClintock, 1965).

The mechanisms that underlie these types of genetic instability are perhaps fewer in number than the names appended to their diverse effects in different organisms. At present they appear to result mainly from three general mechanisms: duplications; short base sequences especially prone to base substitutions, addition, or deletion (point mutational hot spots); and mobile elements.

Although it is unclear how duplications arise, it is clear that they both revert and enlarge to triplications at very high frequencies by recombinational mechanisms, as postulated nearly six decades ago (Sturtevant and Morgan, 1923). So common is this type of instability that it is usually the first explanation that comes to mind when a rapidly reverting mutant is isolated. More recently, however, attention has focused on specific mechanisms responsible for the other two major classes of instability, namely point mutational hot spotting and mobile elements.

HOT SPOTS

The mutational hot spots, first defined by Benzer, are now seen to represent an extreme within a more or less continuous spectrum of point mutation rates extending over some eight or more orders of magnitude. Since the total mutability of a gene tends to be determined by its content of hot spots, it is important to understand their causes. A number of explanations have gradually accumulated, and very recently a new and unexpected association between hot spotting and DNA sequence has become evident.

First, there are now numerous examples of frameshift mutational hot spots located at sequences rich in redundancy, a direct confirmation of the Streisinger theory of frameshift mutagenesis (see Drake, 1981, for references). The two largest T4 rII hot spots, for instance, contribute as many mutations as do the remaining 3600 base pairs of this region and probably arise at AAAAAA sequences; while the Ames frameshift mutagen tester strains respond to mutations in sequences such as GGGG and CGCGCGCG. Not all frameshift mutations, however, so obviously arise within redundant regions. In a striking analysis, Ripley and Shoemaker (1981, and unpublished data) have recently shown that several T4 rII frameshift mutational hot spots reside in a region with the capacity, at those times when the DNA might find itself in a single-stranded configuration (e.g., during replication), to form "hairpin" structures defined by (often imperfect) reverse-complementary base sequences. Such structures should be energetically favored in the extended single-stranded state although disfavored in ordinary double-stranded DNA, unless special driving forces such as superhelicity (Panayotatos and Wells, 1981) are imposed. Hairpin structures would also be expected to exist in competition with the more extended forms of single-stranded DNA that are maintained through the binding of proteins such as those encoded by gene 32 of phage T4 or ssb of E. coli. It is, therefore, presently unclear just how the capacity to form hairpin structures relates to hypermutability, particularly in enzymatic terms. The association is, however, unmistakable, particularly since it arises in very different ways in other situations, as discussed below.

Little is known about the determinants of base substitution rates nor about base substitution hot spots in particular. Koch (1971) was the first to demonstrate experimentally that base substitution rates are fairly strong functions of adjacent and penultimate bases, and more recently it has been shown that certain E. coli hot spots arise as a direct consequence of an oddity of postreplication base modification (Coulondre et al., 1978). The very considerable remainder of base substitution hot spots, however, remain unexplained.

Again, however, some recent results suggest that DNA secondary structure can be a critical determinant. The first example comes from studies on ultraviolet-induced G·C → A·T hot spots in the E. coli lacI gene (Todd and Glickman, 1981). In each of several cases the genetically unstable sites reside within DNA sequences with a moderate to pronounced capacity for forming hairpin structures when single-stranded; and in these examples, the sites are further localized to the unpaired terminal loops of the hairpins. Additional surveys have also suggested a close association between ultraviolet hypermutability and capacity for DNA secondary structure in both the trpA gene of E. coli and the cyc1 gene of S. cerevisiae (B. W. Glickman, unpublished data). In yet another example, this time from phage T4 transition mutation rates at the three positions of a particular rII codon vary markedly (and irregularly: onefold, eightfold and twofold, respectively) as a function of the base composition of a site located, by recombinational measurements, some 20 to 80 nucleotides from the affected region (Conkling et al., 1980). The wild-type DNA sequence of the region is already known (Pribnow et al., 1981), and a preliminary determination has been made of the double-mutant sequence (A. Sugino, unpublished data). However, it remains unclear just how the rate-modulating mutation affects local opportunities for DNA secondary structure.

It should be stressed that the enzymological underpinnings of these several associations between mutability and secondary structure remain completely undetermined. Whatever the mechanism, however, it is now clear that residues, even a score of nucleotides removed from a mutable site, can be expected to play a role in determining that site's specific mutation rate. Thus virtually any particular mutational "temperature" should be possible.

MOBILE ELEMENTS

The final category of genetic instabilities concerns DNA segments of substantial size (from about 0.5 to 50 or more kilobases long) that appear abruptly and more or less randomly within genes. A particular sequence may be present at frequencies from several to several score per genome. First characterized in maize (see McClintock, 1965) and now recognized to be present in bacteria (see Kleckner, 1977), in yeasts, and in fruit flies, such elements may, by virtue of their mobility, generate mutations. Such mutations are very likely to consist of null alleles. They may also display continued instability after insertion, throwing off both back mutations and a wide variety of secondary macrolesions. Among the terms in current use to describe them are movable elements, insertion sequences, translocatable elements, transposons, and nomadic species. The middle-repetitive sequences of higher eukaryotes may also be candidates for one or more of these names.

The evolutionary significance of such sequences is controversial since they may only be examples of molecular parasitism that is more often deleterious than beneficial to the host genome. In at least a few instances they appear to have become involved in legitimate cellular control processes, such as phase variation in Salmonella (Zeig et al., 1977) and mating-type determination in yeast (e.g., Klar et al., 1981). On the other hand, they may contribute importantly to spontaneous mutation.* A substantial fraction, perhaps even a

---

*To date there is no convincing evidence that they play a role in chemically induced mutation.

majority, of spontaneous mutations in bacteria arise from such insertions (see Kleckner, 1977). A classical allele of Drosophila malanogaster, white-apricot, has recently been shown to contain a copy of copia, a common Drosophila mobile element (P. M. Bingham and B. H. Judd, personal communication). Most interestingly, it also appears that a crucial aspect of the phenomenon of hybrid disgenesis in Drosophila involves the mobilization of such sequences. In this situation, certain crosses between long-established laboratory strains and recent wild isolates generate a pattern of instability that, among other things, sharply increases mutation rates.* The first several such mutants to be examined with suitable probes have turned out to contain insertions of characteristic middle-repetitive sequences in association with the mutated locus (P. M. Bingham, G. M. Rubin, and M. Kidwell, personal communication). Furthermore, among a series of 15 mutant alleles of spontaneous origin but not known to have been products of hybrid disgenesis, five contained insertions, and this despite counterselection by the investigators against alleles a priori likely to contain insertions (e.g., complete null alleles and unstable alleles) (P. M. Bingham and Z. Zachar, personal communication). The sudden emergence of genetic instability in response to the injection of a new pool of genomes into an established population is somewhat akin to the induction of lytic reproduction in prophages upon entering a virgin cytoplasm, and an adaptive increase in intracellular mobility on behalf of the molecular parasites themselves might be imagined. Whether this would extent to interorganism infectivity is a completely open question, but the possibility is consistent with patterns of appearance of the capacity for hybrid disgenesis in populations of Drosophila.

REFERENCES

Ahmad, A., Holloman, W. K., and Holliday, R., 1975, Nuclease that preferentially inactivates DNA containing mismatched bases, Nature (London), 258:54.
Auerbach, C., 1976, "Mutation Research. Problems, Results, and Perspectives," John Wiley, New York.

---

*Activation of genetic instability as a result of outcrossing was predicted by Sturtevant (1937). He reasoned that two recently diverged populations would tend to accumulate randomly different modifiers of mutation rate. Upon subsequent crossing between members of the two populations, high mutation rates would segregate. This prediction was then confirmed (Sturtevant, 1939), but whether the effect that he observed is related to that of hybrid disgenesis remains unknown.

Bagci, H., and Stuy, J. H., 1980, Bromouracil-induced mutagenesis in a mismatch-repair-deficient strain of Haemophilus influenzae, Mutat. Res., 73:15.

Baker, M. S., and Topal, M. D., 1981, Cellular DNA precursor triphosphates are targets of N-methyl-N-nitrosourea (MNU), J. Supramol. Struct., Suppl. 5 (Cell. Biochem.), 454.

Baltz, R. H., Bingham, P. M., and Drake, J. W., 1976, Heat mutagenesis in bacteriophage T4: The transition pathway, Proc. Natl. Acad. Sci. U.S.A., 73:1269.

Benzer, S., 1957, The elementary units of heredity, in: "The Chemical Basis of Heredity," W. D. McElroy and B. Glass, eds., The Johns Hopkins Press, Baltimore.

Berger, H., and Pardoll, D., 1976, Evidence that mismatched bases in heteroduplex T4 bacteriophage are recognized in vivo, J. Virol., 20:441.

Bhanot, O. S., Khan, S. A., and Chambers, R. W., 1979, A new system for studying molecular mechanisms of mutation by carcinogens, J. Biol. Chem., 254:12684.

Bingham, P. M., Baltz, R. H., Ripley, L. S., and Drake, J. W., 1976, Heat mutagenesis in bacteriophage T4: The transversion pathway, Proc. Natl. Acad. Sci. U.S.A., 73:4159.

Clayton, L. K., Goodman, M. F., Branscomb, E. W., and Galas, D. J., 1979, Error induction and correction by mutant and wild type T4 DNA polymerases. Kinetic error discrimination mechanisms, J. Biol. Chem., 254:1902.

Conkling, M. A., Grunau, J. A., and Drake, J. W., 1976, Gamma-ray mutagenesis in bacteriophage T4, Genetics, 82:565.

Conkling, M. A., Koch, R. E., and Drake, J. W., 1980, Determination of mutation rates in bacteriophage T4 by unneighborly base pairs: Genetic analysis, J. Mol. Biol., 143:303.

Coulondre, C., Miller, J. H., Farabaugh, P. J., and Gilbert, W., 1978, Molecular basis of base substitution hot spots in Escherichia coli, Nature (London), 274:775.

de Vries, H., 1901, "Die Mutationstheorie," Veit, Leipzig.

Drake, J. W., 1973, The genetic control of spontaneous and induced mutation rates in bacteriophage T4, Genetics Suppl., 73:45.

Drake, J. W., 1981, Neighborly and unneighborly determinants of site-specific mutation rates, in: "The Immune System," Vol. I, C. Steinberg, and I. Lefkovits, eds., Karger, Basel, in press.

Drake, J. W., and Greening, E. O., 1970, Suppression of chemical mutagenesis in bacteriophage T4 by genetically modified DNA polymerases, Proc. Natl. Acad. Sci. U.S.A., 66:823.

Glickman, B. W., 1981, Methylation-instructed mismatch correction as a postreplication error avoidance mechanism in Escherichia coli, this volume.

Hershfield, M. S., 1973, On the role of deoxyribonucleic acid polymerase in determining mutation rates. Characterization of the defect in the T4 deoxyribonucleic acid polymerase caused by the tsL88 mutation, J. Biol. Chem., 248:1417.

Kato, T., and Shinoura, Y., 1977, Isolation and characterization of mutants of Escherichia coli deficient in induction of mutations by ultraviolet light, Mol. Gen. Genet., 156:121.

Klar, A. J. S., Strathern, J. N., Broach, J. R., and Hicks, J. B., 1981, Regulated transcription in expressed and unexpressed mating type casettes of yeast, Nature (London), 289:239.

Kleckner, N., 1977, Translocatable elements in procaryotes, Cell, 11:11.

Koch, R. E., 1971, The influence of neighboring base pairs upon base-pair substitution mutation rates, Proc. Natl. Acad. Sci. U.S.A., 68:773.

Kondo, S., and Ichikawa, H., 1973, Evidence that pretreatment of Escherichia coli cells with N-methyl-N'-nitro-N-nitrosoguanidine enhances mutability of subsequently infecting phage λ, Mol. Gen. Genet., 126:319.

Kondo, S., Ichikawa, H., Iwo, K., and Kato, T., 1970, Base-change mutagenesis and prophage induction in strains of Escherichia coli with different DNA repair capacities, Genetics, 66:187.

Kunkel, T. A., and Loeb, L. A., 1979, On the fidelity of DNA replication. Effect of divalent metal ion activators and deoxyribonucleoside triphosphate pools on in vitro mutagenesis, J. Biol. Chem., 254:5718.

Lemontt, J. F., 1971, Mutants of yeast defective in mutation induced by ultraviolet light, Genetics., 68:21.

Lindstrom, D. M., and Drake, J. W., 1970, The mechanics of frameshift mutagenesis in bacteriophage T4: the role of chromosome tips, Proc. Natl. Acad. Sci. U.S.A., 65:617.

McClintock, B. M., 1965, The control of gene action in maize, Brookhaven Symp. Biol., 18:162.

Meistrich, M. L., and Drake, J. W., 1972, Mutagenic effects of thymine dimers in bacteriophage T4, J. Mol. Biol., 66:107.

Muzyczka, N., Poland, R. L., and Bessman, M. J., 1972, Studies on the biochemical basis of spontaneous mutation. I. A comparison of the deoxyribonucleic acid polymerases of mutator, antimutator, and wild-type strains of bacteriophage T4, J. Biol. Chem., 247:7116.

Panayotatos, N., and Wells, R. D., 1981, Cruciform structures in supercoiled DNA, Nature (London), 289:466.

Prakash, L., 1974, Lack of chemically induced mutation in repair-deficient mutants of yeast, Genetics, 78:1101.

Pribnow, D., Sigurdson, D. C., Gold, L., Singer, B. S., Brosius, J., Dull, T. J., and Noller, H. F., 1981, The rII cistrons of bacteriophage T4: DNA sequence around the intercistronic divide and positions of genetic landmarks, J. Mol. Biol., in press.

Pukkila, P. J., 1978, The recognition of mismatched base pairs in DNA by DNase I from Ustilago maydis, Mol. Gen. Genet., 161:245.

Radman, M., 1974, Phenomenology of an inducible mutagenic DNA repair pathway in Escherichia coli: SOS repair hypothesis, in: "Molecular and Environmental Aspects of Mutagenesis," L. Prakash, F. Sherman, M. W. Miller, C. W. Lawrence, and H. W. Taber, eds., C. C. Thomas, Springfield, Illinois.

Ripley, L. S., 1975, Transversion mutagenesis in bacteriophage T4, Mol. Gen. Genet., 141:23.

Ripley, L. S., and Shoemaker, N. B., 1981, Polymerase infidelity and frameshift mutation, this volume.

Salganik, R. I., Vasjunina, E. A., Poslovina, A. S., and Andreeva, I. S., 1973, Mutagenic action of N^4-hydroxycytidine on Escherichia coli B cyt⁻, Mutat. Res., 70:11.

Schaaper, R. M., and Loeb, L. A., 1981, Depurination causes mutations in SOS-induced cells, Proc. Natl. Acad. Sci. U.S.A., 78:1773.

Sturtevant, A. H., 1937, Essays on evolution. I. On the effects of selection on mutation rate, Q. Rev. Biol., 12:464.

Sturtevant, A. H., 1939, High mutation frequency induced by hybridization, Proc. Natl. Acad. Sci. U.S.A., 25:308.

Sturtevant, A. H., and Morgan, T. H., 1923, Reverse mutation of the bar gene correlated with crossing-over, Science, 57:746.

Tessman, I., Ishiwa, H., and Kumar, S., 1965, Mutagenic effects of hydroxylamine in vivo, Science, 148:507.

Tiraby, J.-G., and Fox, M. S., 1973, Marker discrimination in transformation and mutation of pneumococcus, Proc. Natl. Acad. Sci. U.S.A., 70:3541.

Todd, P. A., Brouwer, J., and Glickman, B. W., 1981, Influence of DNA repair deficiencies on MMS and EMS induced mutagenesis, Mutat. Res., in press.

Todd, P. A., and Glickman, B. W., 1981, Mutagenic specificity of ultraviolet irradiation in the lacI gene of E. coli: Role of DNA secondary structure, Proc. Natl. Acad. Sci. U.S.A., 78: in press.

Volkert, M. R., George, D. L., and Witkin, E. M., 1976, Partial suppression of the LexA phenotype by mutations (rnm) which restore ultraviolet resistance but not ultraviolet mutability to Escherichia coli B/r uvrA lexA, Mutat. Res., 36:17.

Webber, B. B., and de Serres, F. J., 1965, Induction kinetics and genetic analysis of X-ray-induced mutations in the ad-3 region of Neurospora crassa, Proc. Natl. Acad. Sci. U.S.A., 53:430.

Witkin, E. M., and Sicurella, N. A., 1964, Pure clones of lactose-negative mutants obtained in Escherichia coli after treatment with 5-bromouracil, J. Mol. Biol., 8:610.

Zeig, J., Silverman, J., Hilmen, M., and Simon, M., 1977, Recombinational switch for gene expression, Science, 196:170.

CONTRIBUTORS

Anne Bagg, Biology Department, Massachusetts Institute of Technology, Cambridge, Massachusetts 02139, U.S.A.

E. G. Bernstine, Biology Division, Oak Ridge National Laboratory, Oak Ridge, Tennessee 37830, U.S.A.

Maurice J. Bessman, Department of Biology and McCollum-Pratt Institute, The Johns Hopkins University, Baltimore, Maryland 21218, U.S.A.

David Botstein, Department of Biology, Massachusetts Institute of Technology, Cambridge, Massachusetts 02139, U.S.A.

Elbert W. Branscomb, Department of Biological Sciences, Molecular Biology Section, University of Southern California, University Park, Los Angeles, California 90007, and the Biomedical Sciences Division, Lawrence Livermore Laboratory, University of California, Livermore, California 94550, U.S.A.

Douglas Brash, Sidney Farber Cancer Institute, 44 Binney Street, Boston, Massachusetts 02115, U.S.A.

Ahmad I. Bukhari, Cold Spring Harbor Laboratory, Cold Spring Harbor, New York 11794, U.S.A.

Robert W. Chambers, Department of Biochemistry, New York University, School of Medicine, New York 10016, U.S.A.

Roshan Christensen, Department of Radiation Biology and Biophysics, University of Rochester, School of Medicine and Dentistry, Rochester, New York 14642, U.S.A.

Frederic J. de Serres, National Institute of Environmental Health Sciences, P. O. Box 12233, Research Triangle Park, North Carolina, 27709, U.S.A.

CONTRIBUTORS

John W. Drake, Laboratory of Molecular Genetics, National Institute of Environmental Health Sciences, Research Triangle Park, North Carolina 27709, U.S.A.

Stephen J. Elledge, Biology Department, Massachusetts Institute of Technology, Cambridge, Massachusetts 02139, U.S.A.

B. W. Glickman, National Institute of Environmental Health Sciences, P. O. Box 12233, Research Triangle Park, North Carolina 27709, U.S.A.

Myron F. Goodman, Department of Biological Sciences, Molecular Biology Section, University of Southern California, University Park, Los Angeles, California 90007, and the Biomedical Sciences Division, Lawrence Livermore Laboratory, University of California, Livermore, California 94550, U.S.A.

Lynn K. Gordon, Sidney Farber Cancer Institute, 44 Binney Street, Boston, Massachusetts 02115, U.S.A.

William A. Haseltine, Sidney Farber Cancer Institute, 44 Binney Street, Boston, Massachusetts 02115, U.S.A.

Thomas Hjelmgren, Department of Medical Biochemistry, Gothenburg University, Gothenburg, Sweden

M. Heude, Institut Curie-Biologie, Centre Universitaire, Bâtiment 110, 91405 Orsay, France

Robin Holliday, National Institute for Medical Research, The Ridgeway, Mill Hill, London NW7 1AA, England

Anita Jacobsson, Department of Medical Biochemistry, Gothenburg University, Gothenburg, Sweden

Cynthia J. Kenyon, Biology Department, Massachusetts Institute of Technology, Cambridge, Massachusetts 02139, U.S.A.

Hajra Khatoon, Cold Spring Harbor Laboratory, Cold Spring Harbor, New York 11794, U.S.A.

Nancy Kleckner, Department of Biochemistry and Molecular Biology, Harvard University, Cambridge, Massachusetts 02138, U.S.A.

Sohei Kondo, Department of Fundamental Radiology, Faculty of Medicine, Osaka University, Kita-ku, Osaka 530, Japan

Thomas A. Kunkel, The Joseph Gottstein Memorial Cancer Research Laboratory, Department of Pathology, The University of Washington, Seattle, Washington 98195, U.S.A.

CONTRIBUTORS

Christopher W. Lawrence, Department of Radiation Biology and Biophysics, University of Rochester, School of Medicine and Dentistry, Rochester, New York 14642, U.S.A.

Michael W. Lieberman, Department of Pathology, Washington University School of Medicine, St. Louis, Missouri 63110, U.S.A.

Tomas Lindahl, Department of Medical Biochemistry, Gothenburg University, Gothenburg, Sweden

Christina Lindan, Sidney Farber Cancer Institute, 44 Binney Street, Boston, Massachusetts 02115, U.S.A.

Judith Lippke, Sidney Farber Cancer Institute, 44 Binney Street, Boston, Massachusetts 02115, U.S.A.

Kwok Ming Lo, Sidney Farber Cancer Institute, 44 Binney Street, Boston, Massachusetts 02115, U.S.A.

Lawrence Loeb, The Joseph Gottstein Memorial Cancer Research Laboratory, Department of Pathology, The University of Washington, Seattle, Washington 98195, U.S.A.

Victoria Lundblad, Department of Biochemistry and Molecular Biology, The Biological Laboratories, Harvard University, Cambridge, Massachusetts 02138, U.S.A.

Peter D. Moore, Department of Microbiology, The University of Chicago, Chicago, Illinois 60637, U.S.A.

E. Moustacchi, Institut Curie-Biologie, Centre Universitaire, Bâtiment 110, 91405 Orsay, France

Ada L. Olins, University of Tennessee-Oak Ridge Graduate School of Biomedical Sciences and the Biology Division, Oak Ridge National Laboratory, Oak Ridge, Tennessee 37830, U.S.A.

Monica Olsson, Department of Medical Biochemistry, Gothenburg University, Gothenburg, Sweden

Karen L. Perry, Biology Department, Massachusetts Institute of Technology, Cambridge, Massachusetts 02139, U.S.A.

Samuel D. Rabkin, Department of Microbiology, The University of Chicago, Chicago, Illinois 60637, U.S.A.

Lynn S. Ripley, Laboratory of Molecular Genetics, National Institute of Environmental Health Sciences, Research Triangle Park, North Carolina 27709, U.S.A.

Brigitte Royer-Pokora, Sidney Farber Cancer Institute, 44 Binney
    Street, Boston, Massachusetts 02115, U.S.A.

Liane B. Russell, Biology Division, Oak Ridge National Laboratory,
    Oak Ridge, Tennessee 37830, U.S.A.

Björn Rydberg, Department of Medical Biochemistry, Gothenburg
    University, Gothenburg, Sweden

Roeland M. Schaaper, The Joseph Gottstein Memorial Cancer Research
    Laboratory, Department of Pathology, The University of
    Washington, Seattle, Washington 98195, U.S.A.

Ann Schwartz, Department of Radiation Biology and Biophysics,
    University of Rochester, School of Medicine and Dentistry,
    Rochester, New York 14642, U.S.A.

William G. Shanabruch, Biology Department, Massachusetts Institute
    of Technology, Cambridge, Massachusetts 02139, U.S.A.

Nadja B. Shoemaker, Laboratory of Molecular Genetics, National
    Institute of Environmental Health Sciences, Research Triangle
    Park, North Carolina 27709, U.S.A.

David Shortle, Department of Biology, Massachusetts Institute of
    Technology, Cambridge, Massachusetts 02139, U.S.A.

B. Singer, Department of Molecular Biology and Virus Laboratory,
    University of California, Berkeley, California 94720, U.S.A.

Michael Smith, Department of Biochemistry, Faculty of Medicine,
    2146 Health Sciences Mall, University of British Columbia,
    Vancouver, B.C. V6T 1W5, Canada

Bernard S. Strauss, Department of Microbiology, The University of
    Chicago, Chicago, Illinois 60637, U.S.A.

Graham C. Walker, Biology Department, Massachusetts Institute of
    Technology, Cambridge, Massachusetts 02139, U.S.A.

Susan M. Watanabe, Department of Biological Sciences, Molecular
    Biology Section, University of Southern California, University
    Park, Los Angeles, California 90007, and the Biomedical
    Sciences Division, Lawrence Livermore Laboratory, University
    of California, Livermore, California 94550, U.S.A.

INDEX

Adaptive response, 92, 94, 96
Alkylating agents, 2, 74
Alkylation
  background, 91
  DNA, with S-adenosyl-
     methionine, 90
  recurrent, 91
Alkyl sulfates, 2
2-Aminopurine
  base-pairing diagrams, 219
  insertion, 216
  nucleotide pair competition, 216
  tautomeric base mispairing, 216
Antimutators, 366
AP endonucleases, 324
AP sites, 139, 199
Apurinic/apyrimidinic sites
  mutagenic effect, 139, 203, 206
Aromatic amines, 14

Bacteriophage
  φX174, 121, 157, 179, 199, 231
  mu, 235
  T4, 161, 231
Base analogues
  2-aminopurine 67, 74, 213
  5-bromouracil, 74
  6-hydroxyaminopurine, 82
Base mispairs, 214, 216
Base-pairing free energy ladder, 220
Base-substitution mutations, 65, 111, 135, 147, 186, 213

Base tautomers, 216
β-lactamase
  mutagenesis, 151
Bisulfite, sodium
  deamination, 147

Carcinogens, 121
  effect on chromatin, 305
Chromatin, 333
  core vs linker regions, 304
  DNA damage, 307
  excision repair, 303
  pyrimidine dimers, 307
  rearrangement, 307
  unfolding, 306
Chromosomal DNA, 333
Chromosome aberrations, 351
Complementation
  mouse c-locus region, 349
  N. crassa, 337
Crossing over, 260
Cyclic alkylating agents, 10, 12
Cyclobutane pyrimidine dimers, 315, 333

D-loop, 149
Damage-inducible genes
  din regulation, 47, 51, 59
  muc, 56
  umuC, 50, 54
  uvr, 49, 50
Deamination of cytosine, 150
Deletions
  multilocus, 342
  transposon-promoted, 269
Depurination of DNA, 199, 232

Direct-base-pairing models, 366
Direct repeats, 246, 269
DNA
  damage, 315
    in alphoid DNA sequences, 327
  depurination, 199
    consequences, 201
  dissociation, 241
  fusion, 241
  glycosylases, 92
    formamidopyrimidine, 96
    inducible activities, 93
    3-methyladenine, 93
    7-methylguanine, 93
    pyrimidine dimer, 324
  metabolism, 162
  methylation, 66
  polymerase, 163, 179, 180, 201
    copying on φX174 DNA, 179, 202
    influence on fidelity of DNA, 163, 167
    mechanism of action, 231
    passive polymerase model, 225, 234
    pattern of termination, 188
    reactions, 187
    role in frameshift mutagenesis, 231
  rearrangements, 236, 265
    genetic consequences, 241
  sequence
    frameshift mutations, 162
    human β globin genes, 162
    influence on polymerase activity, 169
    mammalian β-like globin, 162
    nonfunctional, 259
    wild-type and frameshift targets, 174
  rearrangements, 269

Ethidium bromide
  mutagenesis of mtDNA, 281
Endonucleases
  AP, 324
  III, IV, 93

Escherichia coli, 20, 43, 65, 89, 103, 157, 231, 245, 265, 315
Excision
  3-methyladenine, 92
  modified purine residues, 92
  repair, short-patch, 92
Exonuclease III, VII, 93

Fidelity
  DNA polymerases, 66, 365
  of DNA replication, 225, 231
    error avoidance, 65
    passive polymerase model, 214
  transcription, 22
  translation, 22
Frameshift mechanism, 162
  DNA metabolism, 163
  on DNA secondary (hairpin) structures, 175
  site specificity, 152, 174
Frameshift mutations
  influence of DNA metabolic protein mutant alleles, 163
  influence of mutant DNA polymerases, 167
  N. crassa, 337
  suppressors, 165
  T4 rII β gene, 164
  yeast, 111

Gaps in DNA
  misrepair, 147
  overlapping
    excision, 180
    daughter-strand, 180
Gene conversion, 71, 259
Genetic instabilities, 371
Genetic risk assessment, 354

Haploid selection in mice, 356
Heteroduplex DNA, 260
Homopolymer fidelity assay, 201
Hot spots, mutational
  lacI, 316
Human cells
  repair-defective, 104

INDEX

Human DNA
  alphoid sequences, 327
  repair in chromatin, 303, 333
Hybrid DNA, 260
Hyper-rec phenotype, 71

Illegitimate recombination, 246, 265
Induced mouse mutations, 348
Insertion mutations, 236
Inverted repeats, 56, 246, 269
Iso-1-cytochrome c, yeast, 113, 157

Junk DNA, 259

Lethal mutations, 127

Macrolesions, 371
Mammalian cells, 20, 91, 104, 303, 327, 333
Map expansion, 71
Manganese ion, 153, 191
Meiosis, 259
Methylation of DNA
  dam mutants, 67
  eukaryotes, 82
  prokaryotes, 67
6-Methylguanine
  mutagenic effect, 138
7-Methylguanine removal, 95
Methyltransferase, 97
Misincorporation, 153
Misrepair reaction, 153
Missense mutations
  N. crassa, 337
Mitochondrial DNA, 333
  excision repair, 282, 286
  petite mutation, 275
  photorepair, 284
  point mutation, 275
  recombination repair, 283
  repair processes, 284, 287
  replication, 276
  versus nuclear DNA, 274
Mobile elements, 372, 374
Mouse specific-locus test, 346
Mouse X-autosome translocations, 348

Multi-lesion problem, 364
Mutagenesis
  2-aminopurine, 216
  bisulfite, 150
  by base analogues, 65
  by chemical carcinogens, 208
  by transposable elements, 240
  direct, 73
  frameshift, yeast, 111
  heat, 364
  in apurinic sites, 206
  in depurinated φX174 DNA, 203
  indirect, 73, 365
  in vitro, 121, 147, 157
  localized, 240
  mtDNA
    cytoplasmic gene control, 279
    genetic control, 276, 282
    inducing agents, 280
    nuclear gene control, 277, 282, 285
    spontaneous, 276
  site-specific, 121, 147, 157
  spontaneous, 65
  SV40 in vitro, 150
  untargeted, 112
  ultraviolet, 368
  UV-induced, 109, 179
  via depurination, 208
Mutagenic DNA alterations, 361
Mutagenic mechanisms, 361
  complexity, 362
Mutants
  adaptive-response-deficient, 104
  bacteriophage φX174
    frameshift, 129-131
    insertion-deletion, 132
    lethal, 127
      missense, 128
    nonsense, 127
    silent, 128
  bacteriophage T4
    DNA polymerases, 167
    frameshift, 162, 171
      recombination tests, 173
  E. coli
    dam, 65

error-prone repair, 181
frameshift, 62, 65
lacI, 74, 80, 157
mismatch-repair deficient, 68
transposon excision, 248
human tumor cells, 104
mice, c-locus, 349
mutator, 68, 80
N. crassa, ad-3, 337
ad-3, 337
frameshift, 337
repair-deficient, 335
specific locus, 335
Salmonella, frameshift, 162
S. cerevisiae
iso-1-cytochrome c deletions, 162
repair-deficient, 110, 282
reversion-deficient, 111
transitions, 152, 338
transversions, 74, 151
Xeroderma pigmentosum cells, 105
Mutational hot spots, 372
Mutation assay sensitivity, 363
Mutation rates in mice, 356

Neurospora crassa, 335
N-nitroso compounds, 2
Nucleic acid conformation
effect on chemical reactivity, 16, 17, 20
Nucleosomes, 333
phasing, 306
Nucleotide pool imbalance, 154

O^6-methylguanine
mutagenic effect, 138
Operon fusion vector, 46
Overlapping
daughter-strand gaps, 180
excision gap, 180

Parthenogenesis
accumulation of selfish DNA, 262
Phenotypic expression of mutations, 356

Plasmids
mutagenesis-enhancing, pKM101, 55, 78
pBR322, 147
pɸXG, 127
Polycyclic hydrocarbons, 14
Polymerase-intervention models, 366
Position effect, 348
Psoralen photoaddition, yeast mtDNA, 281
PyC lesion, 333
in human DNA, 327
in UV-irradiated DNA, 321
mutagenic role, 323
Pyrimidine dimers
distribution in defined DNA sequences, 316
DNA glycosylase, 324
in chromatin, 307

recA protein 45, 149
recA regulation, 45
Recombination
effect of mismatch repair, 71
general, 265
illegitimate, 268
replicative, transposon-promoted, 265, 266
site-specific, 265, 266
Repair
apurinic sites, 200
base-pair mismatches, 254
constitutive, 103
DNA adducts, 20
error-prone, 43, 103, 179, 180
assay procedure, 185
efficiency, 183
excision, 180
chromatin, 306
error-free, 103
mechanism, 323
yeast mtDNA, 282, 286
heteroduplex, 66
inducible, 103
long-patch excision, 181
mismatch, 65, 69, 367
DNA bases, 103
O^6-ethylguanine residues, 99

INDEX

O^6-methylguanine residues, 96
postreplication, 65
SOS functions, 180
Replicating instabilities, 372
Replicon fusion, 242, 266
Reverse transcriptase, 232
Reversionless mutants, 111
Risk estimation
  Neurospora, 342

Saccharomyces cerevisiae, 109, 157, 273
Salmonella, 55, 162, 341
Selfish DNA, 259
Sexual reproduction
  elimination of selfish DNA, 262
Single-stranded gaps, 147
SOS responses, 44, 180, 206, 233
Specific locus mutations
  Neurospora, 335
Suicide inactivation
  of enzyme, 97, 98
Suppression, 136
SV40
  mutagenesis, 150
Synthetic polynucleotides, 26, 201, 232

Target DNA sites, 266
Tautomeric equilibrium, 25
  effects of mutagens, 14, 15, 19
Thy(6-4)Pyo, 321
Transcription
  modified polynucleotides, 22
Transdimer synthesis, 110
Transfection
  heteroduplex DNA, 72
Transition mutations
  mechanism, 233
  models
    base tautomer involvement, 220
    hydrogen bond free energy defenses, 220
  site-specific, 152
Translation
  modified polynucleotides, 22
Transposable elements, 236, excision, 240

Mud(Ap, lac) insertion, 246
Tn5 insertion, 54, 56, 58
Transposition of genetic elements, 259
Transposons, 240, 246, 265,
  excision, 246, 268
Transversion mutations, 74
  site-specific, 151

Ultraviolet light
  damage in chromatin, 303
  damage in defined DNA sequences, 316
  effect on in vitro DNA synthesis, 179, 232
  mutagenesis
    Neurospora, 339, 340
    yeast, 109, 273
Untargeted mutagenesis, 112
UV mutagenesis
  depurinated sites, 179, 233
  E. coli genes, 43, 76, 103
  yeast genes, 110, 282
  yeast mtDNA, 280

X-chromosome inactivation, 348

Yeast, 109, 157, 273, 333